Information Systems Engineering and Management 40

The book series "Information Systems Engineering and Management" (ISEM) publishes innovative and original works in the various areas of planning, development, implementation, and management of information systems and technologies by enterprises, citizens, and society for the improvement of the socio-economic environment.

The series is multidisciplinary, focusing on technological, organizational, and social domains of information systems engineering and management. Manuscripts published in this book series focus on relevant problems and research in the planning, analysis, design, implementation, exploration, and management of all types of information systems and technologies. The series contains monographs, lecture notes, edited volumes, pedagogical and technical books as well as proceedings volumes.

Some topics/keywords to be considered in the ISEM book series are, but not limited to: Information Systems Planning; Information Systems Development; Exploration of Information Systems; Management of Information Systems; Blockchain Technology; Cloud Computing; Artificial Intelligence (AI) and Machine Learning; Big Data Analytics; Multimedia Systems; Computer Networks, Mobility and Pervasive Systems; IT Security, Ethics and Privacy; Cybersecurity; Digital Platforms and Services; Requirements Engineering; Software Engineering; Process and Knowledge Engineering; Security and Privacy Engineering, Autonomous Robotics; Human-Computer Interaction; Marketing and Information; Tourism and Information; Finance and Value; Decisions and Risk; Innovation and Projects; Strategy and People.

Indexed by Google Scholar. All books published in the series are submitted for consideration in the Web of Science.

For book or proceedings proposals please contact Alvaro Rocha (amrrocha@gmail.com).

Ahmed J. Obaid · Zdzisław Polkowski ·
Adriana Burlea-Schiopoiu ·
Muthmainnah Muthmainnah
Editors

Frontiers in AI and Computational Technologies

Proceeding of ICONEST 2024

Springer

Editors
Ahmed J. Obaid
Faculty of Computer Science
and Mathematics
University of Kufa
Najaf, Iraq

Zdzisław Polkowski
Faculty of Technical and Social Sciences
The Jan Wyżykowski University
Polkowice, Poland

Adriana Burlea-Schiopoiu
University of Craiova
Craiova, Romania

Muthmainnah Muthmainnah
Universitas Al Asyariah Mandar
Polewali, Indonesia

ISSN 3004-958X ISSN 3004-9598 (electronic)
Information Systems Engineering and Management
ISBN 978-3-031-89959-1 ISBN 978-3-031-89960-7 (eBook)
https://doi.org/10.1007/978-3-031-89960-7

This Springer imprint is published by the registered company Springer Nature Switzerland AG
The registered company address is: Gewerbestrasse 11, 6330 Cham, Switzerland

Preface

This volume contains the papers presented at the 2nd International Conference on Emerging Trends in AI and Computational Technologies (ICONEST), held at Universitas PGRI Ronggolawe (UNIROW) in Tuban, Indonesia, on November 15–16, 2024. Supporting UNIROW and Honorary Conference Chair: Prof. Dr. Alvaro Rocha ISEG, University of Lisbon, Portugal. The conference has been attended by invited international professors from several countries. ICONEST 2024 has the objective of bringing together few exceptional researchers, engineers, and industry experts in the fields of artificial intelligence and computational technologies to exchange their ideas and discuss their applications. The proceedings cover various aspects and were selected for presentation at the conference after a thorough peer-review process, which resulted in a high quality of accepted papers. Discussed topics spanned the wide range of possibilities enabled by AI techniques, including both language-based models and generative intelligence, as well as advances in machine learning and robotics, as well as the ethical implications of AI, and how to enable AI as a force for good in the world. In addition, it tackled the accelerating speed of AI, its advantages for humanity and the associated potential threats, with abstract sessions on AI use in fields like medicine, entertainment, finance, and creative industries. Other topics of interest in ICONEST 2024 included AI for Climate Change and Biodiversity Conservation, AI in Sustainability and Environmental Impact, and AI in Education and Smart City Planning. The emphasis was on promoting multidisciplinary cooperation and supporting dialogues that might accelerate a paradigm shift for more sustainable and ethical AI approaches. We hope this volume will greatly further engagement with ongoing work on artificial intelligence and provide a window into new state-of-the-art technologies and their transformative possibilities. By investigating the potential pitfalls and possibilities in AI development, this book hopes to provide guidance for future paths toward the sustainable and ethical use of data and technology in our societies. ICONEST 2024 would not have been possible without the hard work and dedication of the organizing committee members, keynote speakers, session chairs, and all attendees. Our heartfelt thanks to all those who contributed to making this conference a venue for innovation and conversation in the realms of AI and computational technologies. We hope these proceedings will help inspire continued growth and "paradigm shifts" within this exciting domain.

About Conference

Conference explores the latest advancements and innovations in artificial intelligence and computational sciences. Topics include contemporary developments in AI, language-based models, generative intelligence, machine learning, robotics, and ethical considerations. We aim to discuss the rapid progress of AI, its benefits for humanity, potential dangers, and ethical dilemmas. The conference will cover diverse AI applications, such as in the medical, entertainment, financial, and creative sectors. By gathering experts from various fields, we strive to foster interdisciplinary collaboration and a comprehensive understanding of AI and computational technologies.

Aims of the Conference

ICONEST 2024 unites scientists, academic researchers, and students to improve communication and dialogue and share experiences, ideas, and knowledge on using artificial intelligence and computational technologies. The conference will act as a collaborative space between many organizations and institutions worldwide to open scientific channels to solve the latest figural problems and challenges in the artificial intelligence domain. The major topics will be the concepts underlying current AI-based applications, the potential exploitation of advances, and trends in machine learning, robotics, and the ethical use of AI. The conference aims to bring together researchers, academicians, technology experts, and students to promote sharing and exchanging their experiences, ideas, and knowledge regarding AI technologies that would affect diverse domains like health, education, and environmental sustainability. ICONEST 2024 aims to act as an interdisciplinary platform for critical discourse across core areas on a wide range of topics, including, for example, the role of AI and climate change in sustainability, the implications for smart city design, and the ethics of technology construction. The conference presents an immense opportunity to publish research articles with closely related AI, recent technologies, and methodologies. Apart from its role in the exchange of knowledge, ICONEST 2024 also aims to propose sustainable solutions to current problems and provide a basis for international institutional bodies to foster scientific suggestions to combat pressing issues. ICONEST 2024 hopes to accelerate sustainable development in society through insightful discussions and cutting-edge presentations on revolutionary AI and computing technologies.

Ahmed J. Obaid
Zdzisław Polkowski
Adriana Burlea-Schiopoiu
Muthmainnah Muthmainnah

Acknowledgments

Successfully held on November 15–16, 2024, at the University of PGRI Ronggolawe (UNIROW), Tuban, Indonesia, the 2nd International Conference on Emerging Trends on AI and Computational Technologies (ICONEST). The focus of this conference was to discover the latest breakthroughs and innovations in artificial intelligence and computational sciences while also discussing the latest developments in AI, language models, generative intelligence, machine learning, robotics, and their respective ethical points of view. We wanted to gather experts from diverse fields, allowing them to interact with each other and foster interdisciplinary collaboration and a comprehensive understanding of AI and its usage in medical, entertainment, financial, and creative fields. These discussions are expected to shape future research and application of critical issues like AI for climate change, sustainability, smart city planning, and agricultural innovations. We want to express our deepest thanks to the head of UNIROW, Prof. Dr. Warli, M. Pd., who strongly supported academic development and a career-oriented approach to interdisciplinary cooperation through his contribution and devotion. Your leadership brought together experts who could help these essential conversations along. Our Honorary Conference Chair, Sir Alvaro Rocha, ISEG, University of Lisbon, Portugal, is given a special acknowledgment. His priceless inputs and counsel were instrumental in bringing quality and scope to the conference, leading to greater heights. We are incredibly grateful to have worked with professors and researchers from different countries who have contributed significantly to the conference discourse. This has promoted the sharing of insights on AI-related topics, including AI for climate change and biodiversity conservation, AI for sustainability and environmental impact, AI in education and learning techniques, and AI application in smart city planning and management, and has enhanced an understanding of the consequences of these technologies. We would also like to thank the Scientific and Organizing Committee members who worked hard to make the conference work without any hitch. Above all, they played a significant role in that, as was their commitment to excellence and meticulous attention to detail. We also thank the esteemed presenters and participants whose research and thoughts are so valuable to share. Your contributions greatly enriched the intellectual discourse and took us forward collectively in the field.

Finally, we'd like to thank the sponsors and partners whose commitment to innovation and sustainability allowed this conference to happen. We have laid a solid foundation for furthering AI and computational technologies. We thank everyone who worked hard to make the 2nd International Conference on Emerging Trends in AI and Computational Technologies successful. Most importantly, your work has also enabled collaboration and knowledge sharing in exciting and growing fields, namely AI and computational sciences, to benefit humanity and solve vital global challenges.

Warm regards

Ahmed J. Obaid
Conference Lead Editor

ICONEST 2024 Committee Members

Honorary Conference Chair

Alvaro Rocha	ISEG, University of Lisbon, Portugal, Iraq

Higher Coordinating Committee

Warli	Rector of UNIROW University, UNIROW University, Indonesia
Ahmed J. Obaid	University of Kufa, Iraq, General Chair of Conference
Christina Innocenti	Head of International Affairs, UNIROW University, Indonesia
Risa Triassanti	Head of International Department, UNIROW University, Indonesia

Conference Editors

Ahmed J. Obaid (Lead of Editor)	University of Kufa, Iraq
Muthmainnah Muthmainnah (Co-editor)	University Al Asyariah, Indonesia
Adriana Burlea-Schiopoiu (Co-editor)	University of Craiova, Romania
Zdzisław Pólkowski (Co-editor)	The UJW University, Poland

Conference Keynote Speakers

Luis Miquel Cardoso, Portugal
Michael Baron, Australia
Agus Wardhono, Indonesia

Steering Committees

Warli	UNIROW University, Indonesia
Anggia Kalista	UNIROW University, Indonesia
Yudi Supiyanto	UNIROW University, Indonesia
Suantoko, S.Pd	UNIROW University, Indonesia

Publicity Chairs

EZdzisław Pólkowski	UJW University, Poland
Bharat Bhushan	Co-Chair of Scientific Committee, Sharda University, India
Hiba A. Ahmed	University of Baghdad, Iraq
Adriana Burlea-Schiopoiu	University of Craiova, Romania

Program Conveners

Erkol Bayram	University of Sinop, Turkey
Luis Miquel Cardoso	University of Coimbra, Portugal
Pastor Reglos Arguelles JR	University of Perpetual Help System DALTA, Philippines
Michael Baron	Analytics & Digital Transformation, Australia
Neyara Radwan	Al Maarefa University, Saudi Arabia
Zdzisław Pólkowski	UJW University, Poland
Bharat Bhushan	Sharda University, India

Program Committee

Abdul Ghofur	Universitas Al Asyariah, Indonesia
Mutmainah	Universitas Al Asyariah, Indonesia
Eka Apriyani	IAIN Curup Bengkulu, Indonesia
Imas Cintamulya	UNIROW University, Indonesia
Mu Jizatin Fadiana	IAIN Curup Bengkulu, Indonesia
Risa Triassanti	UNIROW University, Indonesia

Technical Board

Marita Ika Joesidawati	UNIROW University, Indonesia
Salimullah Tegar sanubarianto	UNIROW University, Indonesia
Mu'jizatin Fadiana	UNIROW University, Indonesia
Asfan Muqtadir	UNIROW University, Indonesia
Amaludin Arifia	UNIROW University, Indonesia
Nibrosu Rohid	UNIROW University, Indonesia

International Advisory Committee

Leonardo Aureliano da Silva	Ilma University, Brazil
Haruna Abubakar Haruna	State Universal Basic Education, State Nigeria
Sherif Mohamed Ismail	University in Cairo, Egypt
Hisham AlSaghbini	Cambridge University, UK
Mohammad Israr	Maryam Abacha American University of Nigeria, Nigeria
Ryo Takahashi	American University of Sovereign Nations, Japan
Shashi Kant Gupta	Chinmay Research Education and Publication Private, India
Subrata Chowdhury	SITAMS (A) Chittoor AP, India
Vasileios Drougas	Ministry of Education and Religious Affairs, Greek
Haitham Alzoubi	Skyline University College, UAE
Christoph Rasche	University of Potsdam, Germany
Van Thanh Tien Nguyen	University of Science and Technology, Taiwan
Alex Khang	Institute of Technology and Engineering, Vietnam
Chin–Shiuh Shieh	National Kaohsiung University of Science, Taiwan
Samantha Curle	University of Bath, UK
Dither Balderas	México University, México
Muhammad Mohsin Waqas	Khwaja Fareed University, Pakistan
Marco Valery	Niccolo Cusano University, Italy
Yew Kee Wong	Hong Kong Hai College, China
Mohammed Zubair Khan	Taibah University, Saudi Arabia
Souvik Ganguli	Thapar Institute of Engineering and Technology, India
Jan Walczak	Uczelnia Jana Wyżykowskiego, Poland
Ali AL-Matari	A'Sharqiyah University, Oman
Lázaro Rodríguez	Universidad de Granada, Spain
Mzili Toufik	Chouaib Doukkali University, Morocco

Contents

AI applications in Smart City Planning and Management

AI for Climate Change and Biodiversity Conservation

AI in Agriculture for Sustainable Food Production

AI in Education and Learning Techniques

Emerging Trends in AI and Computational Technologies

About the Editors

Ahmed J. Obaid is an asst. professor at the Department of Computer Science, Faculty of Computer Science and Mathematics, University of Kufa, Iraq. He holds a bachelor's in Computer Science, a degree in Information Systems from the College of Computers, University of Anbar, Iraq (2001–2005), and a master's degree (M. TECH) in Computer Science Engineering (CSE) from School of Information Technology, Jawaharlal Nehru Technological University, Hyderabad, India (2010–2013), and a Doctor of Philosophy (Ph.D.) in Web Mining from College of Information Technology, University of Babylon, Iraq (2013–2017). He is a Certified Web Mining Consultant with over 14 years of experience working as a Faculty Member at the University of Kufa, Iraq. He has taught courses in Web design, Web Scripting, JavaScript, VB.Net, MATLAB Toolbox, and other courses on PHP, CMC, and DHTML from more than ten international organizations and institutes in the USA and India. He is a member of the Statistical and Information Consultation Center (SICC), at the University of Kufa, Iraq. His main line of research is web mining techniques and applications, image processing in web platforms, image processing, genetic algorithms, information theory, and medical health applications. He is an Associated Editor in the Brazilian Journal of Operations & Production Management (BJO&PM) and an Editorial Board member in the International Journal of Advance Study and Research Work (IJASRW), Journal of Research in Engineering and Applied Sciences(JREAS), GRD Journal for Engineering (GRDJE), International Research Journal of Multidisciplinary Science & Technology (IRJMST), The International Journal of Technology Information and Computer (IJTIC), Career Point International Journal Research (CPIJR). He was an Editor in Many International Conferences such as ISCPS_2020 MAICT_2020 IHICPS_2020, IICESAT_2021, IICPS_2020 ICPAS_2021, etc. (Scopus Indexed Conferences). He has edited some books, such as Advance Material Science and Engineering (ISBN: 9783035736779, Scientific.net publisher), Computational Intelligence Techniques for Combating COVID-19 (ISBN: 978-3-030-68936-0 EAI/Springer), A Fusion of Artificial Intelligence and Internet of Things for Emerging Cyber Systems (ISBN: 978-3-030-76653-5 Springer). He has supervised several final projects for bachelor's and master's in his main line of work and authored and co-authored several scientific publications in journals, books, and conferences with more than 75+ journal research articles, 5+ book chapters, 15+ conference papers, 10+ conference proceedings, 8+ books editing, 2+ patents. He is also a Reviewer in many Scopus, SCI, and ESCI Journals, e.g., CMC, IETE, IJAACS, IJIPM, IJKBD, IJBSR, IET, IJUFKS, and many others. He attended and participated as a keynote speakers (60+ conferences), webinars (10+), and session chair (10+), in many international events in the following countries: India, Turkey, Nepal, Philippines, Vietnam, Thailand, Indonesia, and other countries.

Zdzisław Polkowski is a professor of WSG University Bydgoszcz, professor of UJW at the Faculty of Social and Technical Sciences, and Rector's Representative for International Cooperation and Erasmus+ Program at the Jan Wyzykowski University Polkowice, Poland. Also, since 2019, he has been an adjunct professor in the Department of Business Intelligence in Management at Wroclaw University of Economics and Business and The Karkonosze University of Applied Sciences in Jelenia Góra, Poland. Moreover, he is a visiting professor at the University of Pitesti, Romania, Swami Sahajanand School of Management, Bhavnagar, India, and an adjunct professor at Marwadi University, Swami Sahajanand College Of Commerce & Management, Bhavnagar, India. He is the former dean of the Technical Sciences Faculty during the 2009 -2012 period at UZZM in Lubin. He holds a Ph.D. in Computer Science and Management from Wroclaw University of Technology, a graduate degree in Microcomputer Systems in Management from the University of Economics in Wroclaw, and a Postgraduate IT in Education from Economics University in Katowice. He obtained his Engineering degree in Computer Systems in Industry from the Technical University of Zielona Gora. He has published more than 159 papers in journals 35 conference proceedings, including more than 65 papers in journals indexed in the Web of Science, Scopus, and IEEE. He is co-editor of 29 books which have been published in Springer and CRC Press Taylor & Francis. He served as a member of Technical Program Committee and the member of the editorial board in many International conferences, journals in Poland, India, China, Iran, Iraq, Syria, UK, Romania, Turkey, Indonesia, Philippines, Vietnam, Malaysia, and Bulgaria. Till date he has delivered over 85 invited talks at different international conferences across various countries. He is also the member of the Board of Studies and expert member of the doctoral research committee in many universities in India. His area of interests includes IT in Business, IoT in Business and Education Technology. He has successfully completed research projects on "Developing the innovative methodology of teaching Business Informatics" and "Web Technology & Green Internet of Things" funded by the European Commission. He also owns an IT SME consultancy company in Polkowice and Lubin, Poland.

Adriana Burlea-Schiopoiu is a professor of Management at the University of Craiova, Romania, Romanian ambassador for ARGH (Association Francophone de Gestion des Ressources Humaines; President of the Regional Commission of Experts (CRE)—Central and Eastern Europe, President of the AIS SIG—Cognitive Analytics Management (SIGCAM), Member of the Scientific Council of the University Agency of the Francophonie (AUF); Head of Cyber-Security Branch—Scientific Innovation Research Group (SIRG); Member of the Directors Board of Doctoral School of the University of Craiova. She teaches and researches Human Resources Management, Project Management, Social Audit, Corporate Social Responsibility, and Negotiation Techniques at the University of Craiova, Faculty of Economics and Business Administration in Romania. In addition, she was a visiting professor teaching and researching cross-cultural management, human resources management, and intercultural management at universities in France, Poland, and Latvia. She was invited as the keynote speaker for many international conferences in France, Portugal, Albania, Lebanon, India, Indonesia, Iraq, and Turkey. She is a member of many organizing committees of national and international journals and conferences.

She stimulated and encouraged cooperation between her university and other universities around the world (for example, collaboration with staff and teachers from other international universities in various scientific projects—for example, universities from France, the USA, the UK, Swiss, Belgium, Italy, Portugal, Norway, and Holland). She has published in prestigious journals, for example, the Journal of Cleaner Production, Corporate Social Responsibility, Environmental Management, International Journal of Finance & Economics, Socio-economic Planning Sciences, and International Journal of Environmental Research and Public Health.

Muthmainnah Muthmainnah is a distinguished senior lecturer at Universitas Al Asyariah Mandar, located in West Sulawesi, Indonesia. Global recognition acknowledges her extensive contributions as a lecturer, an international speaker, and an accomplished leader in the academic field. She has published international books with renowned publishers such as Springer, Emerald, Wiley, IGI Global, and Taylor and Francis. She also serves as editor and reviewer in reputable journal, serves an international board member for various scientific innovation research groups (SIRG, Egypt), holds adjunct faculty positions at prominent institutions such as MAAUN Nigeria, Buriram Rajabhat University (BRU) Thailand, and Manipal Global Ntx Malaysia, and is the country director of some international organizations and achieved over 100 international awards. She actively engages in global volunteerism. Her scholarly interests encompass a wide range of topics, including literary development, instructional teaching materials, assurance of learning (ASSURE), Teaching English to Speakers of Other Languages (TESOL), ICT in education, digital education, generative AI (GenAI), and cybergogy. She serves as an external examiner for PhD candidates at A+ accredited universities in Florida (USA), Cyprus (Europe), Australia, India, and Malaysia. Her commitment to international collaboration and teamwork underscores her contributions to academia and leadership. Her exemplary career reflects her dedication to advancing education, fostering collaboration, and shaping the future of learning through innovative and inclusive approaches.

AI applications in Smart City Planning and Management

Blockchain Technology for Collaborative Research – A Qualitative Analysis

Madhumita Das(✉), Pravitha Vijaykumar, and Mamata Bhandar

Manipal GlobalNxt University, Kuala Lumpur, Malaysia
{MD18687,PV18764}@campus.globalnxt.edu.my,
mamata.bhandar@globalnxt.edu.my

Abstract. Blockchain Technology (BCT), introduced with Bitcoin in 2008, has use cases in various organizations such as banking, supply chains, financial markets, cybersecurity, and healthcare. Over time, these use cases have increased due to BCT's decentralized nature, elimination of third-party intermediaries, immutability, and transparency. This study aimed to investigate the possible synergy between Blockchain and Collaborative Research through the lens of researchers (users) working in collaborative research environments to determine how Blockchain can facilitate research collaboration in their respective organizations by addressing challenges associated with protecting intellectual property rights (IPR) and research data. The study used a qualitative methodology to understand researchers' perceptions of how BCT can address the identified challenges by providing a secure, transparent, and decentralized framework. The study revealed that most participants understood BCT and perceived it beneficial for resolving collaborative research's IPR and data challenges. The findings predominantly emphasize the need for increased practical use cases and Blockchain literacy. The study offers valuable insights and recommendations for organizations that intend to adopt BCT for collaborative research.

Keywords: Blockchain Technology · Collaborative Research · Intellectual Property · Security

1 Introduction

1.1 Background

Collaborative research brings together researchers from various fields of expertise and organisations worldwide, thereby creating an environment that enhances the productivity of the research. Collaboration in research is important to solve complex problems that surpass individual capabilities. This can also foster creativity and innovative ways of tackling challenging problems. The more international a project is the greater its potential impact and the greater the number of funding opportunities. The more international a project is, the greater its potential impact, attracting interest from various institutions, governments, companies, and associations for funding. Researchers face many challenges in collaborative projects: cultural and language differences, funding and resource limitations, and geographic distances (for multi-institutional or international collaborative research).

A. J. Obaid et al. (Eds.): ICONEST 2024, ISEM 40, pp. 3–16, 2025.
https://doi.org/10.1007/978-3-031-89960-7_1

Secure storage, research data access, and IPR protection (Lakhan et al., 2008) are notable issues in collaborative research. Cloud service providers charge hefty egress fees for transferring data from the initial cloud storage. The growth of the Internet has made the protection of intellectual property rights even more important since it has raised questions on established notions of copyright. Online tools make it easier for content to be copied at a low cost and shared globally, leading to intellectual property creators' rights violations. Preventing unauthorized replication while maintaining accessibility to these resources is one of the most debatable discussions in Internet regulation today (Alguliyev & Mahmudov, 2015).

Blockchain is a decentralised database with a distributed architecture and provides a cryptographically secure storage solution. The network maintains a chain of ordered records called blocks, linked using cryptographic techniques. Each block in the Blockchain network has a cryptographic hash of the previous block, the timestamp and the transaction data. The transactions are recorded across many computers, ensuring the records are unalterable without the consensus of the entire Blockchain network (Hughes et al., 2019). Academic institutions worldwide are increasingly adopting BCT to manage extensive research data. The technology ensures data integrity for large-scale research data. Many academic institutions have been adopting BCT as a solution for their various requirements, primarily due to the advancements in cost-effective and secure decentralised storage solutions for managing large volumes of critical research data. BCT mitigates vulnerabilities and inefficiencies inherent in centralised storage models, such as the risk of data loss and the high costs associated with restricted access and egress fees. (How Universities Can Use Blockchain to Transform Research, 2024). Considering the developments and trends observed in 2024, it is increasingly evident that BCT is a fundamental element of our digital future because of its transformative potential to offer more secure, transparent, and efficient solutions.

1.2 Problem Statement

Researchers face multiple challenges when they work together on collaborative projects. The significant challenges are ensuring privacy, security, confidentiality, and storage of large volumes of research data and intellectual property (IP) protection (Research Ethics & Data Protection, n.d.). Interoperability, privacy, traceability (reinforcing transparency), and trust should be guaranteed among the collaborators since that allows for effective collaborations among the diverse research community. Protection of IPR is essential as it ensures that the person or the organization is the creator and owner of a particular product. Despite BCT's potential to enhance trust and transparency in collaborative work scenarios (Coelho et al., 2022), limited studies have been conducted to understand how the research community perceives the role of this technology in facilitating collaborative research work and mitigating the current challenges facing them. This study aimed to bridge this gap by shedding light on the role of BCT in supporting secure research collaboration.

Research Objectives (RO) and Research Questions (RQ)

The study has two primary research objectives (RO1 and RO2), with RO1 being mapped to the two research questions (RQ1 and RQ2) and RO2 mapping to RQ3. RO1 explores the challenges researchers face in collaborative research environments in protecting IPR and research data while understanding their perceptions of how BCT can fit into this ecosystem by providing a secure, transparent, decentralized collaboration framework. RO2 expands on the researchers' perceptions to understand the factors facilitating BCT adoption in the research community.

The linking of research objectives (RO1, RO2) to the research questions (RQ1, RQ2 and RQ3) is given in Table 1 below:

Table 1. Linking Research Objectives to Research Questions

RESEARCH OBJECTIVES	RESEARCH QUESTIONS
RO1: To investigate the challenges in the current state of collaborative research, particularly in protecting IPR and research data, and to understand researchers' perceptions on how BCT can address these issues by providing a secure, transparent and decentralised framework for collaboration	**RQ1:** What challenges do researchers face in collaborative research projects in protecting IPR and research data? **RQ2:** What are researchers' perceptions regarding using BCT in facilitating collaborative research projects?
RO2: To understand the enabling factors for BCT adoption in the research community	**RQ3:** What are the enabling factors for adopting BCT in collaborative research projects?

2 Literature Review

2.1 Blockchain Technology (BCT)

BCT is a decentralised and distributed public digital ledger system. It consists of individual records, known as "blocks". Each block consists of a cryptographic hash of the previous block, a timestamp, and transaction data. The validity of each transaction recorded on the network is ensured through a strong verification process. A very useful feature of Blockchain is its ability to create trust between different parties, thereby enabling different parties to engage in various business activities without the need for any intermediaries. BCT was launched in 2008, initially within the cryptography expert community, and now plays a significant role in international trade and businesses. Blockchain is among the various emerging innovative technologies prominent in many business aspects. This technology became famous as the technology behind one of the world's leading Cryptocurrencies, Bitcoin. The idea behind the technology was initially conceptualized by an anonymous individual operating with the pseudonym Satoshi Nakamoto. One year later, in 2009, it became an integral part of the cryptocurrency Bitcoin. Blockchain is the technology that empowers Bitcoin, serving as its virtual infrastructure (Nakamoto, 2008). In 2013, BCT began to gain recognition and was being used in other cryptocurrencies like Ethereum.

Blockchain architecture is the organisational framework of a peer-to-peer (P2P) computer network consisting of computers or servers called "nodes" that act as the backend for applications and other systems. The transactions on the Blockchain network are validated and approved before being added to the ledger (Morkunas et al., 2019). The network does not have a centralised entity to control the interaction between the nodes for authentication, such as banks for transactions involving finance; it is decentralised – although in recent years, 'private' Blockchains have emerged that give a finer grain of control over the system. Based on Blockchain's consensus protocol, data uploaded to Blockchain is shared with all network users and is subject to verification and validation by anybody with the necessary rights.

Data entered into the Blockchain is "hashed" or converted into a new format of a fixed length using a mathematical function to ensure data integrity and confidentiality. If the sender of the transaction does not want other network users to be able to read the message's content, which is the plaintext data contained in the submitted documents, they can encrypt the message to make the data unintelligible to those without authorised access.

Once transactions have been verified, they are kept in "blocks" and "chained" (in chronological order) to one another using cryptographic techniques.

2.2 Current State of Collaborative Research and How Blockchain Might Fit into This Ecosystem

Collaborations within academic organisations and industries or between educational institutions and organisations are commonplace to foster efficient problem-solving. Since collaborative research relies significantly on effective communication, the results and processes must be validated by all those involved in the task. This practice ensures the data's accuracy, validity, and reproducibility (Bote et al., 2012). Collaborative research projects often involve extensive research data being generated, which has increased emphasis on optimal ways of storing this data, sharing and accessing the data quickly, securely and transparently from anywhere in the world and reproducing the data whenever needed. This, in turn, leads organisations and academic institutions to rethink innovative and secure research data management (RDM) practices (Subaveerapandiyan, 2023) that constitute the collection, organisation, storage, and documentation of research data throughout and beyond the duration of a research project. Research organisations are increasingly mandating robust data management plans to ensure that all aspects of data handling are taken care of from the beginning of any research activity (Research Data Management, n.d.).

Cloud storage today relies heavily on large, untrusted third-party providers to handle data storage, transmission, and retrieval. High costs, software quality issues, and data security concerns are notable challenges facing these systems today (Desai et al., 2019). BCT is a viable option in such scenarios because of its decentralized, tamper-proof, transparent and secure nature. Intellectual Property infringement is another common problem in research, especially when multiple collaborators are involved in a specific research project. Disputes may arise related to the ownership of the IP and using it to benefit all the joint collaborators. Protecting the IPR is challenging, especially in today's ongoing technological and Internet developments. Blockchain's impact is gradually felt across various creative industries in its ability to manage and protect IPR from unfair use by others. Blockchain provides an immutable, secure database perfect for tracking IP rights' status and usage in real time and managing them. Companies like Mediachain Lab and BlockAI use Blockchain to safeguard copyrights, allow owners to prove ownership, and detect infringements. (Blockchain Technology: Impact on Intellectual Property, 2024).

Although many studies have been done separately on collaborative ecosystems, researchers' challenges and Blockchains' advantages, limited literature deals with the synergy of "collaborative research" and "Blockchain Technology" from the perspectives of researchers from academic institutions and organisations involved in research collaboration. This study aims to fill this research gap and contribute to the ever-expanding world of digital technology benefits and application areas.

3 Research Methodology

The study adopted a qualitative research design to understand the challenges researchers face in collaborative research projects in protecting IP Rights from unauthorized use, maintaining data integrity and security, and exploring researchers' perceptions of the potential of BCT in addressing the identified challenges. A technology acceptance model for Blockchain in collaborative research has been developed based on Davis's technology acceptance model (TAM). TAM is a three-stage process whereby a system's design features stimulate the cognitive responses (perceived ease of use and perceived usefulness), influencing the behavioural intention of using the system in reality (Davis, 1989). Themes and patterns related to researchers' behavioural intention to use, attitude, perceived usefulness, and perceived ease of use of BCT were gradually uncovered through the inductive coding analysis process.

3.1 Data Collection and Analysis

The authors focussed on collecting primary data through an online survey consisting of open-ended questions because of its openness and flexibility until the saturation point was reached (usually 20–30, though a minimum of 10 is acceptable, as stated in the literature) (InterQ Research, 2018). For this study, data was collected from 10 respondents who met the criteria set for the target audience: years of research experience working on collaborative research projects in organisations or academic institutions. The survey link was sent to the target group that met the study's requirements. Purposive sampling was used to obtain accurate information to meet the study's objectives. A total of 10 respondents participated in the survey. The respondents were Doctoral students involved in various research projects (not necessarily collaborative projects) either in their respective organizations or as part of their doctoral journey. Table 2 provides information on the participants' research areas.

Data from the survey questionnaire was analysed in ATLAS.ti, a qualitative analysis software that uses thematic analysis to identify and code critical themes and patterns in the respondents' data. Table 3 shows the coding framework used.

Table 2. Research Areas of the Participants

Crisis Management and entrepreneurship
Media and Education
Resource Management
Education
Disaster Resilience
Role of leadership in sustainability management
DEIB and AI
Digital Transformation
Supply Chain, finance, management

Table 3. Coding Framework

Research Questions	Categories/Code Groups	Codes	Themes
RQ1: What challenges do researchers face in collaborative research projects in protecting IPR and research data?	Collaborative Research Challenges	IP ownership	Challenges in IP Rights Protection and Data Security
		IP confidentiality	
		Maintaining data confidentiality	
		Joint ownership of patents	
		Maintaining data integrity	
		Indigenous Intellectual Property Management	
		Secure Data Sharing	
RQ2: What are researchers' perceptions regarding using BCT in facilitating collaborative research projects?	Perceived Usefulness	Enhanced data security	Perceptions of Researchers (Users)
		Transparency	
		Data Sharing	
		Increased trust	
		Increased traceability	
	Perceived Ease of Use	Complex	

(continued)

Table 3. (*continued*)

Research Questions	Categories/Code Groups	Codes	Themes
	Perceived Challenges	Scalability	
		Interoperability with legacy systems	
		High Energy Demand	
		High Costs	
RQ3: What are the determinants for adopting BCT in collaborative research projects?	BCT Determinants	Blockchain Literacy	Determinants of Blockchain Adoption (DoBA)
		Cost	
		Complexity	
		Data Security	
		Seamless Integration	
		User-friendly Interface	
		Enhanced Tracking and Traceability	
		Scalability	

4 Findings and Discussions

The findings show that 20% of the respondents had 15–20 years of research experience, while another 20% had 10–15 years. 50% of the participants were between 5 and 10 years, and 10% had less than five years. This was important to know, as one of the study's main objectives was to understand the perceptions of researchers who are or have been involved in research in various stages of their careers and have a good number of years of research experience to answer the survey questions. Researchers with more years of experience can offer deeper insights into their current challenges in securing research data and protecting IP ownership. The main findings of the study are presented in the following sections.

4.1 FINDINGS (for RO1)

The first research objective was *to investigate the challenges in the current state of collaborative research, particularly in protecting IPR and research data, and to understand researchers' perceptions of how BCT can address these issues by providing a secure, transparent and decentralised framework for collaboration.*

RQ1: Researchers' Main Challenges in Collaborative Research

The findings revealed that the primary concerns in collaborative research environments revolved around **Intellectual Property (IP)** and **data issues**. The most significant concerns are in "IP Ownership" and "Maintaining Data Integrity", followed by "IP Confidentiality" and "Maintaining Data Confidentiality". Additional challenges include Indigenous Intellectual Property Management, Joint Ownership of Patents, and Secure Data

Sharing. One participant described his challenge: "*Determining which author/institution owns the IP*". Another participant said," "*My current research is looking into knowledge management theory for indigenous knowledge. The field of indigenous knowledge has many complexities, including the use of knowledge and its IP. It is something that concerns me hugely, especially with AI being able to assimilate indigenous knowledge and seemingly interpret it out of context of deeper indigenous values.*"

These challenges may arise from several factors, mainly because collaborative research involves multiple stakeholders, such as industries, academic institutions, and government organizations, each of whom may have different interests and policies concerning IP ownership, data usage, and confidentiality. There can be several legal and ethical considerations across different jurisdictions regarding IP ownership, making it challenging to establish clear ownership of the IP generated from the collaborative work. This can lead to delays in collaboration and restrict knowledge sharing. Again, when multiple organizations are involved, securely sharing data, mainly sensitive and proprietary information, is a challenge because of acts of data breach, which are commonplace nowadays in our ever-connected world. IPR protection and data security are crucial for ensuring trust and smooth collaboration in research partnerships. Without addressing these challenges, collaborative research efforts can face significant setbacks. Blockchain is based on cryptographic principles, a tamper-resistant platform that can provide indisputable ownership proof of any invention and allow for easy status and usage tracking, thus eliminating the risks of IP infringement such as in copyrights, trademarks or patents (Blockchain Technology: Impact on Intellectual Property, 2024).

The authors felt it necessary to know how much the researchers knew about Blockchain to understand how BCT can significantly mitigate the above-identified challenges. The participants were asked to explain their understanding of BCT in their own words.

Seven of the respondents demonstrated a basic to advanced level understanding of the technology in their own words, while the remaining three respondents' understanding was very limited. It was also found that the participants had limited exposure to BCT. However, they know a certain level of the technology, its history, and its potential applications. One participant said Blockchain is: "*A secure digital database that orders/records data*". Another said, "*Blockchain technology is a decentralized digital ledger that records transactions across multiple computers to ensure security, transparency, and immutability. Each block contains a list of transactions, and once added to the chain, it cannot be altered without altering all subsequent blocks, making it highly secure. Blockchain's decentralized nature eliminates the need for intermediaries, reduces fraud risk, and enhances data integrity. It has applications across various industries, including finance, supply chain, and healthcare, by enabling secure and transparent data sharing and transaction processing.*"

The responses can be interpreted to conclude that 70% of the participants knew what the main essence of BCT is – a secure, transparent, immutable, decentralized digital system for storing and managing data with wide-ranging applications in various industries.

The word cloud generated by the software in Fig. 1 below shows the key concepts in the participants' understanding of BCT: *data, secure, transparent, decentralized, ledger, records, and transaction.*

Fig. 1. Word cloud depicting the frequency of keywords in the respondents' data.

This shows that the participants are all familiar with BCT's technical terms and primary benefits. Terms such as *technology, ledgers, transactions, and records* suggest participants' awareness of Blockchain's role in securely storing information across the network. Words like *encryption, immutability, and safety* indicate participants' recognition of Blockchain's capabilities in enhancing data protection, preventing fraud, and reducing transaction risks. The direct quotes of the participants backed by the word cloud reveal that the participants view Blockchain as a secure, transparent, and decentralized ledger for storing data.

RQ2: Researchers' Perceptions of BCT for Collaborative Research Projects

The researchers' perceptions were analyzed from three perspectives: ***Perceived Usefulness, Perceived Ease of Use and Challenges of Blockchain integration in Collaborative Research.***

Analyzing the qualitative data on the **perceived usefulness of Blockchain for collaborative research**, findings reveal that 80% of the participants perceive that the technology can benefit collaborative research in various impactful ways and can significantly mitigate the challenges previously identified in RQ1 related to IPR and data. Table 4 summarizes and explains the codes generated for the "*perceived usefulness*" of the technology for collaborative research from the respondents' point of view.

Although in the current state of work, they use various mitigation strategies such as data encryption, access controls, role creations, password protections, and legal agreements to address the challenges they face, they perceive that Blockchain will be able to provide enhanced data security and sharing, transparency, better protection of IPR due to its enhanced tracking and traceability features thereby fostering mutual trust

Table 4. Perceived Usefulness Codes Description of BCT for Collaborative Research

"Perceived Usefulness" Codes	Explanation
Enhanced Authenticity, Security and Integrity of Research Data	BCT can ensure that the research data's authenticity, security, and integrity are reliable
Anti-hacking and anti-theft	BCT can protect the research data from hacking and other cyber threats
Enhanced Tracking and Traceability of research information	BCT can track and trace the flow of research data from one point to another on the network
Increased transparency and trust among collaborators	BCT can provide transparency and build mutual trust in a collaborative research network
More streamlined data sharing	BCT can facilitate the smooth sharing of research data among collaborators
Restricted accessibility to authorized collaborators	BCT can ensure that the research data is accessible only to authorized people
Accurate and tamper-resistant data	BCT can guarantee that the research data remains accurate and unaltered over time
Better control of IP rights and ownership	BCT can improve the management and protection of IP rights and ownership
Increased verifiability and credibility of research findings	BCT can strengthen the credibility and validity of the research findings
Efficient and reliable Collaborative Research environment	BCT can support a better, healthier and more trustworthy research environment

among the collaborators. Respondents' **perceptions on the ease of use of Blockchain for collaborative work**, however, reveal that it is currently perceived as 'low' as some participants believe that Blockchain is technically complex. The codes obtained on the **researchers' perceptions of challenges for Blockchain integration in collaborative research ecosystems** are 'scalability', 'interoperability with legacy systems', 'high cost', and 'higher energy demand'.

The study's findings on the perceived usefulness of BCT for collaborative research justify that Blockchain can be a potential solution for the IPR and data challenges faced in collaborative research scenarios. This is mainly due to its inherent security, transparency, and immutability characteristics. As more and more organizations and academic institutions involved in diverse research projects are increasingly shifting their focus towards technology-enabled research data management plans (Subaveerapandiyan, 2023), Blockchain has the potential to revolutionize the collaborative research ecosystem. However, to successfully adopt the technology, it should be made less complicated for the research community, and any perceived challenges should be addressed in the most efficient manner possible as and when they arise.

4.2 FINDINGS (for RO2)

The second research objective was *to determine the enabling factors for adopting BCT in collaborative research projects.*

RQ3: Determinants for Blockchain Adoption in Collaborative Research Projects
Figure 2 depicts the framework for adopting Blockchain in collaborative research focusing on key enablers, perceived usefulness and ease of use, and moderating variables that influence system use.

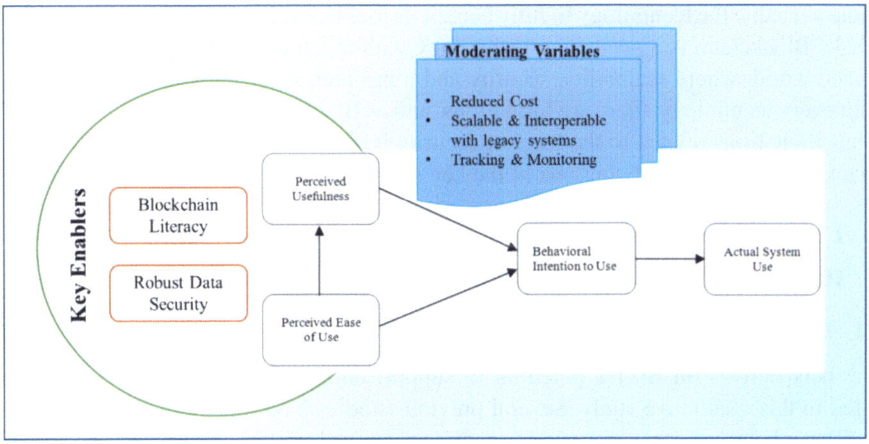

Fig. 2. Framework for Blockchain Adoption in Collaborative Research

Analyzing the data on this aspect indicates that ***Blockchain Literacy*** and a robust ***Data Security*** level are the key enablers for the technology's adoption among the research community. Awareness and understanding of BCT among users are crucial for adopting the technology in the research community. The higher the level of Blockchain literacy, the greater the perceived ease of use and usefulness. The second enabler is the level of security that BCT can provide for the research data. This is also crucial for BCT's adoption in a collaborative environment. The two factors of ***Perceived Usefulness*** and ***Perceived Ease of Use*** represent how users perceive BCT in terms of its *usefulness*, such as efficiency, transparency, security and *ease of use*, meaning how simple or complex they find BCT to implement and use. Both these factors directly contribute to the ***behavioural intention to use*** BCT in the collaborative research ecosystem, which can be linked to the findings of TAM. The moderating variables in this framework are ***reduced cost, scalable and interoperable with legacy systems,*** and ***tracking and monitoring.*** **Cost-effectiveness** can significantly influence the decision to adopt the Blockchain, especially in limited-funding research projects. The ability of Blockchain to scale and integrate with existing systems will increase its chances of adoption, which would create limited disruptions to the existing infrastructure. To what extent can Blockchain efficiently track and monitor research data flow in the network is also a determining factor for its adoption. All these factors, if favourable, will determine the willingness to use BCT (***behavioural***

intention to use) for collaborative research. The outcome is the ***actual system use.*** When the behavioural intention is strong, it is expected to lead to the full implementation and use of Blockchain to secure and manage research data, ensure IPR protection, and facilitate transparent and trusted collaboration.

The study participants have also indicated that although the technology is helpful for collaborative environments, they currently view it as technically complex with high involved costs. While the respondents rated this technology's perceived usefulness highly, as found in the current study, the perceived ease of use remains relatively low among respondents because of its associated complexity factor. The challenges they perceive with Blockchain integration are justified and must be explored further and overcome to enable the technology to fully benefit its users, developers and businesses (Marr, 2023). Blockchain is a promising technology with much potential in today's interconnected world, where scalability, security and transparency are paramount. However, as with every technology, there are limitations, and as Blockchain is still in its early stages, it has limitations related to the level of security it can provide against various forms of attacks despite security being one of the key strengths of the technology.

5 Conclusion and Recommendations for Organizations Looking to Adopt Blockchain for Collaborative Research

5.1 Conclusion

The perspectives on BCT's potential to support collaborative research were investigated in this qualitative study. Several previous studies have shown extensive research on Blockchain use cases in various domains, such as logistics, healthcare, IoT, digital identities, finance and copyrights, to name a few. This paper reviews a new possible use case of Blockchain for Collaborative Research environments through the lens of the study's participants in this field. BCT's potential to enhance data security, trust and transparency aligns with the growing trend of using various emerging technologies, such as Artificial Intelligence and Machine Learning, in collaborative research. Furthermore, the study also discusses the socio-economic drivers behind BCT's adoption, such as IPR protection and facilitating cross-border collaborations. While BCT does have potential, its adoption in this domain will require a thorough evaluation of the scope, number of collaborators, the digital maturity of the participating organizations, decentralization complexities, and finally, the realization of actual business values. The study investigated participants' strategies for mitigating challenges in collaborative research projects through open-ended questions in an online survey distributed to the research community. According to the study, BCT is perceived as promising in improving data security, accessibility, sharing, and IP rights protection among partners. As a result of its perceived complexity, low awareness levels, and people's limited exposure to it, collaborative research environments— one of the technology's promising fields—have not yet seen much success with it.

5.2 Recommendations

As found in the study, researchers perceive Blockchain to be complex. Hence, recommendations for stakeholders planning to adopt Blockchain for collaborative research work should include developing **Blockchain literacy** and **spreading awareness** through workshops, training programs, webinars, and online courses within the research community. These should be customised to the needs of academics and policymakers. Technology providers should allocate more resources to address scalability in large collaborative projects and integration challenges with existing platforms towards developing **easy-to-use** and **intuitive** Blockchain platforms and tools suitable for collaborative research environments. Pilot projects can be initiated and demonstrated to showcase the practical benefits of Blockchain in real-world research scenarios and help build confidence among researchers by providing specific examples of enhanced data sharing, security, and IP rights protection.

6 Limitations, Implications and Scope of Future Research

One of the main limitations of the study was that the participants lacked prior experience with BCT, which may have skewed the findings toward more positive perceptions. As purposive sampling was used for the study, it was ensured that even if the respondents had no experience with this specific technology, they all had good exposure to the world of technology and, more importantly, to research fields. This study contributes significantly to the growing research on technology adoption in the academic and organisational research ecosystem.

Further research can be done on the regulatory and legal implications of using Blockchain, focusing on compliance with existing laws and regulations. Longitudinal studies can be conducted to investigate the long-term impact of BCT on facilitating collaborative research, data security, and IPR protection.

References

Alguliyev, R., Mahmudov, R.: Problems of protection of intellectual property rights in the information society. Informasiya Cəmiyyəti Problemləri **06**(2), 4–12 (2015). https://doi.org/10.25045/jpis.v06.i2.01

Blockchain Technology: Impact on Intellectual Property. (2024, March 27). https://abounaja.com/blogs/blockchain-technology-and-intellectual-property. Retrieved June 21, 2024, from https://abounaja.com/blogs/blockchain-technology-and-intellectual-property

Bote, V.P.G., Olmeda-Gómez, C., De Moya-Anegón, F.: Quantifying the benefits of international scientific collaboration. J. Am. Soc. Inform. Sci. Technol. **64**(2), 392–404 (2012). https://doi.org/10.1002/asi.22754

Coelho, R., Braga, R., David, J.M.N., Stroele, V., Campos, F., Dantas, M.: A blockchain-based architecture for trust in collaborative scientific experimentation. J. Grid Comput. **20**(4) (2022). https://doi.org/10.1007/s10723-022-09626-x

Davis, F.D.: Perceived usefulness, perceived ease of use, and user acceptance of information technology. MIS Q. **13**(3), 319 (1989). https://doi.org/10.2307/249008

Desai, S., Deshmukh, O., Shelke, R., Choudhary, H., Sambhare, S.S., Yadav, A.: Blockchain based Secure Data Storage and Access Control System using Cloud (2019). https://doi.org/10.1109/iccubea47591.2019.9129015

How universities can use Blockchain to transform research. World Economic Forum; Scott Doughman (2024). Retrieved May 31, 2024, from https://www.weforum.org/agenda/2024/03/higher-education-universities-blockchain-transform-research/#:~:text=Blockchain's%20role%20in%20empowering%20scientific%20discovery&text=Moreover%2C%20blockchain%2Dpowered%20data%20storage,peer%20review%20and%20collaborative%20studies
Hughes, L., Dwivedi, Y.K., Misra, S.K., Rana, N.P., Raghavan, V., Akella, V.: Blockchain research, practice and policy: applications, benefits, limitations, emerging research themes and research agenda. Int. J. Inf. Manage. **49**, 114–129 (2019). https://doi.org/10.1016/j.ijinfomgt.2019.02.005
InterQ Research. Determining sample size for qualitative research: What is the magical number? InterQ Research (2018). https://interq-research.com/determining-sample-size-for-qualitative-research-what-is-the-magical-number/
Lakhan, S., Khurana, M., Null, N., Null, N.: Intellectual property, copyright, and fair use in education. Academic Leadership (2008). https://doi.org/10.58809/bxyd1155
Marr, B.: The 5 Biggest Problems With Blockchain Technology Everyone Must Know About. Forbes (2023). https://www.forbes.com/sites/bernardmarr/2023/04/14/the-5-biggest-problems-with-blockchain-technology-everyone-must-know-about/
Morkunas, V.J., Paschen, J., Boon, E.: How blockchain technologies impact your business model. Bus. Horiz. **62**(3), 295–306 (2019). https://doi.org/10.1016/j.bushor.2019.01.009
Nakamoto, S.: Bitcoin: A peer-to-peer electronic cash system (2008). Retrieved from https://bitcoin.org/bitcoin.pdf
Research Data Management. (n.d.). Science Europe. https://scienceeurope.org/our-priorities/research-data/research-data-management/
Research ethics and data protection. (n.d.). https://www.reading.ac.uk/research-services/research-data-management/data-management-planning/research-ethics-and-data-protection
Subaveerapandiyan, A.: Research data management practices and challenges in academic libraries: A comprehensive review (2023). SSRN. https://ssrn.com/abstract=4515473 or https://doi.org/10.2139/ssrn.4515473

Novel Image Quality Index: Integrating UQI and GSI for Enhanced Quality Assessment

Noor Abd Alrazak Shnain(✉) and Mohammed Abdulameer Aljanabi

University of Kufa, Najaf 54000, Iraq
noora.aljanabi@uokufa.edu.iq

Abstract. Assessing image or image quality is extremely crucial in various applications in the fields of image processing and computer vision. Numerous methods have been presented to assess the quality of images by comparing between the reference and distorted images. However, most of the previous metrics presented in this field produced a significant amount of correspondence between the reference image and the distorted images. In view of these challenges, we would like to put forward the proposal for a new comprehensive index that gauges the quality of an image with regard to gradient information and statistical coefficients. This index fuses the Gradient Similarity Index (GSI) with Universal Quality Index (UQI) and the gradient-based edge detection method of Sobel. We evaluated the performance of our approach using two well-known databases the LIVE and CSIQ and the results indicate that the proposed approach outperforms the (GSI) and (UQI) even under distortion and noise.

Keywords: Image Quality · Gradient Similarity Index (GSI) · Universal Quality Index (UQI)

1 Introduction

Image quality Indices are quantifiable tools that describes the quality of digital images. These Indices have significant applications in many fields: computer vision, image processing, and multimedia, and human-computer interaction. Image quality assessment can be done by measuring the perfection of information originally preserved by an image in comparison to its reference or ground-truth image. Image quality Indices have become very significant in view of the growing use of digital images in myriad applications, from simple compression and network transmission to sophisticated image enhancement. Compression artifacts, noise, blurring, and other defects are some of the different distortions these images can undergo during these processes. Characterization and quantification of their effect on perceived image quality form an integral part of creating both robust and efficient image processing algorithms. Image assessment has a lot of useful applications in practice, and the matching metrics would play a very important role in digital image processing. The main purpose of measuring image matching is to develop methods for quality assessment that can go beyond subjectivity in human perception of image quality excellence [1, 2].

A. J. Obaid et al. (Eds.): ICONEST 2024, ISEM 40, pp. 17–26, 2025.
https://doi.org/10.1007/978-3-031-89960-7_2

Comparing digital images for matching is a very well-known task in digital image analysis. These comparisons may, however be restricted to parts of an image. Image quality Indices give quantitative indications of the matching between parts and/or the whole of images. They provide a basis for registration processes, giving information needed to tell if the registration is going in the right direction. Many image matching measures have been put forward for both the medical image field and the computer vision community. Images matching has broad applications in visual search, duplicate product detection, and domain-specific image clustering. It could play an important role in many real-world digital image processing tasks from enhancing to compressing and then restoring them. Ultimately, it means developing objective methods for quality assessment that shall conform with the subjective human judgments [3–5]. Multiple challenges are involved in the analysis of image matching, where some of the more important are reasonably well researched while others are relatively unsolved. Of these particular challenges, designing models of human visual processing that can cope with natural images is, itself a large challenge. Also, processing by higher-level visual areas is essential to be included [6].

Another challenge is the development of algorithms accommodating a few image distortions. Handling perceptually additive versus object-based image distortions requires adaptive visual methodology, which humans use in judging quality and higher-level effects. Simultaneous image distortion is very interesting because when combined together, one must consider possible perceptual interactions between the various kinds of distortions and how these jointly affect the quality of the image. One of the specific challenges is geometric changes in images, which happens to be very difficult for image matching algorithms. Researchers are working continuously to improve and refine their approaches of image matching to overcome such complex problems [7, 8].

This work aims to add a new index to assess image quality that combines the characteristics of conventional indices and relies on more than one feature to solve the shortcomings found in previous indices. The recently proposed index achieves a pleasing balance in performance for similar and different images by combining the advantages of the GSI index and UQI index, these characteristics provide the suggested index remarkable robustness, even in noise or blurred environments.

2 Related Work

Since image quality is essential to both guaranteeing the greatest user experience in a variety of image-related applications and refining image processing algorithms, it is still a subject of current study. To increase the precision and dependability of image quality, new and more complex image quality indices are constantly being developed as technology develops. The goal of image quality indices is to replicate as closely as possible the visual perception of humans. They seek to quantify the degree to which human viewers evaluate a processed or warped image more favourably than the original, unaltered form. Typically, these indices are created using mathematical models that account for a variety of aspects of visual perception, including texture, brightness, contrast, and spatial organization [9]. Over time, scientists have created a variety of image quality Indices, each with its strengths and limitations. A significant advancement in the scope of image quality metrics was introduced in 2004 with Structural Index, this metric relies on statistical resemblances between a pair of images [10].

A revised version of the structural index, known as the feature index for image quality evaluation, was proposed in 2011. The quality assessment of this approach is dependent on the use of low-level of image characteristics [11].

Hedlen et al. presented a novel no-reference image quality technique that extracts features from the frequency and spatial domains using a deep learning model. The model is demonstrated to perform better than current no-reference image quality evaluation techniques after being trained on a sizable dataset of human-rated images [12]. Li et al., presented a new open database for AI-generated image quality evaluation. The database contains over 3,000 images, and it is designed to be used for the development and evaluation of new image quality evaluation methods [13].

A novel scale-guided hypernetwork for blind super-resolution image quality evaluation was presented by Junfu et al. The purpose of the hypernetwork is to figure out how the scale factor and the perceived quality of super-resolved images relate to one another [14]. A novel optimization-based strategy was put out by Ziababnik et al. to enhance the effectiveness of facial image quality evaluation methods. The method's goal is to determine which characteristics and parameters work well together for a particular image quality evaluation procedure [15].

A new large-scale image quality database for artificial intelligence (AI)-generated images (AGIs) was introduced by Wang et al. The nearly 30,000 photos in the collection are intended to convey the caliber, veracity, and correspondence of AGIs as seen from the viewpoints of human observers [16].

This paper is divided into the following sections: The formulation of the suggested approach is described in the next section. After that, we present the experimental results, performance analysis, and discussions. Summarizing the findings from this study is the final step.

3 Image Quality Conventional Indices

3.1 University Quality Index

The Universal Quality Index (UQI) is an index used in assessing the quality of images. It mostly compares any deformed image against a reference image with respect to structural information and pixel intensity. The ranges of the UQI index are between 0 and 1, where 1 refers to perfect quality. In other words, the two images are defiantly similar. [17] The small values close to zero indicate that the two images are very different. The UQI index is mathematically defined by the following formula:

$$UQI(x1, x2) = \frac{(4 * m(x1) * m(x2) * Cov(x1, x2))}{(Var(x1) + Var(x2) + 2 * Cov(x1, x2))} \quad (1)$$

where $x1$ is an original image, $x2$ is distorted version of $x1$, Var is the variance of the original and distorted images, and m is the mean pixel intensities of the original and deformed images. Cov is the covariance of the original and distorted images.

3.2 Gradient Similarity Index (GSI)

The Gradient Similarity Index (GSI) is an index used to benchmark the quality of images. The basic idea behind GSI is that original images' gradient information (which shows variations in color or intensity) should be present in high-quality copies of the original image. In other words, gradation information of image would change as compared to the reference image if it undergoes noise, blur, or compression these are all characteristics synonymous with decreasing image quality. The range of GSI falls between 0 and 1, wherein 1 implies more quality and 0 means greater distortion. [18] The following formula mathematically defines the GSI index where $\times 1$ is an original image, $\times 2$ is distorted version of $\times 1$:

$$GSI(x1, x2) = mean(Csim)*mean(Qsim) \tag{2}$$

where *Csim* (represents the matching of the gradient magnitudes)

$$Csim = \frac{(2*mean(R*D) + c1)}{\left(mean\left(R^2\right) + mean\left(D^2\right) + c1\right)} \tag{3}$$

Qsim (represents the matching of the gradient orientations)

$$Qsim = \frac{(2*mean(\sqrt{(R*D)*\cos(\theta_r - \theta_d)}) + c1}{(mean\left(\sqrt{(R^2)} + mean\left(\sqrt{D^2}\right)\right) + c2} \tag{4}$$

The mean value is indicated by mean (). The gradient magnitude of the reference image is represented by R in the aforementioned equations, while the distorted image gradient is represented by *D*. Additionally, the gradient orientation of the reference image is represented by θ_r, and the small constants $c1$ and $c2$ are added to prevent division by zero.

3.3 Proposed Index

The more general area of image quality evaluation includes images matching because the matching indicates that the test image is of high quality since it contains almost all the information in the reference image, including density, color, and gradient. Different Indices have been formulated and are in practice to cope up with the complexities of image matching and object recognition within the images. All these Indices, despite good attempts, present challenges and demand more work and effort to be put in to formulate dependable algorithms to increase the success rate in image matching testing. One of the major problems in image matching is that high values of matching are often recorded between a reference image and other dissimilar images in a database under certain Indices.

To address this, we propose a novel approach called Universal Gradient Quality Index (UGQI), which is an index used to detect the quality of an image by measuring its gradient information and statistical coefficients. In proposed index several concepts are fused, which includes Gradient Similarity Index (GSI) which benchmarks the matching of distorted image gradient information with that of a reference image, Universal Quality Index (UQI) which relies on statistical coefficients to match between images, and Sobel edge detector that is gradient-based for edge detection of images which applies convolution to specified kernels to calculate the gradient in the horizontal and vertical directions separately, and then combines them to get the edge strength and orientation [19].

The proposed index is comprehensive and well developed because it involves several scales in the operation of finding similarities or identification of images. This brings together ideas, including the gradient-based statistics technique and then combines these gradients to determine the edge strength and orientation. A rational function of the proposed index as follows:

$$UGQI(x1, x2) = \frac{U(x1, x2).S(x1, x2) + \alpha}{G(x1, x2).S(x1, x2) + \beta} \quad (5)$$

where $UGQI(x1, x2)$ represents the proposed index image x and image $x2$; usually, $x1$ represents the reference image and $x2$ represents a corrupted version of $x1$; G represents the Gradient Similarity Index (GSI), U represents Universal quality index. While S represents Sobel correlation; α and β are small constants added to balance the quotient and avoid division by zero. The range of the proposed index $UGQI(x1, x2)$, like the principle of the conventional Indices, its range falls between 0 and 1where 1 indicates the highest image quality, while 0 indicates that the image quality is very low.

4 Result and Discussion

Most image quality systems do the comparison of a reference image with test images, usually stored in memory and forming a database. For our validation experiments, we used matlab r2019a to realize experiments on two well-known databases for performance evaluation of our algorithm.

1. Live database it is one of the most reputed databases used for assessment of image quality and was developed by the laboratory for image and video engineering, live. This database consists of 29 input images; these were distorted using the following distortion types: jpeg2000, jpeg, white noise in the rgb components, gaussian blur in the rgb components, and bit errors in jpeg2000 bit stream when transmitted over a simulated fast-fading rayleigh channel [20] (Fig. 1).

2. The CSIQ database is designed to evaluate how color saliency and spatial information would affect the performance of algorithms in image quality assessment. All the subjective quality scores of the distorted images are provided. The distorted versions include various types of distortions such as noise, blur, contrast changes, etc. [21] (Fig. 2).

In our experiments the new index has been implemented according to Eq. 5, the well-known Universal Quality Indices UQI according to Eq. 1 and GSI according to Eq. 2. All three methods have been implemented simultaneously on two datasets, LIVE and CSIQ.

Fig. 1. Sample of original images and distorted images in LIVE database

Fig. 2. Sample of original images and distorted images in CSIQ database

Figures 3 and 5 show the reference image and test (distorted) images from CSIQ and LIVE database which used for testing.

The performance of the proposed index versus GSI and UQI appears in Figs. 4 and 6. All Indices use the distance between the first and second maximum of the bars, from 0 to 1 where 1 refers to perfect quality. In other words, the distorted image has most features from the reference. While the small values close to zero indicate that the two images are very different. A greater distance thus means the index detected the image that matches the reference image and has the highest quality. Almost in all cases, the proposed index is keeping a larger difference between the reference image and other distorted images; it means obtaining better detection of the highest quality image among the distorted versions (Fig. 6).

The outcomes, as given in Table 1, depict that the average distance achieved by the proposed index is higher compared to the averages achieved by other Indices. From results, it was observed that proposed index performs better than classical GSI, UQI for detecting best quality image. This observation underscores the efficiency of the proposed index in assessing the image quality, demonstrating its superior capability compared to other measures.

Fig. 3. The original image and its distorted versions from CSIQ database

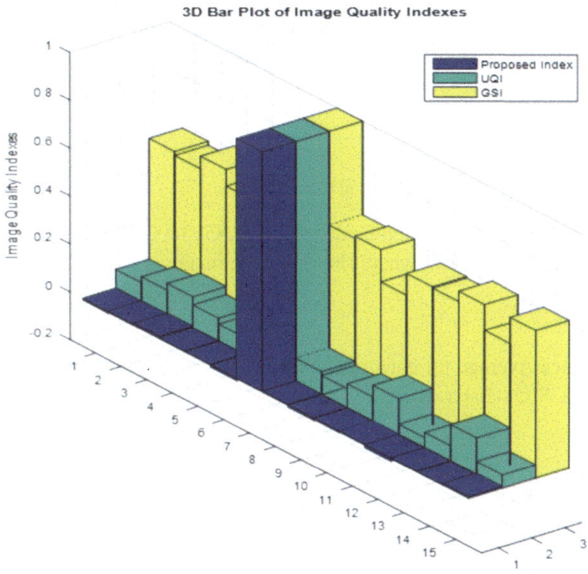

Fig. 4. Performance of UQI with matching distance = 0.70644, GSI = 0.57611 VS proposed index with matching distance = 0.97025 where the reference image is indicated in in Fig. 3 from CSIQ database

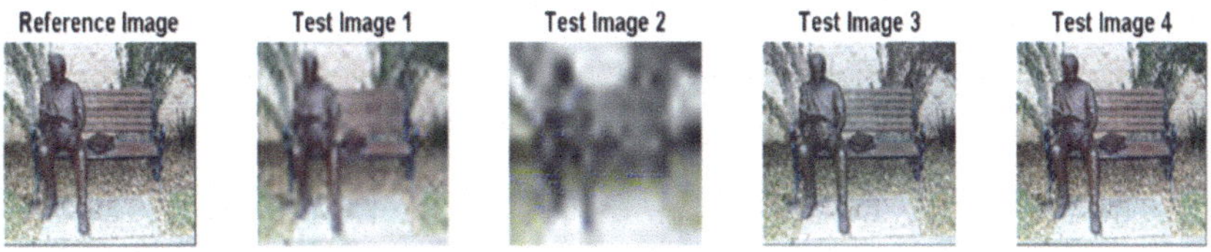

Fig. 5. The original image and its distorted versions from LIVE database

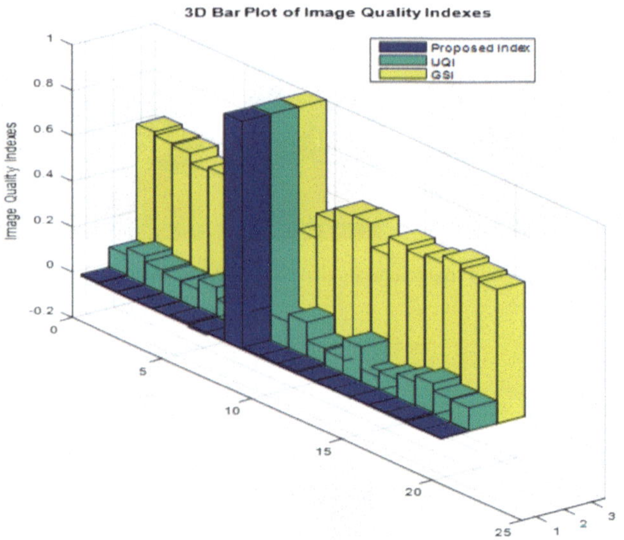

Fig. 6. Performance of UQI with matching distance = 0.85934, GSI = 0.7917 VS proposed index with matching distance = 0.99567 where the reference image is indicated in Fig. 5 from LIVE database

Table 1. The distance Average for each image quality index (UQI, GSI and the proposed index by using LIVE and CSIQ databases).

Distances Average	LIVE database	CSIQ database
Gradiente Index (GSI)	0.0535	0.0873
Universal Index (UQI)	0.7726	0.8851
Proposed Index (***UGQI***)	0.9917	0.9882

5 Conclusion

The proposed image quality assessment index incorporating GSI, UQI, and Sobel gradient-based edge detection methodology outperforms state-of-the-art methods. Our approach uses gradient information and statistical values to measure image quality against original reference image distortion. The results of the evaluation on the highly publicized LIVE and CSIQ databases prove the strength of our approach: our method outperformed existing quality metrics like GSI and UQI even when faced with a plethora of various kinds of distortions and noise. This already opens up a potential direction for our index to provide more reliable and comprehensive image quality assessments in many applications within image processing and computer vision.

Looking ahead, several avenues could be indicated for future work toward the enhancement and enlargement of the capabilities developed under the proposed method. One direction could be toward machine learning techniques specifically, deep learning for refinement in assessment and possible automation in the phase of feature extraction.

Another avenue could be to extend testing to more heterogeneous and challenging image databases, which would validate the generalizability and robustness of the index built for this purpose. Another line of interesting research is the application of our approach to real-time scenarios of image and video quality assessment. After all, computational efficiency and speed in this case become very critical. Finally, the study on the performance of the index with respect to different modalities in images like medical imaging or remote sensing is likely to reveal further practical applications and advantages. It is through the handling of these areas that accuracy, reliability, and versatility in image quality assessment methods can continue to be improved upon, finally leading to real progress in image processing and computer vision.

Acknowledgments. I would like to thank everyone who helped and motivated me to work on this paper.

References

1. Kastryulin, S., Zakirov, J., Pezzotti, N., Dylov, D.V.: Image quality assessment for magnetic resonance imaging. IEEE Access **8**(11), 14154–68 (2023)
2. Thung, K.H., Raveendran, P.: A survey of image quality measures. In: 2009 International Conference for Technical Postgraduates (TECHPOS), pp. 1–4. IEEE (2009)
3. Qaid, G.R.: Encryption and Decryption Image Using Multiobjective Soft computing algorithm
4. Ye, P., Doermann, D.: Document image quality assessment: a brief survey. In: 2013 12th International Conference on Document Analysis and Recognition, pp. 723–727. IEEE (2013)
5. Schlett, T., Rathgeb, C., Henniger, O., Galbally, J., Fierrez, J., Busch, C.: Face image quality assessment: a literature survey. ACM Comput. Surv. (CSUR) **54**(10s), 1–49 (2022)
6. Appalaraju, S., Chaoji, V.: Image similarity using deep CNN and curriculum learning. arXiv preprint arXiv:1709.08761 (2017)
7. Chandler, D.M.: Seven challenges in image quality assessment: past, present, and future research. International Scholarly Research Notices. 2013 (2013)
8. Shnain, N.A.A., et al.: High order statistic-zernike approach for image similarity and face recognition. Int. J. Comput. Sci. Inf. Secur. (IJCSIS) **16**(11) (2018)
9. Zhai, G., Min, X.: Perceptual image quality assessment: a survey. Sci. China Inf. Sci. **63**, 1–52 (2020)
10. Wang, Z., Bovik, A.C., Sheikh, H.R., Simoncelli, E.P.: Image quality assessment: from error visibility to structural similarity. IEEE Trans. Image Process. **13**(4), 600–12 (2004)
11. Zhang, L., Zhang, L., Mou, X., Zhang, D.: FSIM: a feature similarity index for image quality assessment. IEEE Trans. Image Process. **20**(8), 2378–86 (2011)
12. Li, X., Jiang, T., Fan, H., Liu, S.: SAM-IQA: Can Segment Anything Boost Image Quality Assessment?. arXiv preprint arXiv:2307.04455 (2023)
13. Li, C., et al.: AGIQA-3K: An Open Database for AI-Generated Image Quality Assessment. arXiv preprint arXiv:2306.04717 (2023)
14. Fu, J.: Scale Guided Hypernetwork for Blind Super-Resolution Image Quality Assessment. arXiv preprint arXiv:2306.02398 (2023)
15. Babnik, Ž., Damer, N., Štruc, V.: Optimization-based improvement of face image quality assessment techniques. In: 2023 11th International Workshop on Biometrics and Forensics (IWBF), pp. 1–6. IEEE (2023)

16. Wang, J., Duan, H., Liu, J., Chen, S., Min, X., Zhai, G.: AIGCIQA2023: A Large-scale Image Quality Assessment Database for AI Generated Images: from the Perspectives of Quality, Authenticity and Correspondence. arXiv preprint arXiv:2307.00211 (2023)
17. Wang, Z., Bovik, A.C.: A universal image quality index. IEEE Signal Process. Lett. **9**(3), 81–84 (2002)
18. Liu, A., Lin, W., Narwaria, M.: Image quality assessment based on gradient similarity. IEEE Trans. Image Process. **21**(4), 1500–12 (2011)
19. Zhang, C.C., Fang, J.D.: Edge detection based on improved sobel operator. In: 2016 International Conference on Computer Engineering and Information Systems, pp. 129–132. Atlantis Press (2016)
20. Seshadrinathan, K., Soundararajan, R., Bovik, A.C., Cormack, L.K.: Study of subjective and objective quality assessment of video. IEEE Trans. Image Process. **19**(6), 1427–41 (2010)
21. Larson, E.C., Chandler, D.M.: Most apparent distortion: full-reference image quality assessment and the role of strategy. J. Electron. Imaging **19**(1), 011006 (2010)

Detecting Cyber Attacks Using Deep Learning: An Empirical Study on Network Traffic Data

Fatima R. Hamade[1](✉), Ahmed J. Obaid[1], and Parul Agarwal[2]

[1] Faculty of Computer Science and Mathematics, University of Kufa, Najaf, Iraq
pagarwal@jamiahamdard.ac.in
[2] Jamia Hamdard, New Delhi, India
pagarwal@jamiahamdard.ac.in

Abstract. Cyberattacks have become a significant danger to digital data and infrastructure as digital network usage rises rapidly and reliance on technology spreads over all spheres. From distributed denial of service (DDoS) attacks to brute force attacks to long-term targeted attacks (APTs), these attacks call for creating intelligent and strong systems to identify these hazards properly. A hybrid model using LSTM and CNN models with an attention mechanism to improve the performance of detecting cyberattacks has been introduced. CNN model aims to extract spatial features from network data, and LSTM is used to check temporal patterns associated with attack activities. Attention mechanism has a significant role in the proposed model in classifying difficult attacks by improving the model's accuracy, which enhances concentration on the key aspects of the input data. The proposed hybrid model (CNN-LSTM-Attention mechanism) attained an accuracy of up to 98% despite significant difficulties in defining overlapping patterns. The results revealed the model's ability to reduce errors and increase attack detection. The proposed study indicates that increasing performance by employing more varied data and creating analytical methods involves attention to multi-head. This model is developed towards more intelligent and effective cyber security solutions.

Keywords: Cybersecurity · Deep Neural Networks · DDoS Attacks · Networks Analysis of traffic · Convolutional neural networks · Long short term memory

1 Introduction

Cyberattacks have become a serious threat to digital security. From basic approaches to sophisticated tactics aiming at sensitive networks and infrastructure, these attacks have developed from simple means to cause financial losses and a poor image of companies. A significant damage is mainly caused by distributed denial of service (DDoS) attacks, long-range targeted attacks (APTs), and data analysis techniques. Deep learning technology is being applied to find and stop cyberattacks to handle this problem. Promising outcomes have been obtained by using deep learning algorithms straight to network traffic data packets with an eye toward source and destination variables.

A. J. Obaid et al. (Eds.): ICONEST 2024, ISEM 40, pp. 27–41, 2025.
https://doi.org/10.1007/978-3-031-89960-7_3

Predicting and stopping cyberattacks has been found to depend mainly on the Convolutional Long Short-Term Memory Classifier method and further preprocessing including feature scaling (Zhang, et al., 2021). Historically, traditional systems such as statistical analysis and digital signature systems are limited, lack adaptability and flexibility, and have difficulty identifying new threats. Because deep learning techniques can handle large and complex data, cyberattacks have become more common, creating a need for more effective and efficient detection tools (Pang, Shen, Cao, & Hengel, 2021). Cybersecurity faces significant challenges due to the rapid and continuous evolution of cyberattacks such as malware, phishing and denial of service (DDOS). Deep neural networks are essential for identifying sophisticated cyberattacks and threats such as convolutional neural networks(DNNs), long-term memory(LSTMs), and recurrent neural networks (RNNs), where CNNs are used to evaluate conventional network traffic patterns and RNNs and LSTMs to examine temporal data according a recent investigation (Sarker, 2021). DNNs have ability in exact detection of cyberattacks and network traffic data analysis. in spite of LSTMs have the ability to processing temporal data and continues changes, CNNs stand out for their proficiency in gaining insights into fixed patterns. (Macas, Wu, & Fuertes, 2022). With an raise in cyber-attacks on Internet of Things (IoT) systems and traffic networks, the implementation of artificial intelligence, specially neural networks, has become significant for protecting data and reducing susceptibility in digital systems (Mendonça, et al., 2021).

The aim of this research is to demonstrate deep learning models to check network traffic data and evaluate their effectiveness in specifying various attacks.

The quick advancement of cyber threats is diminishing the efficiency of conventional systems. There is a growing demand for models that are capable of dealing with intricate time patterns and unforeseen attacks.

This paper has several important contributions. These include:

- Creating advanced deep neural network models to accurately and efficiently identify cyberattacks.
- It also involves a comparison of the effectiveness of various models, such as CNN and LSTM, based on factors such as accuracy, recall, precision and f1score
- Additionally, the study evaluates these models using actual datasets such as UNSW-NB15.

Utilizing deep learning methods to analyze network traffic data packets is crucial for improving cyber security and minimizing the adverse effects of network and corporate attacks.

The paper is structured as follows: Sect. 2 covers related works that utilize the most recent techniques and offers a comparative analysis based on their proposed approach and goal. Section 3 provides a concise overview of the proposed methodology. Section 4 presents the analysis and discussion of the dataset results. Finally, Sect. 5 delves into the conclusion.

2 Related Works

Recent research has illustrated a sharp rise in the exploration of developed deep learning methods to utilizing of state-of-the-art cyberattack distinguishing systems. These systems are designed to effectively counter different threats, leading investigator to thoroughly scrutinize and assess a multitude of neural models to attain the best possible outcomes. as a result, the following studies have surfaced as the most outstanding and noteworthy in addressing this subject (Smith, 2021) studies focused on assessing the efficiency of convolutional neural networks (CNNs) and recurrent neural networks (RNNs) in detecting cyber threats such as malware, phishing, and network attacks. Findings illustrated that CNNs has ability in analyzing regular network data, while RNNs were particularly effective at identifying temporal patterns, (Sarker, 2021) the studies discover different forms of deep learning such as GANs (Generative Networks) and transfer learning, as they pertain to cyberattacks challenges and how to used such as these techniques to detect the attacks, the results proposed that the deep learning have the ability to improve the accuracy of detection systems.

In the research conducted by (Podder, Bharati, Mondal, Paul, & Kose, 2021), The researchers demonstrated a comprehensive analysis of deep neural network models such as long short-term memory (LSTM) and convolutional networks (CNN) in the Internet of Things (IoT). The study showed that LSTM has ability in detecting attacks when used with KDD Cup99 data. in addition to the importance of developing hybrid systems to improve the general performance, particularly in the ever-changing IoT landscape.

(Ali, et al., 2024) The researchers illustrated a research paper showed the significance of artificial neural networks (ANNs) in securing data against external attacks. The researchers focus attention on the focus attention on to develop systems that based on using artificial intelligence to discover cyberattacks. The resulting showed that the use of neural networks rises the efficiency of systems in discovering sophisticated attacks.

(Khan, et al., 2024) illustrated a research discuss classic machine learning algorithms (SVM, KNN) compared to advanced deep neural network models (CNN, LSTM). The results represented that deep learning surpassed the traditional methods, achieving accuracy rates of 100%, thus affirming its important in analyzing network data.

In their study, (Olaoluwa & Potter, 2024) illustrated how traditional neural networks (CNN) and recurrent neural networks (RNN) can improve intrusion discovery systems. Their research demonstrated that the accuracy to specify new sorts of attacks, such as Zero-Day Attacks, during the analysis of complex patterns. at any rate, the study also noted problems in interpreting the results of these models.

(Dontu, Vallabhaneni, Addula, Pareek, & Balassem, 2024) A research was carried out on using attention mechanisms alongside Auto-encoders to analyze cyberattacks. The results showed that incorporating attention mechanisms increased the accuracy of the models when applied to the BoT-IoT dataset.

The above table helps understand the patterns and difficulties appeared in last studies on deep learning for cyberattacks. It reveals the complexities and nuances in ongoing advancements and research for protecting against cyber threats.

Table 1. Summery of Literature Review

Authors & Year	Proposed Method	Problem Statement	Objective
(Smith, 2021)	Convolutional Neural Networks and Recurrent Neural Networks	Difficulties arise when attempting to identify and analyze temporal patterns in network traffic	Assess the efficiency of CNNs for regular data and RNNs for sequential patterns
(Sarker, 2021)	Generative Adversarial Networks, Knowledge Transfer, and Advanced Reinforcement Learning techniques	Obstacles in mitigating erroneous alerts and improving the efficiency of malware detection systems	Examine advanced learning methods to enhance accuracy in identifying phishing, malware, and cyber intrusions
(Podder, Bharati, Mondal, Paul, & Kose, 2021)	Long Short-Term Memory (LSTM) and Convolutional Neural Network (CNN)	Identification of intricate assaults in IoT networks using dynamic temporal data	Showcasing the ability of LSTM to analyze time-based patterns and achieve an accuracy rate of 99.8%
(Ali, et al., 2024)	ANNs (Artificial neural networks)	Dealing with cyber threats in industrial and academic settings	Create artificial intelligence-based systems for the purpose of examining network traffic and detecting complex cyberattacks
(Khan, et al., 2024)	machine learning algorithms CNN, LSTM, SVM and KNN	Conventional machine learning approaches are unable to attain the desired level of precision in the analysis of network data	Demonstrate the superior performance of deep learning compared to traditional methods with a 100% accuracy rate
(Olaoluwa & Potter, 2024)	Convolutional Neural Networks (CNNs) and Recurrent Neural Networks (RNNs)	Challenges in identifying Zero-Day Attacks and understanding the results of models	Improve intrusion detection systems by utilizing Convolutional Neural Networks and Recurrent Neural Networks to analyze complex patterns
(Dontu, Vallabhaneni, Addula, Pareek, & Balassem, 2024)	Autoencoders equipped with attention mechanisms	Insufficient precision of models in the analysis of IoT datasets such as BoT-ToT	Enhance model precision by 98% by integrating attention mechanisms

3 Proposed Method

This study suggested a new hybrid model depend on using two traditional machine learning CNN and LSTM to detect cyberattacks. Also the combined model supported combined with Attention mechanisms to analyze network attack data and specify patterns related to cyberattack. The proposed method included preprocessing data, build hybrid model using CNNs and LSTMs with attention mechanisms. The performance evaluating using number of measures such as accuracy, recall, confusion matrix. The proposed model can be represented through the layout designed in Fig. 1

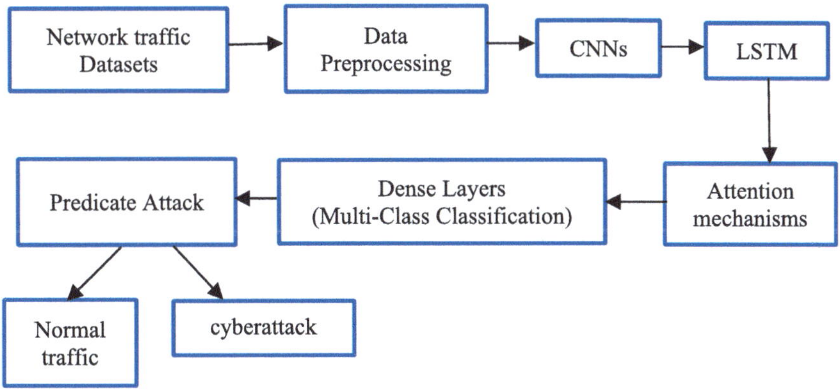

Fig. 1. The general layout of the proposed Method.

3.1 Datasets

In order to build an efficient, cyberattack discovery model, this paper depend on the main dataset: UNSW-NB15. Which is choice for them by coverage of attack sorts and network traffic. The UNSW-NB15 dataset (Moustafa & Slay, 2015), designed by the University of New South Wales, focus to handle the shortcomings of older datasets like KDD Cup 99. It consists of 49 features that provide detailed data about network traffic, including protocols, packet counts, and connection duration. This dataset consist of a variety of attacks such as DoS attacks, reconnaissance, exploits, analysis attacks, and malware and consists of around 2.54 million records, enabling large-scale training of the proposed model.

A dataset can be accessed through a publicly available web platform known as "Kaggle". The dataset is provided in CSV format.

3.2 Data Preprocessing

The preprocessing of data is a significant step in transforming raw network traffic data into form suitable for efficient handling by hybrid model. This operation starts with data cleansing, which includes the deleting of missing values. Remove missing values from record to prevent negative impact on the performance of the proposed model. in addition to outliers and noise values is eliminate to keep away from bias and enhance the model's efficiency. (Abbasi, Shahraki, & Taherkordi, 2021).

After cleaning procedures, the data is undergoes normalized by using Min-Max Scaling method, which transform values to a consistent range through 0 and 1 to enhance the stability of the models and accelerating training speed. The following equation is applied to each feature:

$$V_{normalized} = \frac{V - V_{min}}{V_{max} - V_{min}} \tag{1}$$

After terminate the normalization process of choosing objects features, the process of correlation analysis process as a next step to specify the features that are highly pertinent to discovering cyberattacks. The main features are given special impotent, as previous studies has illustrated the importance effect of these features on the classification of network traffic data (Kshirsagar & Kumar, 2021). The data is ultimately split into a 70% training set and a 30% test set to enhance the model's generalizability. This split helps evaluate the model's performance when exposed to new data. The choice of this ratio is based on recent studies indicating that allocating more data to training improves prediction accuracy. (Oh, Park, & Ye, 2020).

all of these steps enhance the accuracy and increase speed of the model, make sure it processing complex temporal data and cyberattack patterns, crucial for the success of the cyberattack discovering system.

3.3 Convolutional Neural Network (CNNs)

The convolutional neural network (CNN) is an important component of the proposed model in this paper, and it is used to extract spatial features in network traffic data. CNN has ability to evaluates fixed network data to distinguish various distinctive patterns or repetitive occurrences associated to cyberattacks (Adithya & Rajesh, 2020). The elements of CNN as follow:

- **Convolutional Layer**

Within this layer, the data processing using either a filter or kernel in order to detect significant patterns. The convolution between the kernel and the data at each position is computed using Eq. (2).

$$y(x, y) = \sum_{m=0}^{k-1} \sum_{n=0}^{k-1} x(x + m, y + n).w(m, n) \tag{2}$$

where:

- $x(x + m, y + n)$ the value of data at the specified position
- $wm, n)$ the value of kernel
- k size of kernel
- $y(x, y)$ the result values (Feature Map)

- **Pooling Layer**

It is used to reduce the dimensionality of the extracted data (Feature Maps) while retaining important information. Max Pooling technique is implemented using the Eq. (3):

$$y_{pool}(i, j) = max_{p,q \in R}(x(p, q)) \tag{3}$$

This layer contributes to improving efficiency and reducing computational complexity while preserving the most important features.

- **Activation Function**

The ReLU (Rectified Linear Unit) function is used to convert negative values to zero and introduce continuity into the network and calculated by using Eq. (4).

$$f(x) = \max(0, x) \tag{4}$$

- **Flatten Layer**

The result from the convolution and pooling layers is converted into a sequential format to be input into the dense layers.

The proposed model uses CNN for attack discovery, extracting spatial data before LSTM layers analyze temporal patterns. This enhances accuracy for organized attacks such as DDoS. CNN analyzes stable patterns, while Pooling Layers lower data dimensionality and enhance complexity and speed. CNN's flexibility discover multiple attack sorts. (Sahu, Sharma, Tanveer, & Raja, 2021).

3.4 Long Short-Term Memory (LSTM)

LSTM is type of machine learning recurrent neural network (RNN) model which used to handle temporal patterns and long term reliance's in sequential data. LSTMs has ability to handle vanishing gradient difficulties which appear by traditional RNN. Long-short-term memory work by using memory cells which making them to "remember" significant patterns in data over long term of time such as persistent attacks and multi-stage attacks (Laghrissi, Douzi, Douzi, & Hssina, 2021). The LSTM consists of the following components:

1. **Memory Cell:** Each LSTM cell contains three main gates that control the flow of information in and out of the cell:

   - **Input Gate:** Dictates the quantity of fresh data that is to be incorporated into the existing cellular state.

   The input gate is calculated using the Eq. (5):

$$i_t = \sigma(W_i \cdot [h_{t-1}, x_t] + b_i) \tag{5}$$

where:

- i_t : Input gate

- σ : Sigmoid Function
- W_i : represent the weight matrix
- h_{t-1} : Previous hidden state
- x_t : represent the current input
- b_i : represent the Bias

  - **Forget Gate:** It means how much old information should be "forgotten" or ignored.

  The forgetting gate is calculated as follows:

$$f_t = \sigma(W_f \cdot [h_{t-1}, x_t] + b_f) \tag{6}$$

- **Output Gate:** Specifies how much information to pass to the hidden state for the current time point.

  The output gate is calculated as follows:

$$o_t = \sigma(W_o \cdot [h_{t-1}, x_t] + b_o) \tag{7}$$

2. **Cell State:** The cell state is updated using information from the input and forget gates:

$$C_t = f_t \cdot C_{t-1} + i_t \cdot \tilde{C}_t \tag{8}$$

where

- C_t : Current Cell State
- C_{t-1} : Previous Cell State
- $C^{\sim}{}_t$: The new temporary state, which is calculated as follows:

$$C^{\sim}{}_t = \tanh(W_C \cdot [h_{t-1}, x_t] + b_C) \tag{9}$$

3. **Hidden State:** The final result of the cell at time:

$$h_t = O_t + tanh(C_t) \tag{10}$$

LSTMs analyze temporal patterns in network data, detecting cyberattacks like brute force attacks and APTs. They predict emerging attack patterns and can detect new types of cyberattacks (Jony & Arnob, 2024).

In the suggested method, LSTM works as an analysis layer that complements the role of the Convolutional Neural Network (CNN). CNN used to extract the spatial categories that are static in the network data, then these data are passed to LSTM which used to analyze complex temporal patterns. This combination performs to enhances the performance of the model by making it to detect temporal patterns that may not be apparent when analyzing static data alone. Moreover, LSTM helps in analyzing behaviors that span time, which rise the model's ability to generalize and handle multiple sorts of attacks.

LSTM improving detection of continuous and multi-stage attacks. It solves the vanishing gradient problem and efficiently handles long temporal data. Combining LSTM and CNN improves accuracy in discovering complex cyberattacks.

3.5 Attention Mechanism

The Attention Mechanism is a novel method used in neural networks to improve the focusing on the main aspects of the input data. This mechanism is crucial in improving the efficiency of hybrid models such as CNN-LSTM, as it making the model to recognize the relevant sections of the data that have the greatest effect on detecting cyberattacks. Rather than handle each data element equally, attention allows the model to prioritize the most influential information (Jang, Kim, Harerimana, Kang, & Kim, 2020). The components of Attention Mechanism are:

- **The alignment Score Function:** score function is an important component in specify the significant of a certain input associated with the surrounding context. The alignment function calculates the correlation between the decoder's hidden state and the encoder output. The alignment or score function calculated by using the following formula:

$$e_i = f(h_{t-1}, x_i) \tag{11}$$

where:

- e_i : The alignment Score for the input x_i
- h_{t-1} : Previous hidden state
- f: Score Function

  - **Probability Distribution (Softmax):** The Softmax function is used to convert the fit scores into probabilities that express the relative importance of each part:

$$\alpha_i = \frac{exp(e_i)}{\sum_j exp(e_j)} \tag{12}$$

- **Context Vector:** The weighted inputs are combined to generate a context vector that represents the most important information:

$$C_t = \sum_i \alpha_i \cdot x_i \tag{13}$$

where:

- C_t : Context vector which used to enhance the predication
- α_i : Attention weight.

The attention mechanism improves the performance of the hybrid model (CNN-LSTM) for cyberattack detection. It directs the model's focus to influential features in the data, enhancing accuracy by concentrating on attack-relevant features. The mechanism also enhances result interpretability by identifying features or times that impacted the model's decisions. It allows efficient handling of complex data by minimizing irrelevant information and focusing on parts contributing to detecting evolving cyberattacks.

3.6 Dense Layers (Output Layers)

Dense Layers are the final and essential part of the proposed CNN-LSTM-Attention hybrid model, which is used to analyze the features extracted from the previous networks (CNN and LSTM) and the attention mechanism, and then convert them into final outputs that classify the network data into multiple categories, such as "normal traffic" or "cyberattack" (e.g. DDoS, SQL Injection).

After the CNN extracts the spatial features and the LSTM analyzes the temporal patterns and applies the attention mechanism to focus the model on the most important features, all these features are passed to the dense layers for aggregation and full analysis.

The Flatten Layer is used to transform the multi-dimensional features into a single vector) three-dimensional vector) that can be processed in the dense layers. Dense layers typically consist of fully connected layers that take a flat vector as input, and use weights and links between all nodes to compute the final output. The final layer is a classification layer that applies the Softmax function to determine the probability of the data belonging to each class, according to the Eq. (14):

$$P(y_i|x) = \frac{\exp(z_i)}{\sum_{j=1}^{N} \exp(z_j)} \tag{14}$$

where:

- $P(y_i|x)$: The probability which mean the data x belongs to the class y_i
- z_i : output of the final class.
- N : number of classes

Dense layers allow data to be classified into Multi-Class Classification, which is important in detecting various types of cyberattacks.

4 Result Analysis and Discussion

The proposed model (CNN - LSTM - Attention), which integrate the ability of CNN, LSTM, and attention mechanisms, has illustrated powerful performance. When tested the data, the proposed model achieves a great accuracy of 98.7%. This percentage of accuracy represent that the model's ability to effectively classify various sorts of network traffic, characteristic between "normal" and "cyberattack" states. It also has ability in distinguish complex attacks such as Distributed Denial-of-Service (DDoS) and SQL Injection. This model underscores the power of leveraging CNN for extracting spatial features, LSTM for analyzing temporal patterns, and attention to specify the most significant elements. By integrated these techniques, the hybrid model has proven to be a powerful and reliable tool in the field of network traffic classification.

4.1 Evaluation Matrices

There is various performance metrics to evaluate the performance of algorithms based on the specific data and proposed model. These measures represent main points in machine learning field which is accuracy, recall, precision, F1score, true and false positive. These measures plays a significant role in the performance of the models and specify the final results. (Chicco & Jurman, 2020). These measures play an important role in evaluating the performance of the models and determining the final findings (Hicks, et al., 2022). The proposed model will be evaluated for accuracy, precision, and rates of true and false positives using metrics known as the confusion matrix as detailed in the accompanying Table 1.

Table2. Confusion matrix

True negative	False positive
False negative	True positive

Another way to evaluate the performance of hybrid model is by using classification report. Which include number of metrics such as accuracy, precision, recall, and F1 score are fundamental in evaluating model performance is computed by percentage of correct classifications made associated to the total number of classifications in the test data (Foody, 2020), accuracy have calculated as in Eq. (15) (Table 2).

$$accuracy = \frac{TN + TP}{TP + TN + FP + FN} \tag{15}$$

The precision represents the overall number results from true positive divided on true positives added to false positive and computed as in Eq. (16)

$$precision = \frac{TP}{FP + TP} \tag{16}$$

Recall represented as the ability of a model or network to find all the relevant cases within a dataset. The findings of true positive value divided by findings of true positive added to results of false negative and calculated as in Eq. (17).

$$recall = \frac{TP}{TP + FN} \tag{17}$$

The other performance measure is F1-score which is a metric that evaluate the accuracy that considers both the precision and recall of the test. It is the harmonic mean of the precision and recall, with a value between 0 and 1, where 1 is the best possible F1-score. This score is commonly used in fields such as machine learning and information retrieval to compare the performance of different tests (Yacouby & Axman, 2020) and calculated as in Eq. (18)

$$F1_{score} = 2 \cdot \frac{precision \cdot recall}{precision + recall} \tag{18}$$

4.2 Result Discussion

The analysis of the matrix showed that the model accurately categorized most samples with few errors. The primary mistakes were in discerning similar attacks like Reconnaissance and Exploits, which have overlapping characteristics. Nonetheless, these errors were infrequent and made up less than 1% of the samples, indicating the model's high dependability for practical use (Fig. 2).

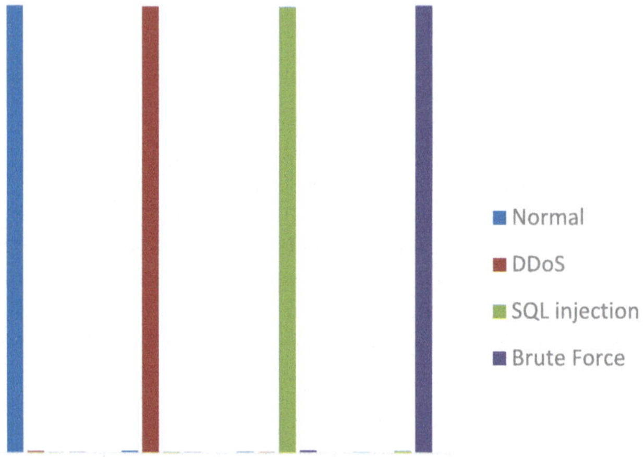

	Normal	DDoS	SQL injection	Brute Force
Normal	98.90%	0.60%	0.40%	0.30%
DDoS	0.50%	98.70%	0.30%	0.20%
SQL injection	0.30%	0.40%	98.60%	0.60%
Brute Force	0.30%	0.30%	0.70%	98.90%

Fig. 2. Confusion Matrix

The accuracy, recall, and precision of the proposed model are displayed in Fig. 4. The hybrid model's performance is evaluated using the performance parameters mentioned above The recall stands out with the highest value compared to the other parameters. As shown in the figure, the (CNN-LSTM-Attention) model achieved a maximum true positive rate of 98.40% and accurately predicted 97.9% of the cases (Fig. 3).

The proposed hybrid model was compared with traditional models such as (KNN, SVM) and with individual models (CNN, LSTM), as shown in Fig. 4.

The proposed model was shown to be superior when compared with traditional models (KNN, SVM) and standalone models (CNN, LSTM).

The attention mechanism helped improve the model's focus on the most important features in the data, which improved the accuracy of attack detection, especially complex attacks such as multi-stage attacks. Despite the high performance, some errors appeared in classifying attacks with overlapping patterns. These errors can be reduced by improving feature extraction or adding additional context to the data. The results indicate that

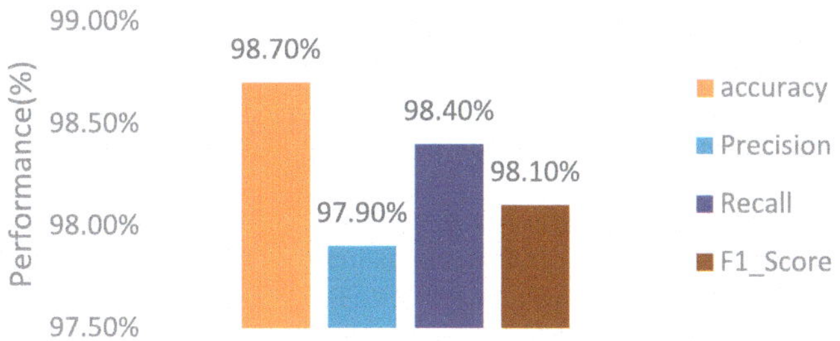

Fig. 3. Performance analysis based on accuracy, precision, recall and F1_score.

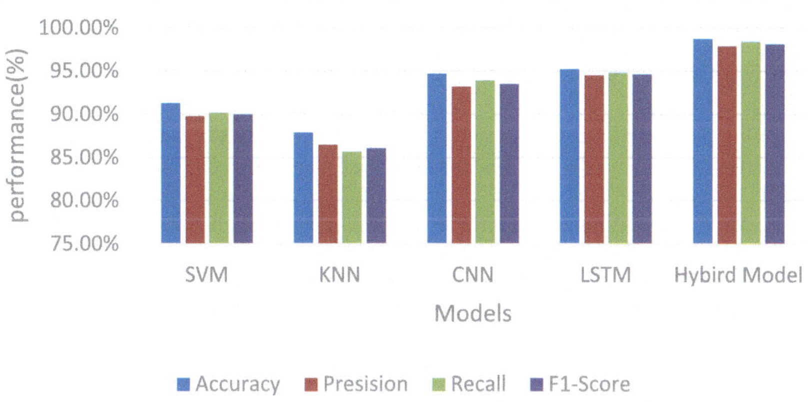

Fig. 4. Comparative analysis with traditional models.

the proposed model can be practically applied in real network environments to detect cyberattacks in real time. With its high accuracy and fast response, it can be useful in protecting sensitive systems such as the Internet of Things (IoT) and cloud infrastructure.

5 Conclusion

The research improves cyberattack detection using an advanced hybrid model based on LSTM and CNN networks with an attention mechanism. Results show efficiency in dealing with diverse data, achieving 98.7% accuracy. LSTM analyzes temporal patterns, CNN extracts fixed features, and attention focuses on important features, improving accuracy and reducing errors. Challenges include distinguishing overlapping attacks and addressing imbalanced data. The proposed model is ideal for modern cybersecurity systems, paving the way for smarter and more effective models in the future.

References

Abbasi, M., Shahraki, A., Taherkordi, A.: Deep learning for network traffic monitoring and analysis (NTMA): a survey. Comput. Commun. **170**, 19–41 (2021). https://doi.org/10.1016/j.comcom.2021.01.021

Adithya, V., Rajesh, R.: A deep convolutional neural network approach for static hand gesture recognition. Procedia Comput. Sci. **171**, 2353–2361 (2020)

Ali, A., et al.: Enhancing cybersecurity with artificial neural networks: a study on threat detection and mitigation strategies. In: 2024 2nd International Conference on Cyber Resilience (ICCR), pp. 1–5 (2024)

Chicco, D., Jurman, G.: The advantages of the Matthews correlation coefficient (MCC) over F1 score and accuracy in binary classification evaluation. BMC Genom. **21**, 1–13 (2020)

Dontu, S., Vallabhaneni, R., Addula, S.R., Pareek, P.K., Balassem, Z.A.: Cybersecurity framework development for BoTNet attack detection using issoa based attention recurrent autoencoder. In: 2024 International Conference on Intelligent Algorithms for Computational Intelligence Systems (IACIS), pp. 1–7 (2024)

Foody, G.M.: Explaining the unsuitability of the kappa coefficient in the assessment and comparison of the accuracy of thematic maps obtained by image classification. Remote Sens. Environ. **239**, 111630 (2020)

Hicks, S.A., et al.: On evaluation metrics for medical applications of artificial intelligence. Sci. Rep. **12**, 5979 (2022)

Jang, B., Kim, M., Harerimana, G., Kang, S.-U., Kim, J.W.: Bi-LSTM model to increase accuracy in text classification: Combining Word2vec CNN and attention mechanism. Appl. Sci. **10**, 5841 (2020)

Jony, A.I., Arnob, A.K.: A long short-term memory based approach for detecting cyber attacks in IoT using CIC-IoT2023 dataset. J. Edge Comput. **3**, 28–42 (2024)

Khan, I., Khan, J., Bangash, S.H., Ahmad, W., Khan, A.I., et al.: Intrusion detection using machine learning and deep learning models on cyber security attacks. VFAST Trans. Softw. Eng. **12**, 95–113 (2024)

Kshirsagar, D., Kumar, S.: An efficient feature reduction method for the detection of DoS attack. ICT Express **7**, 371–375 (2021). https://doi.org/10.1016/j.icte.2020.12.006

Laghrissi, F., Douzi, S., Douzi, K., Hssina, B.: Intrusion detection systems using long short-term memory (LSTM). J. Big Data **8**, 65 (2021)

Macas, M., Wu, C., Fuertes, W.: A survey on deep learning for cybersecurity: progress, challenges, and opportunities. Comput. Netw. **212**, 109032 (2022)

Masum, M., Faruk, M.J., Shahriar, H., Qian, K., Lo, D., Adnan, M.I.: Ransomware classification and detection with machine learning algorithms. In: 2022 IEEE 12th Annual Computing and Communication Workshop and Conference (CCWC), pp. 0316–0322 (2022)

Mendonça, R.V., et al.: Intrusion detection system based on fast hierarchical deep convolutional neural network. IEEE Access **9**, 61024–61034 (2021)

Moustafa, R., & Slay, J. (2015). A comprehensive data set for network intrusion detection systems. School of Engineering and Information Technology University of New South Wales at the Australian Defense Force Academy Canberra, Australia, UNSW-NB15

Oh, Y., Park, S., Ye, J.C.: Deep learning COVID-19 features on CXR using limited training data sets. IEEE Trans. Med. Imaging **39**, 2688–2700 (2020)

Olaoluwa, F., Potter, K.: Deep Learning for Intrusion Detection Systems (IDS) (2024)

Pang, G., Shen, C., Cao, L., Hengel, A.V.: Deep learning for anomaly detection: a review. ACM Comput. Surv. (CSUR) **54**, 1–38 (2021)

Podder, P., Bharati, S., Mondal, M., Paul, P.K., Kose, U.: Artificial neural network for cybersecurity: A comprehensive review (2021). arXiv preprint arXiv:2107.01185

Sahu, A.K., Sharma, S., Tanveer, M., Raja, R.: Internet of things attack detection using hybrid deep learning model. Comput. Commun. **176**, 146–154 (2021)

Sarker, I.H.: Deep cybersecurity: a comprehensive overview from neural network and deep learning perspective. SN Comput. Sci. **2**, 154 (2021)

Smith, D.J.: Deep Learning Models for Cybersecurity: A Comparative Analysis of CNN and RNN Architectures. Universal Research Reports (2021). https://api.semanticscholar.org/CorpusID:273118055

Yacouby, R., Axman, D.: Probabilistic extension of precision, recall, and f1 score for more thorough evaluation of classification models. Proceedings of the first workshop on evaluation and comparison of NLP systems, pp. 79–91 (2020)

Zhang, J., Pan, L., Han, Q.-L., Chen, C., Wen, S., Xiang, Y.: Deep learning based attack detection for cyber-physical system cybersecurity: a survey. IEEE/CAA J. Automatica Sinica **9**, 377–391 (2021)

AI for Climate Change and Biodiversity Conservation

Transitioning to Renewables: Forecasting the U.S. Energy Shift and Its Impact on Emissions and Climate

Zahra Rezaei[1](✉), Hamid Gorji[2], and Mohammad Amin Amini[3]

[1] Judiciary Research Institute, Tehran, Iran
z.rezaei@jri.ac.ir
[2] Department of Computer Engineering and Information Technology, Amirkabir University of Technology (Tehran Polytechnic), Tehran, Iran
[3] Department of Computer Engineering, Islamic Azad Islamic, University of Jasb, Markazi, Iran

Abstract. This research delves into how energy use, CO_2 emissions, and temperature patterns have been intertwined in the United States from 1949 to 2023. By employing a broad methodological design that integrates data visualization, statistical evaluation, and innovative machine learning techniques, the study tracks historical trends and projects forthcoming shifts. Findings point to a marked shift in the nation's energy mix, moving from conventional fossil fuels toward a stronger emphasis on natural gas and renewable options. The analysis uncovers a pronounced positive link between fossil fuel consumption and CO_2 emissions, highlighting fossil energy's central role in driving climate change. Meanwhile, renewable energy appears less strongly tied to temperature deviations, hinting at its promise as a more sustainable path forward. Various machine-learning approaches were utilized to gauge predictive accuracy. Notably, Long Short-Term Memory (LSTM) networks yielded the most accurate forecasts of temperature changes, whereas Facebook Prophet proved especially adept at modeling CO_2 emissions. These results underscore the pressing need to hasten the adoption of renewable sources as a means of mitigating climate impacts. In addition, the study demonstrates the power of machine learning as a tool in climate research, offering valuable predictive insights that can inform both policy-making and strategic planning for a cleaner energy future in the United States.

Keywords: Climate change · CO_2 emissions · per-capita energy use · time-series forecasting · and machine learning models

1 Introduction

Climate change, to a large extent caused by anthropogenic greenhouse gas emissions, is one of the major global problems. During the COP26 summit in Glasgow in 2021, world leaders reaffirmed their commitment to the goals of the Paris Agreement, which aims at limiting temperature increases to below 2°C, with a preference for a target of 1.5°C. [2] A step toward achieving this target requires examining the major variables that shape

A. J. Obaid et al. (Eds.): ICONEST 2024, ISEM 40, pp. 45–60, 2025.
https://doi.org/10.1007/978-3-031-89960-7_4

CO_2 emissions in the top emitters, of which the United States is one. [3] The persistence of fossil fuel sources for energy production and transportation contributes to the rising concentration of CO_2, consequential of huge environmental and climatic impacts. The forces behind these emissions need to be examined, which include energy consumption, economic productivity represented by GDP, financial development, and investments in renewable energy technologies. [4] The excessive use of fossil fuels is a direct driver of rising carbon dioxide concentrations, which in turn feed environmental and climate crises. [5] With its high per capita energy use and heavy dependence on fossil fuels, the United States is considered to be one of the largest greenhouse gas emitters.

[6] Burning of coal, oil, and natural gas for different industries, such as electricity generation, industrial processes, and transport, accounts for a large share of anthropogenic emissions of CO_2, with energy consumption being critical to these activities [7]. Understanding the linkages among different energy types and their implications for carbon and temperature changes is important in developing effective mitigation actions, in particular in countries with high energy consumption, which has important consequences on economic growth and the wellbeing of the population [8] Analyzing the complicated interrelationship between economic growth, energy consumption pattern, and carbon dioxide emission is very essential, especially for those energy-intensive and rapidly industrializing countries such as China, India, and the United States [9] Approaching this "CO_2 emissions–energy use–economic growth trilemma" may provide a strategy to harmonize energy security, economic growth, and environmental sustainability [10]. Anthropogenic greenhouse gas emissions, mostly generated by rising industrialization and the relentless consumption of fossil fuels, increase climatic disorders in developing and developed countries alike. It also comes with quite a few health risks, evidenced by the link between air pollution and enhanced mortality and the dependence of the transportation industry on fossil fuels [11]. In this context, machine learning (ML) methodologies provide powerful tools for the analysis of complex relationships between variables, improvement of emission forecasts, and support to policy formulation [12]. Recent studies highlight the effectiveness of ML algorithms such as Convolutional Neural Networks (CNN), Long Short-Term Memory (LSTM) networks, XGBoost, and Prophet in environmental metric forecasting [13–16]. It is hypothesized first that the ML-based approach with detailed energy consumption data, economic trends, and related metrics can make the model have higher accuracy of forecasting of CO_2 emissions and temperature changes. They capture high potential to advance research into climate change issues in the United States due to the heterogenous impact on carbon emissions and temperature change theses of sources like fossil fuel, nuclear, hydrogen, and renewables [17]. Grounding our methodology in both climate science and energy economics, we selected input factors that are known to influence greenhouse gas emissions. In choosing LSTM networks for time-series data and XGBoost for complex feature interactions, we specifically targeted the temporal and non-linear aspects of these phenomena. By fusing domain knowledge with state-of-the-art ML techniques, we aim to boost both interpretability and accuracy in our models. Following this introduction, Section 2 overviews the relevant literature regarding ML's application in climate modeling and CO_2 emissions projection. Section 3 covers the data sources, detailing energy consumption types (including oil, gas, coal, nuclear, hydrogen, solar, and wind), CO_2 emissions, and surface temperature trends in the U.S. from 1949

to 2023. Section 4 explains our methodology, focusing on descriptive statistics and the use of CNN [14], LSTM [18], XGBoost [15], and Prophet [16]. Section 5 lays out the results, exploring their potential to inform climate policy and shape future investigations. Finally, Section 6 concludes by summarizing the principal insights and offering strategic guidance for decision-makers and industry participants.

2 Related Works

Emerging projections of U.S. emissions, climate dynamics, and energy consumption paint a fluid and ever-changing picture. According to the U.S. Energy Information Administration (EIA), energy-related CO_2 emissions are expected to fall by about 0.7% in 2023—to roughly 5.0 billion metric tons—primarily due to a move toward cleaner energy resources. Even though there are some decreases at the national level, rising global GHG emissions further emphasize the critical need for cooperative international action [19]. Meanwhile, the National Oceanic and Atmospheric Administration registered 2022 as the sixth hottest year ever recorded to further emphasize actions on climate-related impacts [20].

Machine learning, especially methodologies tailored for the forecasting of time series data, is one of the important tools used in estimating past patterns and future changes in energy consumption, greenhouse gas emissions, and metrics relevant to climate change. These approaches provide crucial information for strategic planning, policy formulation, and optimization in energy generation to combat climate change. For instance, studies with the CatBoost Regressor (CBR) have shown high accuracy in modeling global temperature patterns, hence offering highly relevant support to climate adaptation strategies [21]. Moreover, a plethora of research findings underline the flexibility of machine learning approaches in different contexts. In the case of Russia, time series analysis has shown significant relationships between economic growth, energy consumption, and CO_2 emissions.highlighting how renewable energy diversification can mitigate environmental repercussions [10]. The work showed a causal link between rising economic activity and increases in both energy consumption and emissions, underscoring the need for strategic policy reforms.Additional research from Turkey [8] employed ML algorithms—namely MLP, XGBoost, and SVM—to anticipate CO_2 emissions by incorporating data on GDP and vehicle usage. Here, an XGBoost-based scenario (Scenario 4) emerged as the most precise model, indicating the pivotal role AI can play in crafting eco-friendly transportation policies. On a related note, economic modeling of extreme weather events highlighted uneven climate-related economic burdens around the globe, pointing to the necessity of policies that address region-specific vulnerabilities [12]. While the U.S., Europe, and Russia face growing damages, certain energy-exporting nations experience fewer negative effects, illustrating the uneven toll of climate change.Likewise, an LSTM-based study [4] looked at the factors fueling CO_2 emissions in China and India—two of the world's largest coal consumers—and found that energy use exerts the most influence over carbon releases. The projections suggest China's emissions may slow down while India's remain on an upward trajectory, underscoring the imperative for both countries to shift focus toward renewables and energy efficiency. Subsequent studies validate the effectiveness of machine learning methods in understanding and minimizing CO_2 emissions. The comparative analysis of the performance of seven machine-learning models showed that deep-learning models using recurrent neural networks (RNN) are better at making hourly energy consumption predictions, while gradient-boosting methods perform better at making daily predictions [22]. These findings show how important it is to choose models tailored to specific datasets and research objectives in order to arrive at accurate results. By leveraging these methodologies, the research investigation using the Nonlinear Autoregressive Distributed Lag (NARDL) model and machine learning examines the interlinkages of energy consumption with income levels and population growth in China, India, and the United States [9]. It was found that per capita energy consumption is a major driver of emissions in all three countries, hence the need to determine policies to improve energy efficiency and increase the use of clean sources of energy. Another study used Gradient Boosting and Random Forest algorithms for the estimation of CO_2 emissions in lower- and higher-income countries [23]. The study has highlighted important drivers of emissions, including GDP growth, population size, and foreign direct investment, and showed their heterogeneity depending on the income grouping. The results call for complex policy designs tailored to the specific economic and environmental challenges faced by countries at various levels of development.Lastly, an extensive study spanning 1975 to 2015 examined the links between CO_2 emissions, energy usage, economic development, and public health indicators in the top five global emitters—China, the U.S., Russia, India, and Japan [7]. The results reinforce the interconnected nature of environmental, economic, and health factors, suggesting that comprehensive, multisector policies are needed to effectively address climate challenges. Using cointegration techniques and structural break analysis, the researchers found strong correlations between emissions and energy consumption, emphasizing the importance of considering the environmental and health consequences of energy use and economic growth.

This study [1] Also utilizes ML but focuses on China and India, using factors like population growth and GDP. Unlike the article under discussion, it does not provide detailed forecasts for renewable energy transition in the U.S. and relies more on general ML methods without exploring newer models like Prophet. Our research more nuanced view of the U.S. energy sector and compares various ML models, adding depth to its forecasting methodology.

In this research [2] introduces a new machine learning (ML) algorithm to analyze the CO_2 emissions-energy use-economic growth trilemma, focusing on forecasting energy consumption and emission trends in the U.S. Unlike earlier studies, it employs a comprehensive suite of ML models, including LSTM, XGBoost, CNN, and Prophet, to predict CO_2 emissions and temperature changes over time. Mining a most impressively broad data history, the study delineates in full detail U.S. energy transitions and the adoption role being played by renewables toward net emissions reduction. Forward-looking in methodology and markedly valuable in its results, these are indeed findings that will effect wide changes in policy formulation and strategic actions for climate change-researched topics.

3 Methodology

Unveiling Trends and Predictive Insights

This paper conducts an in-depth analysis of changes in carbon dioxide emissions, climatic variables, and energy consumption in the United States from 1965 through 2022. We blend data visualization, statistical methods, and cutting-edge machine learning to spot key trends and interconnections. The insights we gain can support more effective policy-making and guide strategic plans aimed at creating a more sustainable energy landscape.

Dataset Description

This research draws upon multiple datasets spanning from 1910 to 2024, capturing shifts in temperature, CO_2 emissions, and energy use. The temperature records—sourced from NOAA's "Climate at a Glance"—illustrate notable warming trends across North America [25]. Meanwhile, the Global Carbon Budget 2023 provides the CO_2 emissions dataset (1850–2022), giving a broad overview of global carbon outputs and helping us better understand the key drivers of climate change [26]. Lastly, energy consumption data from 1942 to 2023 charts the evolution of primary energy use in the United States, providing crucial insights into the nation's energy trajectory and what it could mean for the future.

Data Preprocessing

In this study, we rely on a carefully assembled dataset that highlights the most significant measures of U.S. energy use and its environmental effects. This dataset encompasses:

- CO2 emissions per capita: Reflecting the country's carbon footprint on a per capita basis.
- Annual surface temperature change: Tracking changes in temperature over time, a key indicator of climate change.
- Per capita energy consumption from various sources: Quantifying the consumption of coal, gas, oil, nuclear, hydrogen, and solar & wind energy per person.

Data Analysis Techniques

The study utilizes a three-pronged approach:

a. Data Visualization: This study uses various graphical methods to show historical trends in U.S. energy consumption, CO_2 emissions, and temperature changes. Line graphs illustrate shifts in energy sources from 1949 to 2023 and the move toward cleaner energy. Scatter plots highlight correlations between energy consumption and emissions, providing insights into their relationships. Comparative visualizations align temperature trends with CO_2 emissions to link greenhouse gas production and climate change. The ETS model forecasts future trends with confidence intervals, illustrating potential future changes and uncertainty in long-term predictions.

Time Series Analysis The ETS (Error, Trend, Seasonality) model forecasts energy consumption and emissions by breaking down data into error, trend, and seasonality components, adapting to diverse patterns. This method is effective for complex climate and energy datasets, providing forecasts with confidence intervals that expand over time.

Machine Learning Models We compare four advanced forecasting models for CO_2 emissions and temperature changes:

XGBoost: A gradient-boosting algorithm for complex datasets.

Bidirectional LSTM: A recurrent neural network for capturing long-term dependencies.

CNN: Adapted from image recognition for time series data.

-Facebook Prophet: An automated model detecting seasonality and trends. This multi-faceted approach aims to enhance understanding of the relationships between energy consumption, emissions, and climate change in the U.S.

4 Discussion

Figure 1 illustrates U.S. energy consumption trends (1949–2023) by energy source as a percentage of total use. Petroleum remains dominant, stabilizing at 37%, while natural gas has risen to 33.5% since the 2000s. Coal has declined sharply from over 30% in the mid-20th century to 10.8% in 2023, reflecting a shift toward cleaner energy sources. Renewable energy, particularly wind (10.2%) and solar (4.9%), has grown significantly in recent decades. Nuclear energy, stable since the 1970s, accounts for 7.8%. Other renewables like biomass (5.6%) and hydroelectric (2.7%) hold smaller shares. The trend indicates a transition from fossil fuels to a more diverse, low-carbon energy mix.

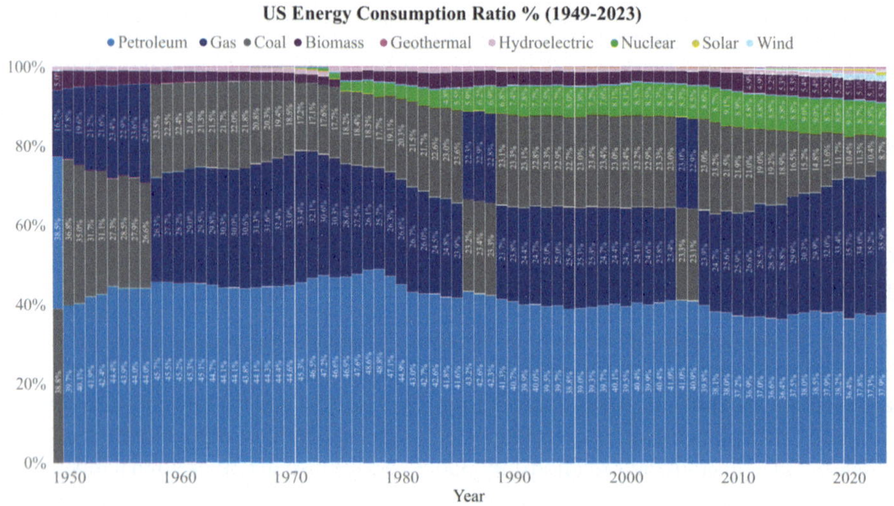

Fig. 1. Evolution of U.S. Energy Consumption by Source (1949–2023): Transition from Fossil Fuels to Renewables

ETS (Error, Trend, Seasonality) Time Series Forecasting Approach
The ETS (Error, Trend, Seasonality) method is a time series forecasting approach that

models data using three components: error, trend, and seasonality. The error represents the difference between actual and forecasted values, while the trend captures the data's underlying direction, and seasonality accounts for recurring patterns. Models like ETS(A, A, N) have specific configurations, such as additive error and trend without seasonality. A 95% confidence interval in ETS forecasting indicates the range within which future values are expected to fall, reflecting the uncertainty of predictions. This method is useful for making forecasts and assessing their reliability.

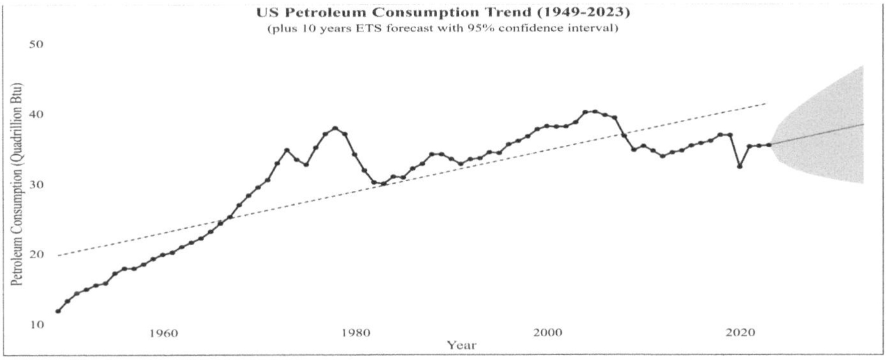

Fig. 2. Historical and Forecasted US Petroleum Consumption (1949–2033): ETS Model with 95% Confidence Interval

Figure 2 shows US petroleum consumption trends from 1949 to 2023 in quadrillion BTUs, along with a 10-year forecast (2024–2033) using an ETS model. The solid black line shows the real-world measurements, while the dashed line reflects the overall trend. The gray band indicates the 95% confidence interval around our forecast. Historically, we see steady growth interrupted by dips during recessions and energy crises. Moving forward, the projections suggest further growth but with greater uncertainty tied to technological progress and the ongoing transition to renewable energy sources.

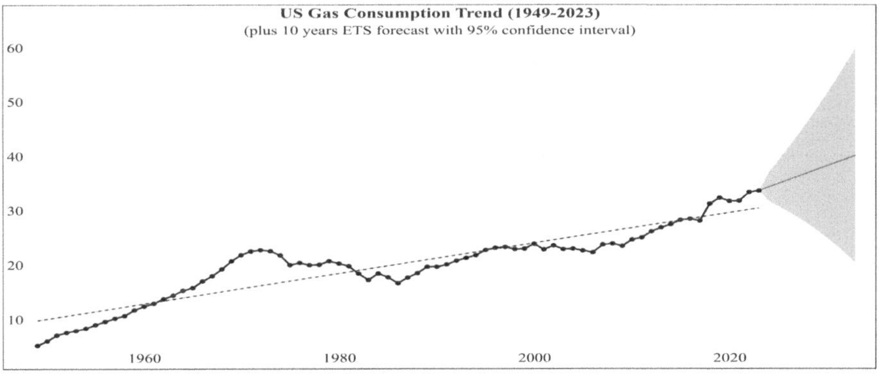

Fig. 3. US Natural Gas Consumption Over Time (1949–2023) and ETS-Based Forecast for the Next Decade

Figure 3 illustrates the historical natural gas consumption in the U.S. from 1949 to 2023 in quadrillion BTUs, with a 10-year projection using the Exponential Smoothing (ETS) method. The data demonstrates consistent growth, with periods of stabilization, particularly from 1980 to 1990, likely due to policy changes or fuel substitution. Recently, consumption has increased, driven by higher energy demand and natural gas's cleaner profile compared to coal. The forecast indicates continued growth, but the wide confidence interval suggests uncertainties from evolving energy policies and market dynamics. The solid line represents the expected path, while the increasing shaded area reflects growing unpredictability in future consumption.

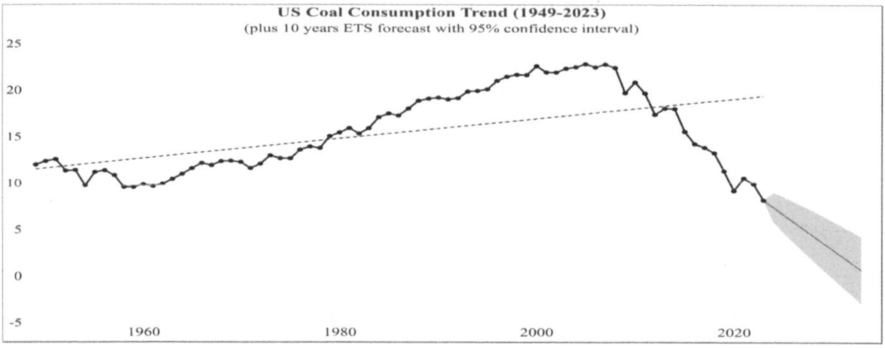

Fig. 4. US Coal Consumption Decline (1949–2023) with a 10-Year Forecast Using ETS Model

Figure 4 presents the trajectory of coal consumption in the United States (1949–2023) alongside a ten-year forecast using the ETS model. After steady growth until the mid-2000s, coal consumption has sharply declined. The forecast, with a 95% confidence interval, predicts a continued decrease, potentially nearing zero by the forecast's end. This trend underscores the transition to cleaner energy sources and the impact of stringent environmental regulations. Despite uncertainty in the rate of decline, the overarching negative trajectory is evident.

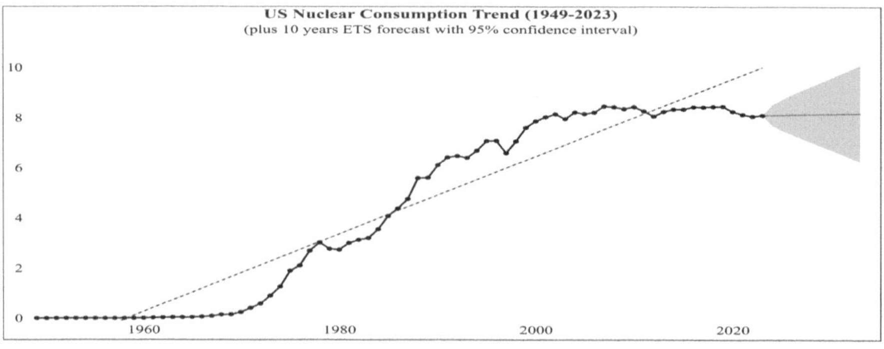

Fig. 5. Historical and Forecasted U.S. Nuclear Consumption Trend (1949–2033): 10-Year ETS Forecast with 95% Confidence Interval.

Figure 5 displays the trend of U.S. nuclear consumption from 1949 to 2023, alongside a 10-year forecast generated using an Exponential Smoothing State Space Model (ETS) with a 95% confidence interval. Historically, nuclear energy consumption in the U.S. showed rapid growth from the 1960s, peaking around the early 2000s, followed by a relatively stable plateau with minor fluctuations. The forecast section (shaded in gray) projects that nuclear consumption will remain relatively stable or slightly increase, with the prediction interval indicating the uncertainty range. The central forecast (solid line) suggests a little deviation from the recent historical trend, while the confidence interval (shaded area) widens over time, reflecting increasing uncertainty about future consumption levels.

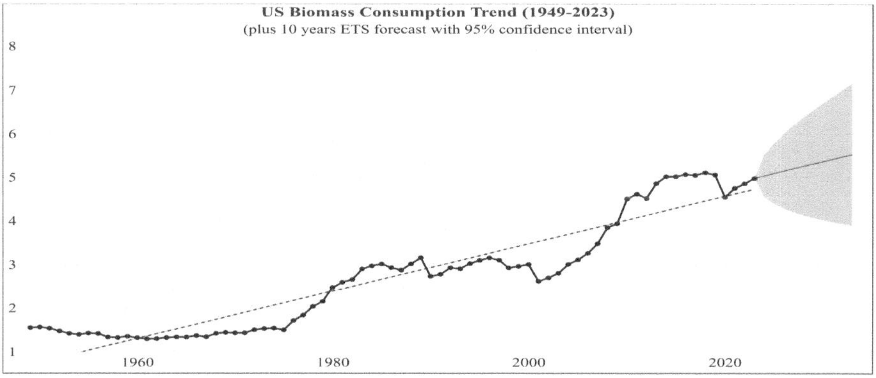

Fig. 6. U.S. Biomass Consumption Forecast (1949–2033): ETS Model with 95% Confidence Interval

Figure 6 depicts the trajectory of U.S. biomass consumption from 1949 to 2023, supplemented by a 10-year forecast using the ETS model with a 95% confidence interval. Biomass consumption exhibited relative stability until the late 1970s, after which it experienced consistent growth, punctuated by occasional declines, notably around the early 2000s. The forecast anticipates a continued upward trend over the next decade, with some variability reflected in the expanding confidence interval, highlighting moderate uncertainty in future projections.

Figure 7 illustrates the progression of U.S. wind energy consumption from 1949 to 2023, accompanied by a 10-year ETS model forecast with a 95% confidence interval. Wind energy use remained minimal until the early 2000s, after which it experienced rapid, exponential growth, particularly in the last decade. The forecast projects further increase, albeit at a slower rate, with expanding confidence intervals signaling rising uncertainty over time. Nonetheless, the overall trend points toward sustained growth in wind energy consumption.

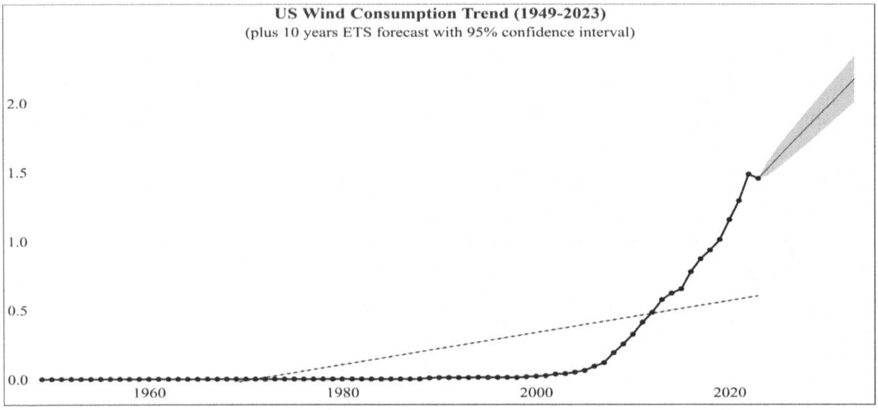

Fig. 7. U.S. Wind Energy Consumption Trend and Forecast (1949–2033): 10-Year ETS Model with 95% Confidence Interval.

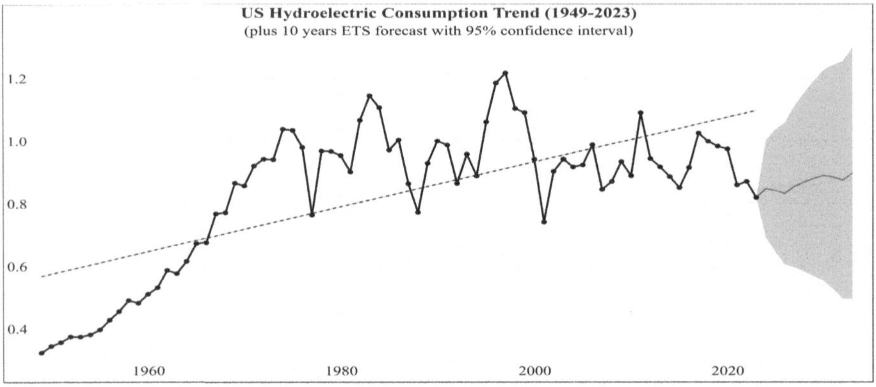

Fig. 8. U.S. Hydroelectric Consumption: Historical Trend and 10-Year Forecast (1949–2033) with 95% Confidence Interval

Figure 8 presents the U.S. hydroelectric consumption from 1949 to 2023, along with a 10-year forecast using the Exponential Smoothing State Space (ETS) model, which includes a 95% confidence interval. The historical data reveal a steady increase in hydroelectric consumption until the early 1980s, after which the trend became more volatile, with noticeable fluctuations over time. Despite the variability, the overall trend points slightly upward. The forecast suggests that hydroelectric consumption will stabilize over the next decade with minimal growth. The shaded region, representing the confidence interval, widens as time progresses, indicating growing uncertainty in the forecast but a generally steady outlook for hydroelectric consumption.

Figure 9 shows U.S. solar energy consumption trends from 1949 to 2023, with a 10-year forecast based on the ETS model and a 95% confidence interval. Consumption was low until the early 2000s, when it began to rise gradually. The past decade saw significant exponential growth. The ETS model predicts continued increases in solar

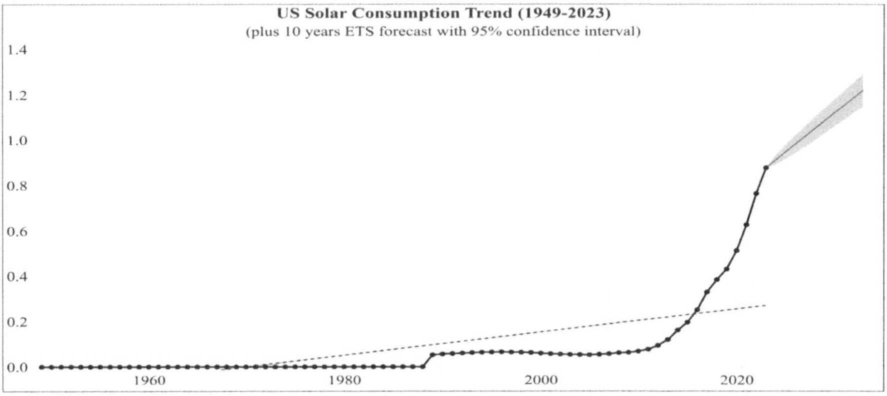

Fig. 9. U.S. Solar Energy Consumption Trend and Forecast (1949–2033): 10-Year ETS Model with 95% Confidence Interval.

consumption, though the growth rate may slow slightly, while the widening confidence interval indicates growing uncertainty but maintains an upward trend.

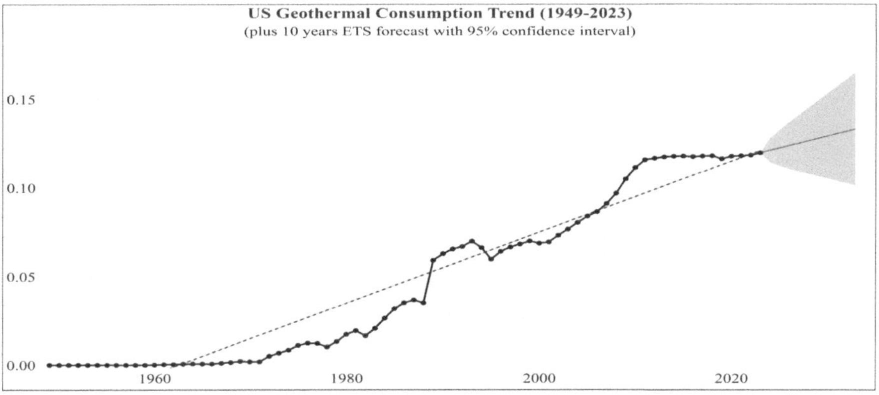

Fig. 10. U.S. Geothermal Energy Consumption (1949–2033): ETS Model Forecast with 95% Confidence Interval.

Figure 10 highlights U.S. geothermal energy consumption trends from 1949 to 2023, with a 10-year ETS model forecast. Consumption saw limited growth until the late 1970s, followed by moderate increases and fluctuations, particularly around the early 2000s. In recent years, usage has stabilized, with the forecast projecting steady levels and a slight potential rise. The widening confidence interval reflects growing uncertainty, but the general stability of the trend persists.

Figure 11 shows the relationship between CO2 emissions (in metric tons) and temperature changes (in degrees Celsius). The scatter plot indicates a positive correlation, with CO2 emissions on the x-axis and temperature change on the y-axis. The upward trend suggests that higher CO2 emissions lead to increased temperatures, aligning with

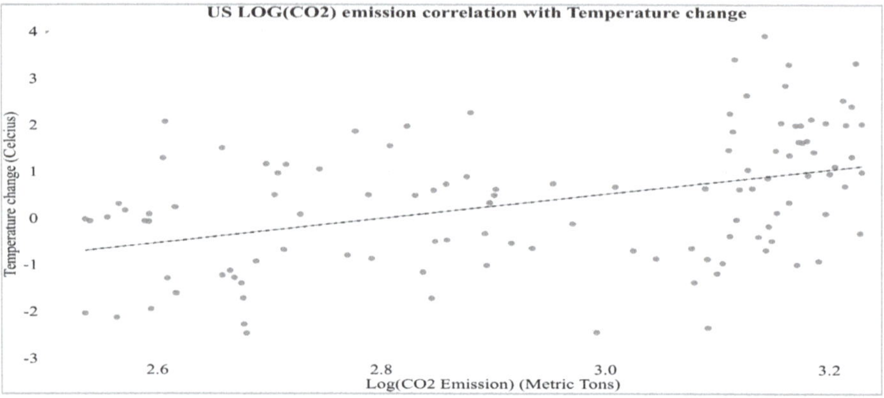

Fig. 11. Correlation Between U.S. CO2 Emissions and Temperature Change: Logarithmic Analysis

the scientific understanding of the greenhouse effect and climate change. Despite some dispersion at higher emission levels, the overall trend highlights the connection between rising CO2 levels and global warming.

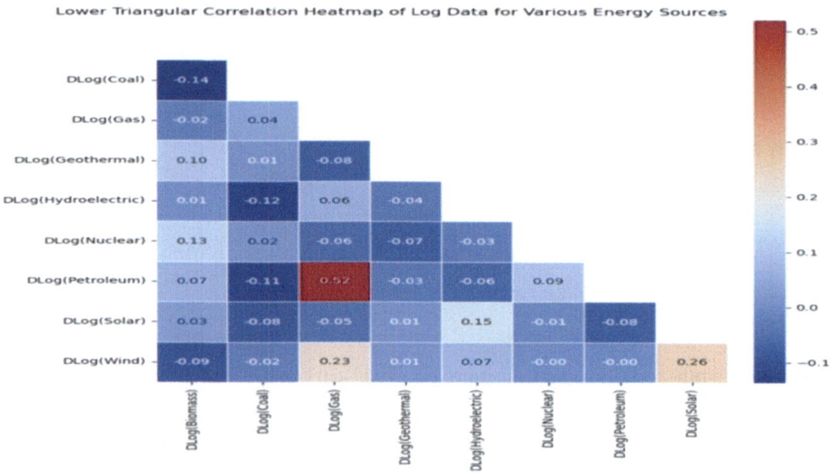

Fig. 12. Correlation Matrix of Log-Transformed Data for Various Energy Sources

Figure 12 illustrates U.S. solar energy consumption trends from 1949 to 2023, accompanied by a 10-year ETS model forecast with a 95% confidence interval. Solar energy consumption remained minimal until the early 2000s, followed by gradual growth and significant exponential increases in the past decade. The forecast projects continued growth, albeit at a slightly slower rate, with a widening confidence interval reflecting increased uncertainty but sustaining an upward trajectory. Correlation analysis reveals weak negative correlations between Biomass and Coal (-0.14) and Gas and Nuclear (-0.07), indicating slight inverse relationships. Positive correlations are strongest between Petroleum and Nuclear (0.52), with moderate associations for Coal and Petroleum (0.18) and Solar and Wind (0.26). Most energy source correlations are near zero, reflecting limited interdependence.

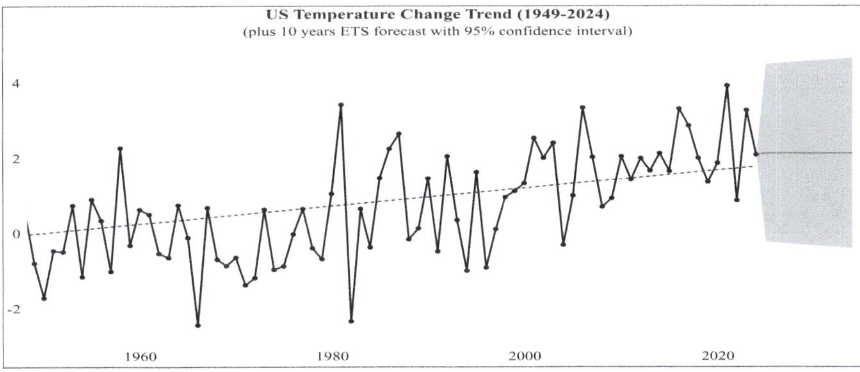

Fig. 13. Historical and Forecasted U.S. Temperature Change (1949–2024): ETS Model with 10-Year Projection and Confidence Interval

Figure 13 shows U.S. temperature trends from 1949 to 2024, including a 10-year forecast with a 95% confidence interval. While historical data shows significant year-to-year fluctuations, the overall trend is an upward increase in temperature. The forecast suggests continued slight rises in temperature over the next decade, but the growing uncertainty in the confidence interval indicates that short-term predictions remain less certain.

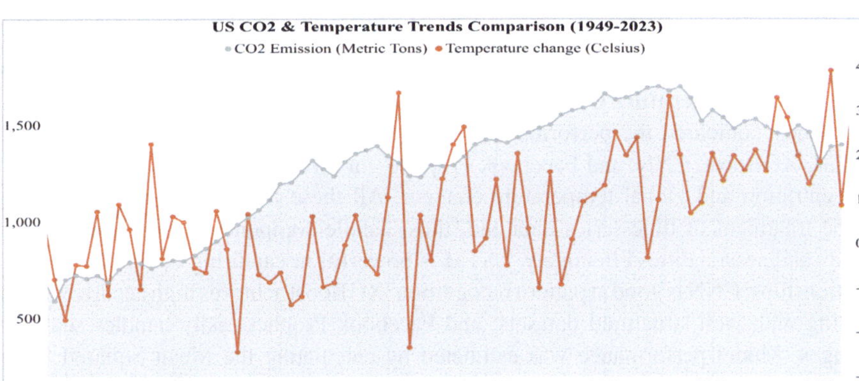

Fig. 14. U.S. CO2 Emissions and Temperature Trends (1949–2023): A Comparative Analysis

Figure 14 compares U.S. CO2 emissions and temperature changes from 1949 to 2023. The gray area represents a steady rise in CO2 emissions since the 1950s, peaking in the early 2000s and slightly declining recently. The orange line shows temperature changes, which have significant variability but an overall upward trend, especially after the 1970s. This contrast highlights the complex relationship between greenhouse gas emissions and temperature, reinforcing CO2's role in climate change.

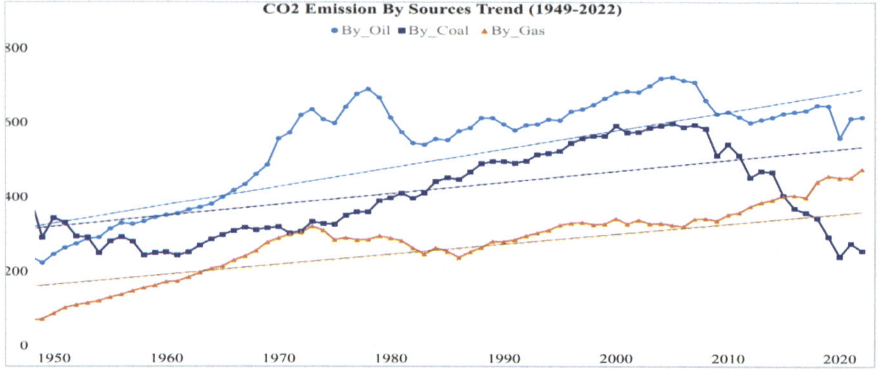

Fig. 15. U.S. CO2 Emissions by Source (1949–2022): Long-Term Trends for Oil, Coal, and Gas

Figure 15 shows trends of CO_2 emissions from oil, coal, and gas in the U.S. (1949–2022) are illustrated with distinct lines: blue for oil, dark blue for coal, and orange for gas. Oil and coal emissions increased similarly until the early 2000s, with oil peaking in 2005 before declining. Coal emissions significantly dropped after 2007 due to a shift away from coal. Meanwhile, gas emissions have risen steadily since the 1950s, with rapid growth after 2005 as the energy mix transitioned toward natural gas, reducing coal reliance. Long-term trends are indicated by dashed lines, showing decreases in coal and oil emissions but persistent increases in gas emissions, further emphasizing the current energy transition taking place within the United States.

Comparative Performance of Time Series Forecasting Models for CO_2 Emissions and Global Temperature Changes

This study compared the performances of four different machine learning models—LSTM, XGBoost, CNN, and Facebook Prophet—in forecasting carbon dioxide (CO_2) concentration and global temperature changes. All these frameworks are specialized in the treatment of time-series data and, thus, handle components such as long-term trends and seasonalities effectively. LSTM is powerful in capturing complex temporal relationships; CNN is good at pattern recognition; XGBoost achieves high accuracy when dealing with well-structured datasets; and Facebook Prophet easily handles seasonal changes. Model performance was estimated by calculating the Mean Squared Error (MSE) for CO_2 and temperature predictions separately, so that each method's strengths in contributing to climate forecast accuracy could be appreciated.

Table 1 shows the comparison of the predictive performance of four machine learning models: LSTM, XGBoost, CNN, and Prophet, with respect to CO_2 emissions and temperature fluctuations, evaluated through the Mean Squared Error (MSE) metric. Notably, LSTM has the highest accuracy in predicting temperature anomalies with an

Table 1. Evaluating Multiple Approaches to CO_2 Emissions and Their Influence on Global Temperature Anomalies.

Model	CO_2 Emissions (MSE)	Temperature change (MSE)
LSTM	2.00437	**0.006371**
XGBoost	1.90656	0.01123
CNN	1.03471	0.01532
Prophet	**0.33435**	0.03780

MSE of 0.006371, as it can capture complex temporal dependencies. On the other hand, Prophet gives better results in predicting CO_2 emissions with an MSE of 0.33435, hence showing its proficiency in modeling seasonal and trend components effectively. CNN and XGBoost, though effective, exhibit higher MSE values, suggesting limitations in capturing temporal or seasonal complexities compared to LSTM and Prophet.

5 Conclusion

The study highlights a shift in the U.S. energy landscape from fossil fuels to renewables and their implications for CO_2 emissions and climate. Fossil fuels strongly correlate with emissions, driving climate change, while renewables like wind, solar, and biomass show weaker correlations with temperature anomalies, emphasizing their sustainability potential. Advanced models like LSTM and Prophet effectively capture energy-climate dynamics but face limitations with non-linear and interdependent climate phenomena. Future research should integrate socio-economic data and physics-based models with machine learning for enhanced predictive accuracy, supporting policy-making and a sustainable energy transition.

References

1. 2. C. C. 2. IPCC: The Physical Science Basis. Contribution of Working Group. ixth Assessment Report of the Intergovernmental Panel on Climate Change (2021)
2. United Nations Framework Convention on Climate Change (UNFCCC). The Paris Agreement (2015)
3. U. E. P. Agency: Inventory of U.S. Greenhouse Gas Emissions and Sinks (2021)
4. Ahmed, M., Shuai, C., Ahmed, M.: Influencing factors of carbon emissions and their trends in China and India: a machine learning method. Environ. Sci. Pollution Res. **29**(32), 48424–48437 (2022). https://doi.org/10.1007/s11356-022-18711-3
5. I. E. A. (IEA): World Energy Outlook (2020)
6. U. E. I. A. (EIA): Annual Energy Review (2021)
7. Li, J., et al.: The relationship between energy consumption, CO2 emissions, economic growth, and health indicators. Int. J. Environ. Res. Public Health **20**(3), 2325 (2023). https://doi.org/10.3390/ijerph20032325
8. Çınarer, G., Yeşilyurt, M.K., Ağbulut, Ü., Yılbaşı, Z., Kılıç, K.: Application of various machine learning algorithms in view of predicting the CO2 emissions in the transportation sector. Sci. Technol. Energy Transit. **79**, 15 (2024). https://doi.org/10.2516/stet/2024014

9. Ahmed, M., et al.: The Effect of energy consumption, income, and population growth on CO2 emissions: evidence from NARDL and machine learning models. Sustainability **15**(15), 11956 (2023). https://doi.org/10.3390/su151511956
10. Magazzino, C., Mele, M.: A new machine learning algorithm to explore the CO2 emissions-energy use-economic growth trilemma. Ann. Oper. Res. **345**(2–3), 665–683 (2022). https://doi.org/10.1007/s10479-022-04787-0
11. Goodfellow, C.A.: Deep Learning. MIT Press, Cambridge (2016)
12. Reichstein, M.: Deep learning and process understanding for data-driven Earth system science. Nature **566**(7743), 195–204 (2019)
13. Li, X., et al.: Long short-term memory neural network for air pollutant concentration predictions: Method development and evaluation. Environ. Pollut. **231**, 997–1004 (2017). https://doi.org/10.1016/j.envpol.2017.08.114
14. Keiron O'Shea, R.N.: An Introduction to Convolutional Neural Networks. arXiv:1511.08458 (2015)
15. Tianqi Chen, C.G.: XGBoost: a scalable tree boosting system. In: KDD 2016, San Francisco, CA, USA, 13-17 August 2016.https://doi.org/10.1145/2939672.2939785
16. Taylor, S.J., Letham, B.: Forecasting at Scale. Am. Statist. **72**, 37–45 (2019). https://doi.org/10.1080/00031305.2017.1380080
17. Rolnick, D.: Tackling climate change with machine learning. arXiv preprint arXiv:1906.05433 (2019)
18. Staudemeyer, R.C., Morris, E.R.: Understanding LSTM -- a tutorial into Long Short-Term Memory Recurrent Neural Networks. arXiv:1909.09586 (2019)
19. E. (2023) Short-Term Energy Outlook. https://www.eia.gov/outlooks/steo/
20. NOAA: State of the Climate: Global Climate Report for 2022 (2023). https://www.ncdc.noaa.gov/sotc/global/202213
21. Shams, M.Y., et al.: A machine learning-based model for predicting temperature under the effects of climate change. In: Hassanien, A.E., Darwish, A. (eds.) The Power of Data: Driving Climate Change with Data Science and Artificial Intelligence Innovations, pp. 61–81. Springer Nature Switzerland, Cham (2023). https://doi.org/10.1007/978-3-031-22456-0_4
22. AZafeirioua, A., et al.: Smart energy strategy - a comparative study of energy consumption forecasting machine learning models. Chem. Eng. Trans. **103**, 691–696 (2023)
23. El-Aal, M.F.A.: The relationship between CO2 emissions and macroeconomics indicators in low and high-income countries: using artificial intelligence," Environment. Development and Sustainability (2024)
24. Ahmed, M., Shuai, C., Ahmed, M.: Influencing factors of carbon emissions and their trends in China and India: a machine learning method. Environ. Sci. Pollut. Res. **29**(32), 48424–48437 (2022). https://doi.org/10.1007/s11356-022-18711-3
25. Climate at a Glance Global Time Series. National Oceanic and Atmospheric Administration (NOAA). https://www.ncei.noaa.gov/access/monitoring/climate-at-a-glance/global/time-series/northAmerica/land/1/1/1850-2024. Accessed 12 July 2024
26. Exeter, U.: The Global Carbon Budget 2023 (Friedlingstein et al., 2023b, ESSD) for all data. GCB data. https://explore.globalcarbonbudgetdata.org/timeseries.html. Accessed 11 June 2024
27. EIA: Total Energy, U.S. Energy Information Administration. https://www.eia.gov/totalenergy/data/browser/?tbl=T01.03#/?f=A. Accessed 19 Apr 2024

Anomaly Detection on Abrupt Climate Change Using AI Techniques

Mohammed Hatem Jawad[1](✉), Ahmad Al Yakin[2], Luis Cardoso[3], and Suantoko[4]

[1] Ministry of Education, General Directorate of Education of Najaf, Najaf, Iraq
stsuantoko109@gmail.com
[2] Universitas Al Asyariah Mandar, Polewali, Indonesia
[3] Polytechnic Institute of Portalegre and Centre for Comparative Studies, University of Lisbon, Lisbon, Portugal
[4] Universitas PGRI Ronggolawe Tuban, Tuban, Indonesia

Abstract. The pattern in data that do not conform to expected behavior is the definition of anomaly detection, where an anomalous pattern is the non-conformant pattern. It is also known as outliers, abnormalities, exceptions, deviants, or surprises. The data are gathered in a variety of ways, and they may be measurements for entity observations. If the data adopt an unexpected form, it is called an outlier. In general, handling outliers involves two steps: first, if the outlier is not interested in analysis, it will be identified as noise and eliminated from the data set; second, analyzing the outlier will provide valuable information about the system's abnormality, many studies have been established on ACC give the evidence that the abrupt climate change is real, where Several methods were proposed in the literature for the change point detection, a point that massively changed in the data pattern is a change-point so this paper will focus on the abnormality in climate.

Keywords: Anomaly · Anomaly Detection Techniques · Abrupt Climate Change (ACC)

1 Introduction

Climate change can characterize as the increasing or be decreasing in the average value during the given period of time, this changing can happen gradually(liner) or it can happen rapidly (non-linear, abrupt) the observation on the earth warmth in recent decades and the effect on climate system show unexpected behavior in climate. This unexpected climatic changing known as Abrupt climate change (ACC) which is lead to disasters occurs in different places all over the world flood, drought and more can be as an example for sudden catastrophe that causes many losses in life and expenses around the world. ACC is a changing from one state to another state or it can define as change happen when the system of climate is compelled to cross limit [1].

Many studies have given the evidence that the abrupt climate change is real. Moreover, these studies can be classified depending on the climate variable that needs to detect the abrupt change in it. For example, detecting abrupt temperature change was

A. J. Obaid et al. (Eds.): ICONEST 2024, ISEM 40, pp. 61–71, 2025.
https://doi.org/10.1007/978-3-031-89960-7_5

the main concern in numerous climate studies test the trend of temperature and ACC the result show that abrupt temperature change happens in the year 1990 and there is also uptrend by the temperature average rather was annual or seasonal, similarly used quantile regression to detect the extreme value in temperature the result show that the trend of temperature highly increased in a nonlinear way. While precipitation was the major concern for another numerous studies [2], while, using Bayesian method to detect the changing point on precipitation data show that there are three abrupt changes happen in (1981,1986,2006) [3], in the same way Man-Kendall on precipitation used in China and the result shows that the rainfall had significant reduction from the year 1965 which is evidence that abrupt change happens [4].

Other researcher tries to detect the abrupt in temperature and precipitation at the same time, using cumulative sum test to detect change point in temperature and precipitation show three change point in (86–87,97–98,2003–2004), [5], detecting abrupt dynamic using moving cut data approximate method on daily maximum temperature, daily minimum temperature, and daily precipitation from north of China show an abrupt dynamic change in precipitation in 1980 [6].

Detection methods for ACC can be classify into two major types, first, change in dynamic structure for climate system and shift change. Dynamic structure methods are based on the change in space and time while detection shift change (change point) methods detect the abrupt climate change through the time series at one-time point.

1.1 A Change-Point Detection for Abrupt Climate Change

A point that massively changed in the data pattern is a change-point. It is can be considered as a point where the parameters of distribution or the model parameters describe time series AC here a brief summarizing on the development of the theoretical aspect and systematic studies of AC in the climate system. Focusing on the climate regime shift DM.

The moving *t*-test (MTT) which testing the distinction the mean for two sequences, while The Man-Kendall test method assume that the components sequence to be analyzed is a fixed time series, which included typical dispersion, and individual samples are largely autonomous. M-K method general use is to test the pattern of AC in a time series. The main idea is the null value for $\alpha 1 < \alpha 0$, which demonstrates that there is a substantial pattern change in this series, If the result has been in the confidence average, that mean the trend changing in the authentic sequence; otherwise, the changing trend is substantial. The MK additionally has vast use in for a time series abrupt trend detecting Using Man-Kendall to detect abrupt climate change and moving t-test and Yamamoto method, this study showed that the temperature rate raised by of 0.36 °C10a − 1 during 1961–2010.

Additionally, a rank-sum test, was proposed by Mann and Whitney in 1947. Where assumption that two samples took from two non-similar groups for which the mean value identical excepted from all of the statistical features. Basis on this, these two groups are tested whether it have substantial non-similarity in the mean values.

Another type of DM is Bayesian analysis which used to detect sudden changing in value, which was development of Bayes theory [7], however used Segments number (Bayesian information criterion and DP algorithm depend on linear regression where first temporal analysis and then spatial analysis the result Show that temperature anomaly has an uptrend.

The cumulative sum method is the other type of detection method for the abrupt mean value change and it mainly includes the cumulative sum test and the cumulative frequency test 10 detect abrupt changes in climate indicators also analyze the trend and spatial extent and investigate the relationship with the characteristics of atmospheric circulation patterns cusum with bootstrapping cusum give us the information where abrupt happen in direction and will be divided in to part depended on this point and it will repeat until no significant change found using bootstrapping to determine the confidence for each point the result show increase in the mean of tmean by 05c tmax by 03c t90 by 87d and pdry by 91d the main idea of cusum is to obtain the evolutionary trend of a time series then we can identify whether there is a significant trend change in the time series.

2 Data Mining

The practice of extracting novel patterns from massive data sets is known as data mining [8]. The goal of a data mining effort is normally either to create a descriptive model or a predictive model Many data mining algorithms have been developed and typically fall into categories: classification, association analysis, clustering, prediction and anomaly detection (outlier).

2.1 Clustering

Clustering is grouping data in clusters each clusters objects are similar, each cluster represent a pattern clustering is unsupervised learning which means there is no prior information about the construction of data set, it used in several fields to extract the useful information form data, many algorithms had been proposed in literatures in this subfield which is meet the requirement of real-life data mining problem. Clustering technique majorly can be divided into hierarchical method (agglomerative and divisive) algorithms and partitioning method (k-mean, k-medoids, density- based, fuzzy c –mean) algorithm.

2.2 Classification

Classification is to build a classifier base on labeled data by training sub of dataset and then test the classifier, classification is a supervised mode, classification used in many applications and it has a wide attention in literature many algorithms developed in this field (nyve base, decision tree, supervised neural network, k-nearest neighbor (K-NN).

2.3 Association Rule

Association rule which designs to find the relationship among variables, It aims to find interesting correlation or association among attributes/items in large datasets, association rules predict the occurrence of an object based on the occurrence of other objects the most known algorithm (Apriori, FP growth tree).

2.4 Prediction

Using present records to forecasting future event is a prediction: neural networks, regression, (SVM), and discriminant analysis.

2.5 Anomaly Detection

Anomaly detection can be defined as "finding the pattern in data that do not conform to expected behavior" [9].

2.5.1 Anomaly Detection Application

Anomaly detection applied in deferent application and domain for example of these applications. Firstly, intrusion detection in the network, sometimes the traffic and operating system calls show abnormal behavior due to malicious activity. Using anomaly detection for intrusion have been used widely in the literature, [10] used clustering to detect intrusion while combined between clustering and classification [11], using classification intrusion detection [12]. Secondly, credit card Freud where the Freud behavior can consider as anomalous, anomaly detection has been applied to other different application like medical, sensors and also on climate data [13].

Anomalies in climate data hide the changing in mechanism for the climate system where this changing cannot be expected. Using outlier detection to identify the abnormality in temperature data the result shows how effective to use outlier detection with climate data. Similarly, [14], using anomaly detection on climate data where the anomalies can consider as an abrupt change [15], detecting anomalous from temperature and precipitation data where is the anomalous is that significantly different from the other point similarly [16], detect useful anomaly from temperature and precipitation data the result show how important to specify the kind of outlier (Basic Element) [17].

3 Anomaly Detection Technique

Base on the mode anomaly detection technique can be divided into two main categories.

3.1 Supervised Anomaly Detection

Supervised mode which also known as classification which required a labeled data to train the model which build for predicting normal and anomalous. The most common algorithm and technique for anomaly supervised are support vector machine supervised SOM, rule-based, k-nearest neighbors.

3.2 K- Nearest Neighbor (K-NN)

K-nearest neighbor (K-NN) classifying sample base on nonparametric techniques it calculates the distance between different points and then assigns the new instance to the class. Introduced k-NN to detect anomalous in global climate system and find the day and location that different from the other first the data arranged to reduce the high dimensionally and then apply the K-NN where the more distance more anomalous [16], combined genetic algorithm with K-NN for feature selection and weighting. All the features are weighted and the feature that gets the highest is selected [18].

3.3 Rule-Based

Decision tree (DT) One the powerful rule-based algorithm for classification and prediction, main three component for decision tree: arcs and nodes of leaves where ach node is given attribute, which is useful from the attributes that aren't yet considered along the path from the root. Each arc that emerges from the node is given a feature value. Each leaf is designated by a class or category., each instance will cheek and classify base on path from root to leave node that the instance trace it through the tree, using decision tree to detect anomaly where the first it scan from root to leave to find the probability then scan the tree from the leave to the root to find the probability for each node, combined Naive Bayes and decision tree to build an adaptive anomaly detection model [19].

3.4 Support Vector Machine (SVM)

Proposed support vector machine first maps the input vector into a higher –dimensional feature space and then obtain the optimal separating,moreover a decision boundary the separating hyperplane calculate by support vector rather than training all the data if the test instance falls within learned class it will concede as normal else it is anomaly, SVM is a binary classification which mean that there are two classes, Detecting Anomalous Sensor Events in Smart Home Data for Enhancing to detect anomalous by using Support vector machine to evaluate using precision, recall and F-measure where the data point that near to the region before transfer consider as anomaly the result show SVMs provide a good out-of-sample generalization data linearly separable however It assume that the ground truth available while this not general in all cases and does not mention to the error from sensor measure [20].

3.5 Bayesian Network Based

Naive Bayes basic technique of Bayesian network estimate the posterior probability of observing a class label, the test data instance will be predicted to the largest class posterior, Bayes theorem is useful because it provides ways to calculate the posterior probability, combined between -k-medied and naïve Bayes classification to distinguish between normal and attack behavioral [11].

3.6 Unsupervised Anomaly Detection

Clustering base technique is grouping data in clusters each clusters objects are similar, each cluster represent a pattern, clustering is unsupervised learning which mean there is no prior knowledge about the structure of data set.

3.6.1 Unsupervised Neural Network

The main Unsupervised neural network is Unsupervised Self-organizing map (SOM) the output of (SOM) are clusters in a grid, the network moves the winning node, which is the closest neuron, after determining which node is nearest to each training set. in the training course, Using extreme learning machine to build a model base on regression and unsupervised classification (SOM) to anomalies in large aviation data sets [21].

3.6.2 K-mean

K-mean divided data in cluster each similar object will be in the same cluster K represented the number of clusters for anomaly the small cluster will be an anomaly cluster, Clustering K-mean for Anomaly detection method using network pattern analysis process [22], using two algorithms to determine the first cluster and centered for k-mean clustering for Improved K-Means Using in Anomaly Detection [23]. To accurate the label class for test point by used Clustering k-mean with the vertical data format. Many developed versions of K –mean proposed [24], using genetic k-means instead of traditional k-mean [25].

To group numerous uncertainties in ocean surface temperature, clustering method with uncertainty must be utilized. Type 2 fuzzy hypothesis considers the instability of membership grade while fuzzy C mean (FCM) not. Based on the analysis of interval type-2 fuzzy C means (IT2FCM), the researchers utilize relationship as similarity estimation in time series clustering If relation be tween's objects is extensive, the principles of SST change in the comparing districts are noteworthy relative, at that point these regions are grouped into a similar cluster, the authors use two ordinary cloud models to express fuzzifiers m1 and m2, and utilizations two cloud drops to substitute them.

3.6.3 DBSCAN (Density-Based Spatial Clustering of Application of Noise)

DBSCAN can locate the random forms the concept behind DBSCAN is that each cluster item must have a certain amount of data objects in its neighborhood within the specified radius. Two parameters need to determine for DBSCAN which is Epsilon (radius) and Mints (minimum number of points as a neighborhood point). The anomaly point that is does not fit to any cluster (depend on the threshold) climate data arranged as for season so it necessary to remove it before applying anomaly technique by using DBSCAN base cluster (outlier point that does not core or border) Show the effective of using DBSCAN to detect an anomaly in climate data.

4 Ensemble Clustering

Cluster ensembles mean combine multiple clustering of a set of objects into a single consolidated clustering [26]. There are two parts in cluster ensemble Generation Mechanisms, Consensus Functions [27] Consensus Function, which computes a new partition by integrating all of the partitions acquired in the generating step, and Generation, which deals with creating a set of partitions of these objects [28]. The supervised learning field was first presented and well-studied for the ensemble method. Since this paradigm has been successfully used to classification tasks in the past few decades, especially to clustering problems, academics have tried to adapt it to the field of unsupervised learning. This could be problematic, though, for the following clear reason. This is because with unsupervised learning, there is no prior knowledge of the hidden structure, specific characteristics we are looking for, or what we believe to be sound data solutions [35] Clustering ensembles used for enhancing robustness, stability, and accuracy of unsupervised learning arrangements. The major problem of clustering ensemble is the consensus function [27].

5 Discussion

Several methods were proposed in the literature for the change point detection in one single variable, the advantage of these approaches that it can easily detect the high variability of value in any statistical method. For example, detected point anomaly by using density based scan to identify the anomaly point from temperature data where the anomaly that point does not fit any clusters [14], detected abrupt climate change in temperature data by applying Man-Kendall method, [29], using cusum to detect abrupt change temperature or precipitation data the result states that abrupt change happened in mean temperature for the years (1986,1996) [5], applied Bayesian information criterion with DP algorithm to detect change point in temperature data [7], also detected abrupt climate change in temperature time series [30].

However, detect the shift point in rainfall data need to concern to the relationship between that point and the other event in temperature and relative humidity at the same time stamp where abrupt rainfall happens, because of there is strong relationship between temperature and precipitation where the increasing or decreasing temperature can cause drought or flood [31].

The above studies focusing on detect change point in a single variable. While this change point which has higher value not necessary to be an abrupt change but it can be just an outlier [32] under those circumstances and from computer science view this outlier effect the accuracy of detection approach, also help to initiate suitable measure for adaption at any period of time [33].

Moreover, to find the information behind this abrupt change this might only be inferred when taking the combination of a various variable into account [38], more work needs to be done to combine different weather elements to perform clustering and interpret the synoptic relations among scenarios [34].

However finding change point in a multivariate environment still difficult and big challenging for earth observation data.To reduce the high dimensionality by feature selection, feature are selected temperature, precipitation [17], use the data of temperature and precipitation [35], mean temperature, maximum temperature, warm day, total precipitation, 5-day maximum precipitation, dry day, wind [5], use Land surface temperature, precipitation rate [33], using longitude, latitude, mer-wind, zone wind, humidity, air temperature, sea surface temperature [36], However, reduction in the number of features may lead to the loss in the information content and thereby lower the accuracy of the resulting [37].

Numerous studies that used supervised algorithm for anomaly detection, for instance, used Decision tree to detect anomalous where tree scanned from root to leave to find the probability then scan the tree from the leave to the root then find the probability for each nodes [38], while Detecting Anomalous Sensor Events in Smart Home Data for by using Support vector machine where the data point that near to the region before transfer consider an anomaly [39], also used Naive Bayes and decision tree to detect intrusion detection [38] used som to detect credit attack [19].

used K-Mean with vertical data format using pca for confirming weather the test instant belongs to the nearest cluster [24], using unsupervised clustering aggregation, including threshold aggregation (TA) and spatial aggregation technique to detect anomaly [40], dense clustering technique Used to classify the data in to normal and anomaly in network [41], using moving window on time series and count the fuzzy c mean for each subsequences [25], using genetic K-Means instead of traditional K-Mean to improve the outlier detection technique [42], using DBSCAN base cluster to detect outlier in temperature data,however, using clustering is useful for grouping data in to normal and anomaly but when new instance need cheek clustering will be not useful due to way which only grouping data [43], however, the main aim to use clustering is to group similar point in one cluster while the aim of this study is to detect the new instance that comes and cheek it if it anomaly for that many researcher's choices the use combination between clustering and classification to detect anomaly.

The ground truth is not required for an unsupervised mode which is most widely applicable when dealing with the unlabeled data set. For instance, detecting anomaly based similar data instances are grouped by K- Medoids technique and resulting clusters are classified into using SVM classifiers [44], anomaly detection method using network pattern analysis process clustering K-mean [22], to detect intrusion detection by used RF model to identify known attack and remove it then apply unsupervised SOM to have normal and unknown attack, to detect anomaly in wind turbine [45], using (SOM) the result show efficiency bycatch anomalies [46], An Improved K-Means Using in Anomaly Detection using two algorithms to determine the first cluster and centered for K-Mean clustering [23].

To distinguish between normal and attack behavioral by using k-medied and naïve bayes classification [11], companied between the cluster center and nearest neighbor, [47] combined Fuzzy C-Means and Artificial Neural Network [48], BIRCH hierarchical clustering algorithm used with SVM classifier [49], used Fuzzy C Means clustering with Artificial Neural [50], combined between of K-means clustering and Sequential Minimal Optimization (SMO) classification, Network all the above studies show up that the combined technique outperforms the classic algorithm in detection accuracy, however, as mention before the climate data is correlated data which mean each variable can affect the other in deferent way which leads to search for better technique can deal with this manner in literature the technique to deal with correlated data is rule-base technique which design to find the relationship among variables [51], Its aim is to which finds interesting correlation or association among attributes/items in large datasets [52].

When using unlabeled data this lead to be unsupervised learning (clustering) where there is no ground truth about data and it difficult to determine which clustering algorithm is suitable due to the different result that can obtain when applying different algorithm to the same dataset or change the parameter value to same algorithm [28]. Therefore, using ensemble clustering which mean combined many clustering partition in to one result which is also referred to as consensus ensemble or clustering aggregation to enhance the clustering performance, however there are many challenging when using ensemble clustering because it is not straightforward to transfer from supervised to unsupervised, the clustering ensembles technique need to improve their accuracy [27].

On the other hand, the abrupt climate is rare and for this, it is difficult to get a concrete example of it occurrence for that these change point can consider as an anomaly which has a high value compared with the other values in time series.

6 Conclusion

In this paper we try to give a brief view about anomaly detection technique that used to detect abrupt climate change, including Abrupt climate change and Anomaly detection application, in an attempt to support potential avenues for anomaly detection research.

References

1. Alley, R.B. et al.: Abrupt Climate Change, p. 11 (2003)
2. Franzke, C.: A novel method to test for significant trends in extreme values in serially dependent time series. Geophys. Res. Lett. **40**(7), 1391–1395 (2013). https://doi.org/10.1002/grl.50301
3. Chen, S., Li, Y., Kim, J., Kim, S.W.: Bayesian change point analysis for extreme daily precipitation. Int. J. Climatol. **37**(7), 3123–3137 (2016). https://doi.org/10.1002/joc.4904
4. Hao, L., Min, J.: Why precipitation has reduced in North China, pp. 940–943 (2011)
5. Fischer, T., Gemmer, M., Liu, L., Buda, S.: Change-points in climate extremes in the Zhujiang River Basin, South China, 1961–2007. Clim. Change **110**(3–4), 783–799 (2011). https://doi.org/10.1007/s10584-011-0123-8
6. Jin, H., He, W., Liu, Q., Wang, J., Feng, G.: The applicability of research on moving cut data-approximate entropy on abrupt climate change detection. Theor. Appl. Climatol. **124**(1–2), 475–486 (2016). https://doi.org/10.1007/s00704-015-1428-8
7. Wang, N.: Quantifying temporal-spatial variability of land-surface temperature anomalies using DP-based algorithm, pp. 1078–1084 (2015)
8. Kalyankar, M.A., Alaspurkar, S.J.: Data mining technique to analyse the metrological data. Int. J. Adv. Res. Comput. Sci. Softw. Eng. **3**(2), 114–118 (2013)
9. Chandola, V., Banerjee, A., Kumar, V.: Anomaly detection: a survey. ACM Comput. Surv. **41**(3), 1–58 (2009). https://doi.org/10.1145/1541880.1541882
10. Ranjan, R., Sahoo, G.: A new clustering approach for anomaly intrusion detection. Int. J. Data Mining Knowl. Manage. Process **4**(2), 10 (2014). https://doi.org/10.5121/ijdkp.2014.4203
11. Chitrakar, R., Chuanhe, H.: Anomaly based Intrusion detection using hybrid learning approach of combining k-medoids clustering and Naïve Bayes classification (2012)
12. Pajouh, H.H., Javidan, R., Khaymi, R., Dehghantanha, A., Raymond, K.: A two-layer dimension reduction and two-tier classification model for anomaly-based intrusion detection in IoT backbone networks, vol. 6750, no. c, pp. 1–11 (2016). https://doi.org/10.1109/TETC.2016.2633228
13. Aggarwal, C.C., Heights, Y.: Outlier Analysis. Second Edition (2016). https://doi.org/10.1007/978-3-319-47578-3
14. Dada, F., ÇELİK, M.: Anomaly detection in temperature data using DBSCAN algorithm. 91–95 (2011)

15. Panigrahi, S., Verma, K., Tripathi, P., Sharma, R.: Knowledge discovery from earth science data. In: 2014 Fourth International Conference on Communication Systems and Network Technologies (CSNT), pp. 398–403 (2014). https://doi.org/10.1109/CSNT.2014.85
16. Das, M.: Anomaly Detection and Spatio-Temporal Analysis of global climate system, pp. 142–150 (2009)
17. Sun, Y., Xie, K., Ma, X., Jin, X., Pu, W., Gao, X.: Detecting spatio-temporal outliers in climate dataset: a method study, no. 40235056, pp. 760–763 (2005)
18. Su, M.Y.: Real-time anomaly detection systems for Denial-of-Service attacks by weighted k-nearest-neighbor classifiers. Expert Syst. Appl. **38**(4), 3492–3498 (2011). https://doi.org/10.1016/j.eswa.2010.08.137
19. Harbi, N., Rahman, M.Z.: Combining naive Bayes and decision tree. Int. J. Netw. Secur. App. **2**(2), 12–25 (2010)
20. Vapnik, V.N.: An overview of statistical learning theory. IEEE Trans. Neural Netw. Publ. IEEE Neural Netw. Council **10**(5), 988–999 (1999). https://doi.org/10.1109/72.788640
21. Janakiraman, V.M., Nielsen, D.: Anomaly detection in aviation data using extreme learning machines (2016)
22. Minho Han, I.K.: Anomaly detection method using network pattern analysis of process, pp. 159–163 (2015)
23. Yin, C., Zhang, I.K., Wang, S.J., Kim, J.: An improved k-means using in anomaly detection, pp. 3–5 (2015). https://doi.org/10.1109/CCITSA.2015.11
24. Prathibhamol, C.P., Amala, G.S., Kapadia, M.: Anomaly detection based multi label classification using association rule mining (ADMLCAR), pp. 2703–2707 (2016)
25. Izakian, H., Member, S., Pedrycz, W.: Anomaly detection and characterization in spatial time series data: a cluster-centric approach, vol. 6706, no. c, pp. 1–13 (2014). https://doi.org/10.1109/TFUZZ.2014.2302456
26. Ghosh, J., Acharya, A.: Cluster ensembles. WIREs Data Mining Knowl. Disc. **1**(4), 305–315 (2011). https://doi.org/10.1002/widm.32
27. Mekala, M., Elango, P.: A survey: clustering ensemble techniques with consensus function. IJSRSET **2**(4), 597–601 (2016)
28. Vega-pons, S., Shulcloper, J.R.: A survey of clustering ensemble algorithms. Int. J. Pattern Recogn. Artif. Intelligence **25**(03), 337–372 (2011). https://doi.org/10.1142/S0218001411008683
29. Yi, X., Li, G., Yin, Yu.: Temperature variation and abrupt change analysis in the Three-River Headwaters Region during 1961–2010. J. Veograph.l Sc. **22**(3), 451–469 (2012). https://doi.org/10.1007/s11442-012-0939-9
30. Yi, Xi., Li, G., Yin, Y.: Temperature variation and abrupt change analysis in the Three-River Headwaters Region during 1961–2010. J. Geogr. Sci. **22**(3), 451–469 (2012). https://doi.org/10.1007/s11442-012-0939-9
31. IPCC, Climate Change 2007 Synthesis Report (2007). https://doi.org/10.1256/004316502320517344
32. Rodionov, S.N.: A comparison of two methods for detecting abrupt changes in the variance of climatic time series, no. 2005, pp. 1–32 (2016)
33. M. Das, "Detection of Climate Zones using Multifractal Detrended Cross-Correlation Analysis : A Spatio-temporal Data Mining Approach," pp. 1–6, 2015
34. Minghua Zheng, B.A.C., Edmund, K., Chang, M.: Applying Fuzzy Clustering to a Multi-model Ensemble for U . S . East Coast Winter Storms: Scenario Identification and Forecast Verification, pp. 881–903 (2017). https://doi.org/10.1175/WAF-D-16-0112.1
35. Fan, L.: A novel frequent patterns mining method of unusual climate events in data of East Asian monsoon Zone, pp. 3–6 (2010
36. Kolhe, P.N., Ugale, R.M., Shingne, V.: Detection of rare patterns in climate change using data mining techniques. vol. 8, no. 2, pp. 25–31 (2015)

37. Jain, A., Duin, R., Mao, J.: Statistical pattern recognition: a review. IEEE Trans. Pattern Anal. Mach. Intell. **22**(1), 4–37 (2000). https://doi.org/10.1109/34.824819
38. Delibalta, I., Gokcesu, K., Simsek, M., Baruh, L., Kozat, S.S., Member, S.: Online anomaly detection with nested trees. IEEE Signal Process. Lett. **23**(12), 1867–1871 (2016)
39. Jakkula, V., Cook, D.J.: Detecting Anomalous Sensor Events in Smart Home Data for Enhancing the Living Experience (2011)
40. Yang, D., Miao, D., Qin, X., Wei, G.: A Novel Anomaly Detection with Temporal and Spatial Aggregation in Mobile Networks," no. 1 (2016)
41. Zhao, C., Cui, Y., Zhou, X., Wang, Y.: Evaluation of performance of different methods in detecting abrupt climate changes. Discrete Dyn. Nat. Soc. **2016**, 5898697 (2016)
42. Marghny, M.H., Taloba, A.I.: Outlier detection using improved genetic k-means. Int. J. Comput. Appl. **28**(11), 975–8887 (2011). https://doi.org/10.5120/3458-4723
43. ÇELİK, M.: Anomaly detection in temperature data using DBSCAN Algorithm, June 2011
44. Chitrakar, R., Chuanhe, H.: Anomaly detection using support vector machine classification with k-Medoids clustering, pp. 1–5 (2012)
45. Kim, E.: A Novel Hierarchical Detection Method for Enhancing Anomaly Detection Efficiency, pp. 1018–1022 (2015). https://doi.org/10.1109/CICN.2015.202
46. Du, M., Ma, S., He, Q.: A SCADA Data based Anomaly Detection Method for Wind Turbines, no. Ciced, pp. 10–13 (2016)
47. Zainaddin, D.A.A., Hanapi, Z.M.: Hybrid of fuzzy clustering neural network over NSL dataset for intrusion detection system. J. Comput. Sci. **9**(3), 391–403 (2013). https://doi.org/10.3844/jcssp.2013.391.403
48. Lin, W.C., Ke, S.W., Tsai, C.F.: CANN: An intrusion detection system based on combining cluster centers and nearest neighbors. Knowl Based Syst **78**(1), 13–21 (2015). https://doi.org/10.1016/j.knosys.2015.01.009
49. Horng, S.J., et al.: A novel intrusion detection system based on hierarchical clustering and support vector machines. Expert Syst. Appl. **38**(1), 306–313 (2011). https://doi.org/10.1016/j.eswa.2010.06.066
50. Pandeeswari, N., Kumar, G.: Anomaly detection system in cloud environment using fuzzy clustering based ANN. Mobile Netw. App. **21**(3), 494–505 (2016). https://doi.org/10.1007/s11036-015-0644-x
51. Mohamed, S., Mohamed, A., Mokhtar, R.A.: Anomaly detection approach using hybrid algorithm of data mining technique (2017)
52. Manikandan, M., Mala, R.: Survey on analysis of meteorological condition based on data mining techniques. Int. J. Adv. Eng. Res. Sci. **3**(7), 70–75 (2016)

Sustainable Finance Echelon Futurism: Vanguard Visionaries' Financial Mastery for Organizational Excellence and Business Prosperity

Enkeleda Lulaj(✉)

Faculty of Business, University Haxhi Zeka, Peja, Kosovo
enkeleda.lulaj@unhz.eu

Abstract. The study investigates the financial mastery of Vanguard visionaries by focusing on sustainable finance factors: organizational optimization and resource capital (OORC), financial innovation and investment (FII), and optimizing business performance (OBP). The goal was to identify correlations between these factors and provide insights into best practices for sustainable finance and business prosperity. Data from 200 companies in 2023 were analyzed by applying exploratory and confirmatory factor analyses using SPSS and AMOS software. The results show a strong relationship between OORC, FII, and OBP, highlighting the importance of organizational optimization, financial innovation, and business performance for sustainable financial success. OORC variables, such as resource capital management and financial stability, play a crucial role. The impact of FII on innovation and liquidity and the focus of OBP on talent management contribute significantly to business prosperity. The study provides valuable insights for developing strategies for sustainable finance and business prosperity and highlights the importance of Vanguard's financial mastery for developing strategies for sustainable finance futurism, organizational excellence, and business prosperity.

Keywords: Sustainable Finance · Financial Mastery · Futurism Finance · Vanguard Visionaries' · Business Prosperity

1 Introduction

In this dynamic period of financial evolution, there is a growing need for a deep understanding of sustainable finance, financial mastery, and strategic vision within businesses of all types. This growing demand is due to the challenges presented by rapid technological change and increasing expectations for sustainability in the business environment. Therefore, in this research, the contributions of other authors have been analyzed in terms of sustainable finance echelon futurism through vanguard visionaries' financial mastery for organizational excellence and business prosperity based on the three main factors included in OORC (Organizational Optimization and Resource Capital), FII (Financial and Investment Innovation), and OBP (Optimizing Business Performance). According

A. J. Obaid et al. (Eds.): ICONEST 2024, ISEM 40, pp. 72–89, 2025.
https://doi.org/10.1007/978-3-031-89960-7_6

to Sadiq et al. (2023), the conceptual framework constructed experiences changes in organizational performance by comparing the types of diversification identified in the research of (Boz et al., 2013). The analysis of the time orientation of managers, according to the study of (Chen et al., 2023), highlights the impact of this orientation on the allocation of financial assets in companies. The findings of Tseng et al. (2021) highlight the importance of blockchain technology, cash flow scarcity, return factoring, risk assessment, and the triple bottom line (TBL) in sustainable finance. According to Lulaj (2023), it is emphasized that the key factors for achieving sustainable business profit through customer relationships include employee behavior, efficient handling of requests, strong pre- and post-purchase support, clear communication of product details (such as discounts, usage, expiration dates, and payment methods) and providing transportation to customers. However, another contribution by Lulaj et al. (2024), emphasizes that businesses will struggle to maintain strategic profit and competitive market survival if they fail to adapt quickly to environmental changes, lack strong managerial skills and effective leadership, or neglect the implementation of innovative strategies, debt repayment, and employee performance evaluation.

The synthesis of research in recent decades is organized around two main themes: sustainable finance and FinTech/digitalization, according to the studies by Saidane and Ben Abdallah (2021). In the panel data analysis of Anifowose et al. (2020), (IR) capital has a positive effect on income growth, while human and natural capital indirectly affects the cost of financing. According to Azadda et al. (2023), sustainable finance has an impact on firms' risk resilience and risk management infrastructure, including the moderating effect of managerial capability and innovation capability. On financiers and cost of capital, Willem van Gelder et al. (2021) and Tanjung (2023) provide perspectives on the criteria and their impact on firm performance and market value. The results of the study by Ahmad et al. (2023) show that intellectual capital and corporate social responsibility have a positive impact on the sustainable competitive performance of SMEs. In the analysis by Neher et al. (2018), organizations with embedded management values show better financial performance, including customer focus, innovation, and sustainability. Sustainable finance concludes with an analysis of the effectiveness of the flow of operations according to Afy-Shararah and Rich (2018). Furthermore, according to Abbas et al. (2024), the results show a positive impact of competition on firm performance, while digital financial innovation has a partially positive mediating effect, deepening the literature on digital innovation.

The open business model of the circular economy is analyzed as an innovative variant that offers greater flexibility and social benefits widely distributed among different actors (Fedele and Formisano, 2023). Private equity investments show a significant impact on fostering firm innovation, increasing the role of innovation support compared to separate investments (Liang et al., 2023). These findings highlight differences in how firms respond to pressures for growth and survival, causing innovation to shift from a cooperative to a passive state (Su et al., 2023). Lulaj et al. (2023) emphasize that to achieve sustainable profit, businesses should focus on managing total liabilities, improving total asset performance, increasing net financial income, increasing total business income, and investing in employee skills and technology (e.g., equipment and machinery). The developed model shows that changes in private risk can turn a previously stable institution into a source of conflict, especially for firms with weak internal monitoring (Farrer et al., 2017). This leads to poor performance and the intervention of corporate raiders (Markides and Singh, 1997). According to Lulaj and Iseni (2018), the great importance of cost-volume-profit analysis is emphasized for achieving a sustainable profit in terms of sustainable finances and business prosperity, moreover (Lulaj, 2021) underlines that large corporations usually have a stronger financial position than small businesses, which are vulnerable to the risk of bankruptcy. Through a quantitative approach, the results show that human resource capacity and business diversification have an impact on the financial performance of SMEs, while Sharia financial technology has a positive effect on their sustainability (Menne et al., 2022). The study presents a new approach to the simulation of business processes, using event logs to discover accurate simulation models and provide a better approach to the performance of processes (López-Pintado et al., 2024).

Finally, it is emphasized that the study will build a theoretical and research perspective on the financial mastery of the vanguard visionaries and the sustainable finance echelon futurism based on the main factors of this study OORC, FII, and OBP. Therefore, the purpose of this study is to identify and demonstrate the correlations between these factors, bringing knowledge and guidelines for best practices that contribute to sustainable finance, organizational excellence, and business prosperity, also within this section the hypotheses of this research have been built based on the literature exciting and the purpose of the research. The constructed hypotheses are as follows, which will be elaborated in detail and through the conceptual model presented in Fig. 1.

H: There is a statistically significant and positive relationship between OORC, FII and OBP factors.

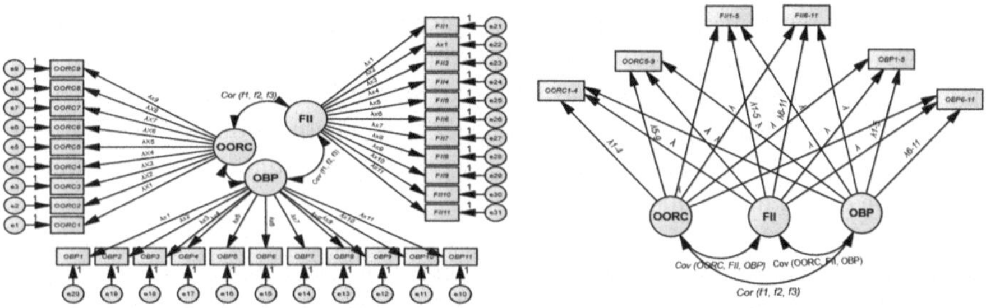

Fig. 1. Design model of OORC, FII, and OBP factors *Source: Prepared by the author.*

Figure 1 shows the design model that explores the relationships between organizational optimization and resource capital (OORC), financial innovation and investment (FII), and optimizing business performance (OBP) within the realm of sustainable finance futurism. This framework is guided by the financial expertise of pioneering visionaries and promotes organizational excellence and business prosperity. To accurately design the model, the use of exploratory (EFA) and confirmatory (CFA) factorial analyses for hypothesis building and validation is essential. Therefore, hypothesis (H) represents a statistically significant and positive relationship between the factors: OORC, FII, and OBP. Essentially, it examines whether there is a strong and positive relationship between efficient organizational practices, resource utilization, financial innovation, investment, and overall firm performance. In summary, it will be concluded that the research will be a valuable guide for companies and finance professionals, illuminating the complex dynamics of how optimizing organizational and resource capital, fostering financial innovation, and making strategic investments will shape overall business performance. Particularly in the context of sustainable finance, it will serve as a beacon for making informed strategic decisions and building resilient financial plans for the future.

2 Methodology

The research focuses on the financial mastery of Vanguard visionaries and factors of sustainable finance futurism, such as organizational optimization and capital resources (OORC), financial innovation and investment (FII), and business performance optimization (OBP). Thus, the objective is to analyze and illustrate the relationships between these factors, providing valuable insights and recommendations for best practices that promote sustainable finances and business prosperity.

2.1 Data Analysis

To assess the significance of the model and validate the proposed hypothesis, this research's data were analyzed using exploratory factor analysis (EFA), reliability analysis (Cronbach's Alpha), and confirmatory factor analysis (CFA) through SPSS and Amos software. The econometric model is presented below in graphical form.

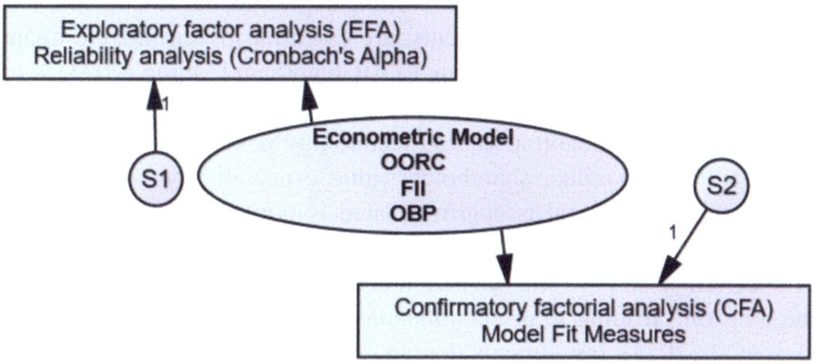

Fig. 2. Econometric Model. Source: Prepared by the author.

Figure 2 shows the econometric model for the three factors of this research (OORC, FIII, and OBP) elaborated in two steps. Therefore, in the first step (S1) of this research, EFA was applied, as highlighted by Spearman (1927), noting that EFA is one of the most widely used statistical analyses in various fields, including economics. Therefore, based on the arguments of (Floyd and Widaman, 1995) that EFA is employed through questionnaires in the assessment of instruments across multiple factors, a questionnaire was developed for the factors (OORC, FII, and OBP). Hence, in the second step (S2), CFA was utilized to determine the model structure by applying regression (β) and correlation (r) in predicting items from latent factors (OORC, FII, and OBP), as well as multiple regression/correlation analysis (Cohen et al., 2003).

2.2 Data Collection

The dataset was gathered from 200 businesses in the year 2023 through a questionnaire completed by financial leaders. Participants were made aware that the data collected would be used for research purposes only and would be kept confidential. They participated voluntarily to support understanding of the future of sustainable finance and business growth. The questionnaire was designed using a Likert scale, ranging from 1 (strongly disagree) to 5 (strongly agree).

3 Results and Discussion

The study examined the main factors (OORC, FII, and OBP) highlighted in the literature introduction and methodology sections. This was followed by a rigorous hypothesis (H) testing through econometric analyses (EFA, CFA). These analyses were based on the variables related to the identified factors, providing a strong and comprehensive assessment of the proposed hypothesis. In this section, an attempt is made to improve the understanding of the paper by establishing links between the findings of other authors and the results of this research for each factor and its variable. According to Schilling and Seuring (2023), it is emphasized that companies providing online services have a positive impact on the information, financial, and physical flows of local supply chains. Lulaj and Minguez-Vera (2024) highlight that key factors that drive positive cash flow, such as clear optimization strategies, effective working capital management, sound financial decisions, and technological improvements, are essential for sustainable finances and business prosperity. In addition, according to AlGhamdi and Durug (2023), a model of strategies for intra-and inter-organizational levels is proposed that includes communication skills, recruitment, negotiation, and technology development. In addition, firm diversification is found to reduce shareholder value, especially for single business firms that diversify once, with internal asset growth rated as more effective than diversification through acquisitions (Panayiotis et al., 2016). The impact of return predictability and parameter uncertainty on portfolio allocation is associated with the degree of hopelessness in the long run, leading to a recommendation to hold zero units of a risky asset (Kontosakos et al., 2024). Investments in renewable resources and capital development are found to have a positive impact on sustainability (Mao et al., 2023). Lulaj et al. (Lulaj et al., 2024) explore how financial accounting digitalization (FAD) impacts advanced financial accounting systems (AFAS) and accounting 2.0/FinTech (AFT). The study shows that FAD drives the adoption of AFAS and AFT, increasing financial sustainability and operational efficiency. The strong links between FAD and AFT highlight the transformative potential of digitalization in improving financial practices and supporting financial sustainability and business prosperity. Additionally, Lulaj et al. (2024)

emphasize that sales velocity and profit growth help in sustainable business growth. This section also examines megaprojects, organizational diversity, revenue management efficiency, and the impact of personal protective equipment (PPE) in a specific context, providing insights into business and economic challenges and opportunities.

In line with the contribution of Sawarni et al. (2023), their research focuses on the impact of earnings management (EM) on the effectiveness of working capital management (WCM) in businesses. The findings highlight that EM has the potential to negatively impact WCM efficiency, leading to changes in the cash conversion cycle and inventory management. In addition, research by Rasouli et al. (2024) explores the impact of advanced personal protective equipment (PPE), examining both the benefits and challenges associated with these technologies in protecting personnel and mitigating workplace risks. A key conclusion by Moon (2022) is that the uptake of research and development can affect the external financing of technological innovation by firms. In addition, it is found that explosive product cycles can cause a twofold advance in the equation rate of stocks, which has implications for business prosperity, organizational excellence, and the future of sustainable finance (Jinnai, 2015). Furthermore, according to Ikeda et al. (2023) emphasized that this research contributes to the understanding and development of pricing strategies in the context of trade and service industries, addressing the challenges of market trends.

From the perspective of other authors, the results of this research will now be elaborated and elaborated in detail, coming to very important conclusions and recommendations for companies.

Table 1 shows the pattern matrix according to (EFA) which emphasizes the creation of three factors (Organizational Optimization and Resource Capital- OORC, Financial and Investment Innovation- FII, and Optimizing Business Performance- OBP) with their subfactors (OORC1-8, 10, FII1 -11, and OBP1-11) regarding the sustainable finance echelon futurism through Vanguard visionaries' financial mastery for organizational excellence and business prosperity. All the subfactors of the factors are important as their value is above 0.50. According to the KMO test (Kaiser and Rice, 1974) for all three factors, it is confirmed that the data have a reliable fit to the model as its value is above 0.80 (OORC, 0.868; FII, 0.812; and OBP, 0.882). Moreover, according to the reliability analysis (Cronbach's Alpha) (Cronbach, 1951), it is emphasized that the factors have a high value of reliability (OORC, $0.81 \geq \alpha \leq 0.80$; FII, $0.84 \geq \alpha \leq 0.80$; OBP, $0.85 \geq \alpha \leq 0.80$).

Table 2 presents the findings of the confirmatory factorial analysis (CFA) regarding the sustainable finance echelon futurism through Vanguard visionaries' financial mastery for organizational excellence and business prosperity among the factors: OORC (1–8, 10), FII (1–11), and OBP (1–11). In this analysis according to (Bollen, 1989), the results show a significant and statistically reliable influence of each observed variable (OORC1–10, OORC10), (FII1–11), and (OBP1–11) on the latent variables OORC, FII, and OBP. According to the estimation, it is emphasized that all latent variables (OORC, FII, OBP) are statistically significant as they have a standardized regression weight greater than 0.5, and $p < 0.001$ (***), (Kline, 2016). Therefore, in the organizational optimization and resource capital factor (OORC), the most important variables are (OORC10 < ---0.607) or RMC_Efficiency (resource management and capital), (OORC5 < ---0.600)

Table 1. EFA and Cronbach's Alpha

Item	Construct	Factor Loading λ	KMO and Bartlett's Test	Cronbach's Alpha	Interpretation
Organizational Optimization and Resource Capital- OORC					
OORC8	Financial stability and financial value creation	0.542	KMO = 0.868 Approx. Chi-Square = 403.501 df = 36 Sig. = 0.000	α = 0.812 N of Items = 9	Substantial adequacy (Kaiser, 1970, 1974) Good internal consistency (Cronbach, 1951, 2004)
OORC7	Technology transfer and value creation in entities	0.581			
OORC1	Efficient diversification in the organization	0.585			
OORC2	Determine assets and value in the context of loss	0.527			
OORC3	Efficient management of limited resources	0.584			
OORC4	Efficient management of organizational structure	0.585			
OORC5	Organizational diversity and value creation	0.617			
OORC10	RMC_Efficiency (resource management and capital)	0.588			
OORC6	Effective Equipment Utilization and Inter-Unit Equipment Collaboration	0.517			

(*continued*)

Table 1. (*continued*)

Item	Construct	Factor Loading λ	KMO and Bartlett's Test	Cronbach's Alpha	Interpretation
Financial and Investment Innovation- FII					
FII1	Focus on innovation in main activities	0.644	KMO = 0.882 Approx. Chi-Square = 556.540 df = 55 Sig. = 0.000	α = 0.841 N of Items = 11	Substantial adequacy Good internal consistency
FII2	Increasing liquidity and business innovation	0.660			
FII3	Innovation in taking advantage of good prices offered for assets	0.540			
FII4	Shifting innovation resources to high-growth or high-opportunity businesses	0.559			
FII5	Innovating in response to adverse economic conditions	0.509			
FII6	Funding innovation in other investments	0.586			
FII7	Innovation to improve business underperformance	0.554			
FII8	Innovation in compulsory investment in loss-making units	0.537			
FII9	Innovation to correct past investment mistakes	0.514			

(*continued*)

Table 1. (*continued*)

Item	Construct	Factor Loading λ	KMO and Bartlett's Test	Cronbach's Alpha	Interpretation
FII10	Innovation to reduce the number and variety of operating units	0.564			
FII11	Innovation to curate a successful business	0.602			
Optimizing Business Performance- OBP					
OBP1	Improve business structure to facilitate the growth of independent units	0.513	KMO = 0.885 Approx. Chi-Square = 599.767 df = 55 Sig. = 0.000	α = 0.849 N of Items = 11	Substantial adequacy Good internal consistency
OBP2	Modernization and elimination of obsolete technology or know-how	0.629			
OBP3	Select segments that do not create an unfavorable public image	0.565			
OBP4	Exiting segments where customers are strong	0.661			
OBP5	Exit segments where suppliers are strong	0.532			
OBP6	Reduce the risk of over-diversification of business	0.684			
OBP7	Manage talent shortages	0.555			
OBP8	Free up management talent for other areas	0.536			

(*continued*)

Table 1. (*continued*)

Item	Construct	Factor Loading λ	KMO and Bartlett's Test	Cronbach's Alpha	Interpretation
OBP9	Prevent the loss of competitive advantage due to new regulations	0.607			
OBP10	Manage new regulations to minimize waste	0.579			
OBP11	Avoid loss of economies of scale in production or distribution	0.536			

Note: *** $p < .001$. *Source: Table prepared by the author.*

or organizational diversity and value creation, and (OORC8 < ---0.593) or financial stability and financial value creation. In the financial and investment innovation factor (FII), the most important variables are (FII2 < ---0.662) or focus on innovation in main activities, and (FII1 < ---0.638) or increase liquidity and business innovation. In the optimizing business performance factor (OBP), the most important variables are (OBP5 < ---0.671) or exit segments where suppliers are strong, (OBP9 < ---0.668) or prevent the loss of competitive advantage due to new regulations, and (OBP7 < ---0.646) or manage talent shortages. A 99.9% confidence level confirms that these results are highly reliable. This confirmatory analysis provides a crucial enhancement to our grasp the complex dynamics of financial mastery in the context of Vanguard's vision and reflects a solid foundation for developing strategies for the future of sustainable finance and business prosperity.

Table 3 shows the results of the covariance analysis, which emphasizes the correctness of the relationships between the latent variables (OORC, FII, and OBP). The estimated values present statistically significant results (***) for all three factors at the 1% level: the relationship between (OORC < -- > FII) has a positive covariance in the value of 0.300, the relationship (OORC < -- > OBP) has a positive covariance in the value of 0.285, as well as the relationship (FII < -- > OBP) has a covariance in the value of 0.282. These positive interactions suggest that firms that successfully manage resources and foster innovation are more likely to secure stable finances and experience prosperity. Results further underscore the impact of organizational excellence strategies in achieving a successful combination of financial sustainability and business growth (Anwar, 2018).

Table 4 shows the correlation analysis, which shows strong and positive relationships between the latent variables (OORC, FII, and OBP) regarding the sustainable finance echelon futurism through Vanguard visionaries' financial mastery for organizational excellence and business prosperity. Regarding (OORC < -- > FII), there is a high and

Table 2. CFA

O.V	L.V	S. R. W	Estimate	S. E	C. R	C.I of 99.9%
OORC1	OORC	0.568	1.000	0.147	5.987	Statistically Significant
OORC2		0.518	0.880	0.160	6.256	
OORC3		0.549	1.001	0.154	6.215	
OORC4		0.545	0.955	0.158	6.663	
OORC5		0.600	1.050	0.161	6.390	
OORC6		0.566	1.026	0.158	6.430	
OORC7		0.570	1.015	0.153	6.607	
OORC8		0.593	1.008	0.162	6.722	
OORC10		0.607	1.086			
FII1	FII	0.638	1.000	0.146	7.903	Statistically Significant
FII2		0.662	1.156	0.135	6.768	
FII3		0.550	0.914	0.121	6.693	
FII4		0.543	0.813	0.126	6.669	
FII5		0.541	0.842	0.130	6.874	
FII6		0.560	0.890	0.125	6.703	
FII7		0.544	0.839	0.132	6.802	
FII8		0.553	0.896	0.122	6.319	
FII9		0.508	0.773	0.137	7.116	
FII10		0.583	0.976	0.124	7.163	
FII11		0.588	0.886			
OBP11	OBP	0.550	1.000	0.154	6.249	Statistically Significant
OBP10		0.560	0.960	0.169	7.026	
OBP9		0.668	1.189	0.159	6.310	
OBP8		0.567	1.006	0.170	6.878	
OBP7		0.646	1.171	0.159	5.980	
OBP6		0.526	0.951	0.176	7.046	
OBP5		0.671	1.242	0.165	6.111	
OBP4		0.542	1.009	0.149	6.066	
OBP3		0.537	0.905	0.168	6.588	
OBP2		0.604	1.105	0.147	5.930	
OBP1		0.520	0.871			

*Note: Standard Error (S.E.), Critical Ratios (C.R.), *** $p < .001$ indicates statistical significance. The confidence interval is set at 99.9% (CI). Standardized Regression Weights (S. R. W.). Observed variable (O.V), Latent Variable (L.V). Source: Table prepared by the author.*

positive correlation between OORC and FII with a value of 0.860, with confidence interval of (0.810, 0.905). This result confirms a very strong and positive correlation between OORC and FII (r = 0.86). Furthermore, regarding (OORC < -- > OBP), there is a high and positive correlation between OORC and OBP with a value of 0.887, and a confidence interval of (0.842, 0.918). This result confirms a strong and positive correlation between OORC and OBP (r = 0.89). As for (FII < -- > OBP), there is a high and positive correlation between FII and OBP with a value of 0.830, a confidence interval of (0.780, 0.876), proving the importance of these factors (r = 0.83). Therefore,

Table 3. Covariance

Covariance's	Est	S.E	C.R	Interpretation
OORC < -- > FII	0.300	0.055	5.487	Statistically significant
OORC < -- > OBP	0.285	0.055	5.184	and positively
FII < -- > OBP	0.282	0.053	5.367	covariance

*Note: *** $p < 0.001$ indicates statistical significance, Estimate (Est.). Source: Table prepared by the author.*

Table 4. Correlations

Correlations	Estimate	Confidence Interval (95%)	p-value	Asterisk	Interpretation
OORC < -- > FII	0.860	(0.810, 0.905)	p < 0.001	***	Highly strong
OORC < -- > OBP	0.887	(0.842, 0.918)	p < 0.001	***	and positive
FII < -- > OBP	0.830	(0.780, 0.876)	p < 0.001	***	correlation

*Note: *** $p < 0.001$ indicates statistical significance. The confidence interval is set at 99.9% (CI). Source: Table prepared by the author.*

these findings suggest that the financial mastery of Vanguard visionaries, through the management of operational risk and financial information, has a positive and lasting impact on organizational excellence and business prosperity in the future of sustainable finance.

Table 5 shows the results of the model fit for the factors (OORC, FII, OBP). Based on the tests performed, the results are as follows: CMIN/DF ($p = 1.197 \approx 1.2$), RMR ($p = .049 \approx 0.5$), GFI ($p = .870 \approx 87\%$), CFI ($0.959 \geq 0.95$), PClose ($0.999 \geq 0.05$), and RMSEA ($0.031 \leq 0.06$). These values indicate excellent model. Overall, the interpretation of the model fit based on all tests emphasizes the significance and accuracy of the data related to the (OORC, FII, OBP) factors, their factors (OORC1–8, 10; FII1–11, and OBP1–11), and the subfactors within each other.

Figure 3 presents the relationship regarding the sustainable finance echelon futurism through Vanguard visionaries' financial mastery for organizational excellence and business prosperity based on the factors (OORC, FII, OBP) as well as these factors with their variables (OORC1–8, 10; FII1–11; and OBP1–11). Therefore, according to the path diagram, there is a high and positive relationship between OORC (Organizational Optimization and Capital Resources) and FII (Financial and Investment Innovation) with a value of ($r = .86$), therefore it is emphasized that companies that focus on optimizing their organizational structure and capital resources have a tendency to use financial innovation and investment in an efficient way that contributes to the security and stability of finances. Furthermore, the OORC factor has a statistically high and positive relationship with the OBP (Optimizing Business Performance) factor at a value of ($r = .89$), it is emphasized that efforts towards improving business performance have a positive impact on achieving financial prosperity of sustainable, deepening business capabilities

Table 5. Model Fit

Model Fit Measures	Value of Model OORC & FII & OBP	Cutoff Criteria	Description	Interpretation
CMIN ($\chi 2$)	489.757	--	$\chi^2 - \chi'^2 = \sum_{i=1}^{k} \frac{\chi_i^2}{m_i} - \sum_{i=1}^{k} \frac{\chi^2}{m_i'}$ (Fisher, 1922)	--
DF	409	--	(Eisenhauer, 2008) (X2/df)	--
CMIN/DF	1.197	Between 1 and 3	Chi-square Minimization per Degree of Freedom	Excellent
CFI	0.959	> 0.95	Comparative Fit Index (Hu and Bentler, 1999)	Excellent
SRMR	0.051	< 0.08	Standardized Root Mean Squared Residual (Diamantopoulos and Siguaw, 2000)	Excellent
RMSEA	0.031	< 0.06	(MacCallum et al., 1996)	Excellent
PClose	0.999	> 0.05	P-value of the null hypothesis (Gaskin and Lim, 2016)	Excellent
RMR GFI	.049 .870	≤ 0.07 ≥ 0.9	Goodness of Fit Index (Hu and Bentler, 1998) Root Mean Squared Residual (Steiger, 2007)	Excellent

Note: Degrees of freedom (DF), Root Mean Square Error of Approximation (RMSEA). Source: Table prepared by the author.

to adapt and develop successfully in a changing economic environment. Then, the factor (FII) has a high and positive correlation with the factor (OBP) with a value of (r =.83), therefore it is emphasized that this positive relationship can be interpreted as a signal that investments and strategies that promote financial innovation have the potential to have a positive impact on business prosperity.

Table 6 shows that the findings confirm a strong and positive relationship between the OORC, FII, and OBP factors, indicating a strong fit of the model and validating the proposed hypothesis regarding sustainable finances and business prosperity.

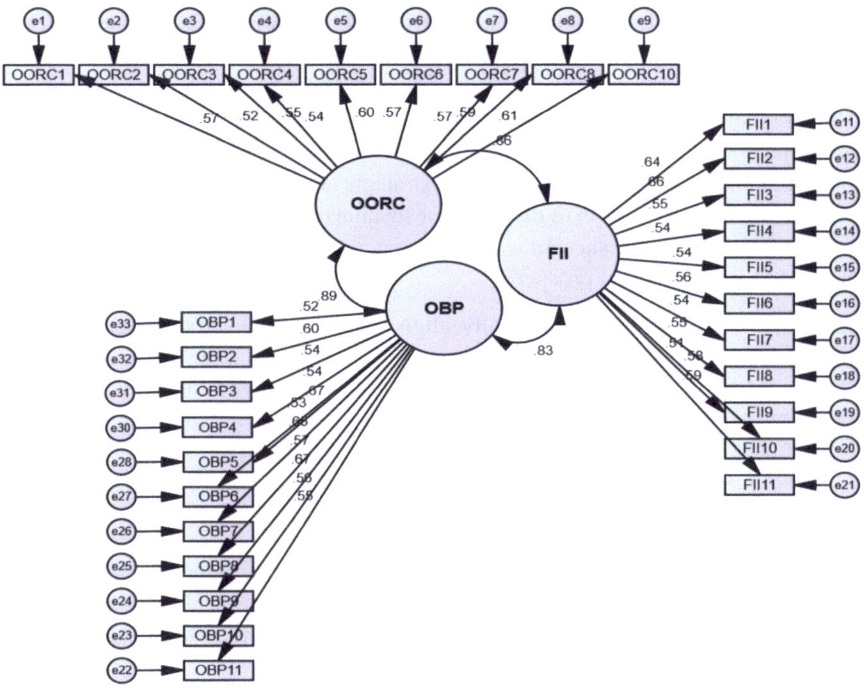

Fig. 3. Path diagram. Source: Prepared by the author.

Table 6. Hypothesis Assessment

Hypotheses	Elaboration	Tests	Rejected/Accepted
H	There is a statistically significant and positive relationship between OORC, FII and OBP factors	Excellent Model Fit	Accepted

Source: Table prepared by the author.

4 Conclusions and Future Studies

In summary, this research explores the complex dynamics of sustainable finance and organizational prosperity, focusing on the financial mastery demonstrated by Vanguard visionaries. The key success factors identified, namely Organizational Optimization and Resource Capital (OORC), Financial and Investment Innovation (FII), and Optimization Business Performance (OBP), emerge as critical elements for companies navigating the complexities of the financial landscape. Vanguard's leadership model serves as a beacon, highlighting the central role of organizational excellence, financial innovation, and strategic business performance in achieving long-term success. The study not only provides practical guidance for financial leaders, with particular attention to factors such as RMC efficiency and innovation in key activities but also makes a significant contribution to the academic discourse by validating models that predict sustainable finance. The robust econometrical analyses validate the reliability of the identified factors and underscore their importance in shaping the future of financial practices. As companies strive for prosperity in an evolving economic environment, the study's findings provide a valuable roadmap, emphasizing the need for strategic alignment, risk mitigation through innovation, and the promotion of diversity for sustained financial stability and organizational excellence. Achieve success with these concise recommendations for sustainable financial and organizational prosperity:

- Integration for Success: Strategically align organizational improvement, financial innovation, and business performance.
- Prioritize Efficiency and Innovation: Foster a culture of innovation while prioritizing resource management and capital efficiency.
- Drive financial innovation: Drive success by encouraging investment in cash and core business innovation.
- Diversity for Excellence: Promote diversity and inclusion for organizational excellence and financial stability.
- Proactive Risk Management: Stay ahead of the curve by managing risk through innovative financial practices.
- Adaptability: Keep pace with market changes by regularly monitoring and adjusting financial strategies.
- Invest in talent: Ensure adaptability and address talent shortages by investing in talent development.
- Collaborative Success: Achieve sustainable finance by fostering collaboration between organizational optimization, financial innovation, and business performance.

References

Abbas, J., Balsalobre-Lorente, D., Amjid, A.M., Al-Sulaiti, K., Al-Sulaiti, I., Aldereai, O.: Financial innovation and digitalization promote business growth: the interplay of green technology innovation, product market competition and firm performance. Innovation and Green Development (2024). https://doi.org/10.1016/j.igd.2023.100111

Afy-Shararah, M., Rich, N.: Operations flow effectiveness: a systems approach to measuring flow performance. Int. J. Oper. Prod. Manag. **38**(11), 2096–2123 (2018). https://doi.org/10.1108/IJOPM-09-2016-0575

Ahmad, M., Qiang, W., Khattak, M.S.: Intellectual capital, corporate social responsibility and sustainable competitive performance of small and medium-sized enterprises: mediating effects of organizational innovation. Kybernetes **52**(10), 4014–4040 (2023). https://doi.org/10.1108/K-02-2022-0234

AlGhamdi, M., Durug, C.M.: Business-to-business co-creation management practices for intellectual property value: Insights from Saudi Arabia. World Patent Inf. **72**, 102180 (2023). https://doi.org/10.1016/j.wpi.2023.102180

Andreou, P., LoucaPetrou, C.: Organizational learning and corporate diversification performance. J. Bus. Res. **69**(9), 3270–3284 (2016). https://doi.org/10.1016/j.jbusres.2016.02.022

Anifowose, M., Abang, S., Zakari, M.A.: Integrated capitals reporting and companies' sustainable value: evidence from the Asian continent. Asian Rev. Account. **28**(4), 567–589 (2020). https://doi.org/10.1108/ARA-10-2019-0184

Anwar, M.: Business model innovation and SMEs performance — does competitive advantage mediate? Int. J. Innov. Manage. **22**(07), 1850057 (2018). https://doi.org/10.1142/S1363919618500573

Azadda, W.N., Koomson, S., Klutse, S.K.: Sustainable finance and business risk resilience: a conceptual perspective and suggestions for upcoming research. Vilakshan - XIMB Journal of Management **21**, 66–78 (2023). https://doi.org/10.1108/XJM-02-2023-0034

Bollen, K.A.: Structural Equations with Latent Variables. Wiley, New York (1989)

Boz, I., YiğitAnıl, T.: The relationship between diversification strategy and organizational performance: a research intented for comparing Belgium and Turkey. Procedia Soc. Behav. Sci. **99**(6), 997–1006 (2013). https://doi.org/10.1016/j.sbspro.2013.10.573

Chen, C.H., Wang, T., Jia, X.: Short-termism in financial decision-making: Uncovering the influence of managerial myopia on corporate financial asset allocation through MD&A textual analysis. Int. Rev. Financ. Anal. **90**, 102900 (2023). https://doi.org/10.1016/j.irfa.2023.102900

Cohen, J., Cohen, P., West, S.G., Aiken, L.S.: Applied multiple regression/correlation analysis for the behavioral sciences (3rd ed.). Mahwah: NJ: Lawrence Erlbaum (2003)

Cronbach, L.J.: Coefficient alpha and the internal structure of tests. Psychometrika **16**(3), 297–334 (1951)

Cronbach, L.J.: My current thoughts on coefficient alpha and successor procedures. Educ. Psychol. Measur. **64**(3), 391–418 (2004)

Diamantopoulos, A., Siguaw, J.A.: Introducing LISREL. Sage Publications, London (2000). https://doi.org/10.4135/9781849209359

Eisenhauer, J.G.: Degrees of freedom. Teach. Stat. **30**(3), 75–78 (2008)

Farrer, B., Holahan, R., Shvetsova, O.: Accounting for heterogeneous private risks in the provision of collective goods: controversial compulsory contracting institutions in horizontal hydrofracturing. J. Econ. Behav. Organ. **133**, 138–150 (2017). https://doi.org/10.1016/j.jebo.2016.11.008

Fedele, M., Formisano, V.: Waste from criticality to resource through an innovative circular business model: a case study in the manufacturing industry. J. Clean. Prod. **407**, 137143 (2023). https://doi.org/10.1016/j.jclepro.2023.137143

Fisher, R.A.: On the interpretation of $\chi 2$ from contingency tables, and the calculation of P. J. Roy. Stat. Soc. **85**(1), 87–94 (1922). https://doi.org/10.2307/2340521 1922

Floyd, F.J., Widaman, K.F.: Factor analysis in the development and refinement of clinical assessment instruments. Psychol. Assess. **7**, 286–299 (1995)

Gaskin, J., Lim, J.: Model Fit Measures AMOS Plugin. Gaskination's StatWiki. AMOS Plugin, Gaskination's StatWiki

Glémain, P.: The strategy and fundamentals of sustainable finance serving sustainable development. In: Sun, W., Louche, C., Pérez, R. (eds.) Finance and Sustainability: Towards a New Paradigm? A Post-Crisis Agenda, pp. 187–209. Emerald Group Publishing Limited (2011). https://doi.org/10.1108/S2043-9059(2011)0000002015

Hu, L.T., Bentler, P.M.: Cutoff Criteria for Fit Indexes in Covariance Structure Analysis: Conventional Criteria versus New Alternatives. Struct. Equ. Model. **6**, 1–55 (1999)

Hu, L.-T., Bentler, P.M.: Fit indices in covariance structure modeling: sensitivity to under parameterized model misspecification. Psychol. Methods **3**(4), 424–453 (1998). https://doi.org/10.1037/1082-989X.3.4.424

Ikeda, S., Nishimura, N., Sukegawa, N., Takano, Y.: Prescriptive price optimization using optimal regression trees. Oper. Res. Perspect. **11**, 100290 (2023). https://doi.org/10.1016/j.orp.2023.100290

Jinnai, R.: Innovation, product cycle, and asset prices. Rev. Econ. Dyn. **18**(3), 484–504 (2015). https://doi.org/10.1016/j.red.2014.10.002

Joreskog, K.G.: A General Approach to Confirmatory Maximum Likelihood Factor Analysis. Psychometrika **34**, 183–202 (1969)

Kaiser, H.F.: An index of factorial simplicity. Psychometrika **39**, 31–36 (1974). https://doi.org/10.1007/BF02291575

Kaiser, H.F. (1970). A second generation little jiffy. Psychometrika, 35, 401–415 (1970). https://doi.org/10.1007/BF02291817

Kaiser, H.F., John, R.: Little Jiffy, Mark Iv. Educ. Psychol. Measur. **34**, 111–117 (1974). https://doi.org/10.1177/001316447403400115

Kim, H., Seung-June, H., Yoon, W.: Industry cluster, organizational diversity, and innovation. International Journal of Innovation Studies **7**(3), 187–195 (2023). https://doi.org/10.1016/j.ijis.2023.03.002

Kline, R. B. (2016). Principles and practice of structural equation modeling (4th ed.). Guilford Press

Kontosakos, V., Hwang, S., Kallinterakis, V., Pantelous, A., A.: Long-term dynamic asset allocation under asymmetric risk preferences. Eur. J. Oper. Res. **312**(2), 765–782 (2024). https://doi.org/10.1016/j.ejor.2023.07.038

Lulaj, E., Dragusha, B., Hysa, E.: Investigating accounting factors through audited financial statements in businesses toward a circular economy: why a sustainable profit through qualified staff and investment in technology? Admin. Sci. **13**(3), 72 (2023). https://doi.org/10.3390/admsci13030072

Lulaj, E., Dragusha, B., Lulaj, D.: Market mavericks in emerging economies: redefining sales velocity and profit surge in today's dynamic business environment. J. Risk Financ. Manage. **17**(9), 395 (2024). https://doi.org/10.3390/jrfm17090395

Lulaj, E., Minguez-Vera, A.: Cash flow dynamics: amplifying swing models in a volatile economic climate for financial resilience and outcomes. Sci. Ann. Econ. Bus. **71**(3), 315–336 (2024). https://doi.org/10.47743/saeb-2024-0022

Lulaj, E.: Quality and reflecting of financial position: an enterprises model through logistic regression and natural logarithm. J. Econ. Dev. Environ. People **10**(1), 26–50 (2021). https://doi.org/10.26458/jedep.v10i1.690

Liang, W., Wang, Y., Hou, L.: Does private equity investment influence enterprise innovation strategy? Financ. Res. Lett. **58**, 104593 (2023). https://doi.org/10.1016/j.frl.2023.104593

López-Pintado, O., Dumas, M., Berx, J.: Discovery, simulation, and optimization of business processes with differentiated resources. Inf. Syst. **120**, 102289 (2024). https://doi.org/10.1016/j.is.2023.102289

Lulaj, E.: A sustainable business profit through customers and its impacts on three key business domains: technology, innovation, and service (TIS). Bus. Manage. Econ. Eng. **21**(1), 19–47 (2023). https://doi.org/10.3846/bmee.2023.18618

Lulaj, E., Iseni, E.: Role of analysis CVP (cost-volume-profit) as important indicator for planning and making decisions in the business environment. Eur. J. Econ. Bus. Stud. **4**(2), 104 (2018). https://doi.org/10.26417/ejes.v4i2.p104-120

Lulaj, E., Gopalakrishnan, A., Kehinde Lamidi, K.: Financing, Investing in Women-led Businesses: Understanding Strategic Profits, Entrepreneurial Expectations by Analysing the Factors that Determine Their Company Success. Periodica Polytechnica Social, Management Sciences (2024). https://doi.org/10.3311/PPso.22532

Lulaj, E., Hysa, E., Panait, M.: Does digitalization drive sustainable transformation in finance and accounting? Kybernetes, Vol. ahead-of-print No. ahead-of-print (2024). https://doi.org/10.1108/K-04-2024-1006

MacCallum, R.C., Browne, M.W., Sugawara, H.M.: Power analysis and determination of sample size for covariance structure modeling. Psychol. Methods **1**, 130–149 (1996)

Mao, Z., Li, Y., Guan, Z., Uktamov, K., Ageli, M.M.: COP27 perspective of resources management: from conflict to COVID-19 of emerging countries. Resour. Policy **83**, 103708 (2023). https://doi.org/10.1016/j.resourpol.2023.103708

Markides, C., Singh, H.: Corporate restructuring: a symptom of poor governance or a solution to past managerial mistakes? Eur. Manage. J. **15**(3), 213–219 (1997). https://doi.org/10.1016/S0263-2373(97)00002-9

Menne, F., Surya, B., Yusuf, M., Suriani, S., Ruslan, M., Iskandar, I.: Optimizing the financial performance of SMEs based on sharia economy: perspective of economic business sustainability and open innovation. J. Open Innov. Technol. Market Compl. **8**(1), 18 (2022). https://doi.org/10.3390/joitmc8010018

Mitchell-Ketzes, S.: Optimising business performance through innovative workplace strategies. J. Facil. Manage. **2**(3), 258–275 (2003). https://doi.org/10.1108/14725960410808249

Moon, B.: Unleash liquidity constraints or competitiveness potential: The impact of R&D grant on external financing on innovation. Eur. Res. Manage. Bus. Econ. **28**(3), 100195 (2022). https://doi.org/10.1016/j.iedeen.2022.100195

Neher, A., Jungmeister, A., Wang, C., Burmeister, O.: Chapter 10 the effect of embedded managerial values on corporate financial outcomes. In: Schwartz, M., Harris, H., Comer, D.R. (eds.) Visual Ethics, pp. 165–204. Emerald Publishing Limited (2018). https://doi.org/10.1108/S1529-209620180000019010

Rasouli, S., Alipouri, Y., Chamanzad, S.: Smart personal protective equipment (PPE) for construction safety: a literature review. Saf. Sci. **170**, 106368 (2024). https://doi.org/10.1016/j.ssci.2023.106368

Sadiq, M., Aysan, A.F., Kayani, U.N.: Digital currency and blockchain security in accelerating financial stability: a mediating role of credit supply. Bursa Istanbul Rev. **25**, 1251–1262 (2023). https://doi.org/10.1016/j.bir.2023.09.009

Saidane, D., Abdallah, S.B.: Sustainable finance: concepts, analyses and perspectives. In: Bourghelle, D., Pérez, R., Rozin, P. (eds.) Rethinking Finance in the Face of New Challenges, pp. 181–192. Emerald Publishing Limited (2021). https://doi.org/10.1108/S2043-905920210000015028

Sawarni, K., NarayanasamyPathan, K.: Impact of earnings management on working capital management efficiency. Financ. Res. Lett. **54**, 103778 (2023). https://doi.org/10.1016/j.frl.2023.103778

Schilling, L., Seuring, S.: Mobile financial service-enabled micro-businesses driving sustainable value creation in emerging markets. Technol. Forecast. Soc. Chang. **192**, 122596 (2023). https://doi.org/10.1016/j.techfore.2023.122596

Spearman, C.: The Abilities of Man. MacMillan, New York (1927)

Steiger, J.H.: Understanding the limitations of global fit assessment in structural equation modeling. Personal. Individ. Differ. **42**(5), 893–898 (2007). https://doi.org/10.1016/j.paid.2006.09.017

Su, T., Yu, Y., Chen, Y., Hou, W.: On or off: the triggering effect of underperformance duration on cooperative innovation. Technovation **126**, 102817 (2023). https://doi.org/10.1016/j.technovation.2023.102817

Tanjung, M.: Cost of capital and firm performance of ESG companies: what can we infer from COVID-19 pandemic? Sustainab. Account. Manage. Policy J. **14**(6), 1242–1267 (2023). https://doi.org/10.1108/SAMPJ-07-2022-039

Tseng, M.-L., Bui, T.-D., Lim, M.K., Tsai, F.M., Tan, R.R.: Comparing world regional sustainable supply chain finance using big data analytics: a bibliometric analysis. Ind. Manag. Data Syst. **121**(3), 657–700 (2021). https://doi.org/10.1108/IMDS-09-2020-0521

Willem, J., van Gelder, L., German, R.B.: Biofuels investments in tropical forest-rich countries: implications for responsible finance. Sustainab. Account. Manage. Policy J. **3**(2), 134–160 (2012). https://doi.org/10.1108/20408021211282296

AI in Agriculture for Sustainable Food Production

The Ecological Footprint as an Indicator of Living Standards: The Relationship Between Consumption Patterns and Environmental Sustainability

Nadia Stezhko[1](✉), Ahmed J. Obaid[2], and Yevgen Kotukh[3]

[1] Kyiv National Economic University named after Vadym Hetman, Kyiv, Ukraine
nadijastezhko@gmail.com
[2] University of Kufa, Kufa, Iraq
ahmedj.aljanaby@uokufa.edu.iq
[3] Dnipro University of Technology, Dnipro, Ukraine

Abstract. The article examines the ecological footprint as a key indicator of the standard of living and its influence on the formation of sustainable development policy. Analyzing data from the Environmental Performance Index (EPI), high-scoring countries such as Canada, Australia and Luxembourg demonstrate successful environmental policy strategies. The use of correlation analysis confirms that the growth of the gross domestic product and the level of human development is closely related to the increase of the ecological footprint. At the same time, the results show that health care costs have a slight negative impact on the ecological footprint, which emphasizes the importance of an integrated approach in assessing environmental indicators.

Keywords: ecological footprint · standard of living · ecological efficiency · ecological sustainability · sustainable development

1 Introduction

In today's conditions of global economic and environmental changes, the analysis of the relationship between the standard of living and the environmental load becomes particularly relevant. The increase in the standard of living, which is expressed in the increase in the income of the population, the improvement of living conditions and the increase in access to goods and services, is inevitably accompanied by the increase in the consumption of natural resources. Higher incomes lead to increased demand for energy, water, food and physical goods. The expansion of consumption leads to the growth of greenhouse gas emissions, depletion of natural resources and environmental degradation. These processes are interdependent, since the increase in economic activity and well-being is accompanied by the intensification of the use of resources. In turn, the increased environmental burden resulting from increased consumption can lead to a deterioration in the quality of life due to environmental problems such as air pollution, land degradation and climate change.

A. J. Obaid et al. (Eds.): ICONEST 2024, ISEM 40, pp. 93–109, 2025.
https://doi.org/10.1007/978-3-031-89960-7_7

It should also be noted that greater life and quality standards in developed countries are largely building on the back of internationalization of use of natural resources. The ecological footprint of such countries is, therefore, usually higher than their own carrying capacity hence leading to a shift of the ecological burden to supply producing developing countries. This leads to unequal distribution of environmental burden that another advantage whereby developed countries make their own living standards at the expense of other regions' ecological losses.

Hence the need to consider ecological footprint analysis as a leverage to the extent of new economic models in addressing natural resource conservation. It helps you look at how much resources were used to the opportunities of restoring those and see what needs to be done to ensure that economic growth is achieved while minimizing the rate of environmental degradation.

2 Related Works

The analysis of the changes in living standards and their reflection in consumption rates and environmental sensitiveness is now a domain of numerous scientists. Ecological footprint was defined by William Rees and Matisse Wackernagel [1] and has become the initial index for measuring the impact of human activity on environment. They looked at how the contemporary consumer behavior in different countries outstrips the Earth's carrying capacity. According to Tim Jackson [2] economic growth does not necessarily increase the standard of living and there is a need to begin to seek out new models of consumption that are not damaging to the environment. Jackson also asked questions about income growth welfare, filling the gap and listed the threats that came with everything mentioned above, even labeling the current status as a dangerous imbalance. This has made his work to be an invaluable contribution towards the discussion of environmental sustainability in the emerging economy. As Kate Revers in [3] notes, the problem of scarcity and consequent lack of supply of necessary resources, and the necessity of the redistribution of capital for social equity and sustainable ecology. Her model of the donut economy also holds the idea that there are inconvenient ecological niches that can not be crossed in order to preserve the planet's stability, and the living standards cannot be below a certain social level. Nikola Stern [4] considered the economic consequences of climate change and emphasized the importance of understanding the relationship between economic growth, resource consumption, and climate change. He proves that environmental crises can negatively affect the standard of living, and therefore it is necessary to look for ways to sustainable economic development. Herman Daly [5] in his works focused on the need to take into account limited natural resources in the process of economic growth. Paul Ehrlich [6] emphasizes that the excess of natural resources due to excessive consumption and population growth can lead to catastrophic environmental consequences. At the same time, the analysis of the mentioned sources shows that a number of issues related to the study of the ecological footprint remain unresolved, which led to the choice of the topic and the substantive focus of the study.

3 Methodology

Research methodology is formed on the basis of the provisions of modern economic theory, scientific works of leading domestic and foreign scientists in the field of research on the relationship between the standard of living, consumer models and environmental sustainability. In the research process, the dialectical method of scientific knowledge was used, as well as a system of general scientific methods and approaches: system analysis and theoretical generalization (for researching the theoretical foundations of the ecological footprint and standard of living); logical generalization (to ensure the sequence of performance of assigned tasks), methods of statistical analysis (to study factual material that characterizes the current state and ecological efficiency of consumption models); system analysis (to identify dialectical differences in the formation of organizational and economic stability of consumer models and environmental sustainability), etc.

In order to identify the relationship between the standard of living and the ecological footprint, one of the most common methods of economic analysis was used - correlation analysis. This type of analysis is a method of studying the interdependence of characteristics in the general population, which are random variables with a normal distribution. When conducting such an analysis, the main requirements are a sufficient number of observations, a set of factor and result indicators, as well as their quantitative measurement and reflection in information sources. The main tasks are:

– determination of the form of communication;

– measurement of connection density (strength);

- identification of the influence of factors on the effective sign.

To carry out a correlation analysis, it is necessary to perform the following successive stages:

1) establishment of cause-and-effect relationships between the investigated features (identification of factors and selection among them of those that have the greatest impact on the performance indicator);
2) formation of a correlation-regression model (information provision of the analysis, selection of the type and form of communication, assembly of the model);
3) determination of correlation characteristics (indicators of communication);
4) statistical assessment of communication parameters (economic interpretation, assessment of the significance of correlation coefficients (to what extent the selected factors explain the variation of the performance indicator) and their use for solving practical problems).

At the first stage of the analysis, the relationships between the resulting and factor characteristics are determined. Such connections can be different depending on the nature of the dependence, direction of action and analytical expression.

At the second stage, the initial information collected for the study is evaluated, various statistical criteria are also used (mean square deviation, coefficient of variation, etc.), based on the results, a model of stochastic communication is formed.

The following formula is used to determine the correlation coefficient:

$$\rho = M(\mathrm{xa})/\sigma \mathrm{x}^{*}(\mathrm{yb}) \quad (1)$$

where a = Mx; b = Mu – centers of distribution of x and y values;

σ2x and σ2y are their variances.

Under the condition of correct application of correlation methods, it is possible to determine the deep essence of the processes of interrelationships. Correlations are not found in every single case, but on average for most cases. These relationships between cause and effect do not have a complete correspondence, but there is a certain correlation. Peculiarities of correlation relationships give rise to two tasks in correlation theory: to determine the theoretical form of the relationship (regression analysis) and to measure the density of the relationship (correlation analysis). Determining the theoretical form of the relationship (regression analysis) consists in finding the form of the functional relationship that best corresponds to the essence of correlational dependence. The measurement of connection density (correlation analysis) involves measuring with the help of special indicators to what extent the correlation connection approaches the functional connection [7].

To analyze the main indicators of the standard of living and well-being of developed and underdeveloped countries, key indicators reflecting economic, social and environmental aspects were used, namely, gross domestic product per capita (GDP per capita), Human Development Index (Human Development Index, HDI), Poverty Index (Poverty rate), ecological footprint per capita (Ecological Footprint per capita). This study is based on Eurostat and World Bank data.

4 Result Analysis and Discussion

Today, the ecological footprint plays an important role in the development of sustainable development policies at the national and international levels. It is also used to raise environmental awareness among the public and to motivate actions aimed at reducing resource consumption. Various organizations and governments around the world are actively using ecological footprint data to evaluate the effectiveness of their environmental programs and plans.

In order to compare the environmental achievements of countries and monitor their progress in achieving environmental goals, we will analyze the Environmental Efficiency Index (Environmental Performance Index, EPI). The EPI is an indicator that assesses how effectively countries manage their environmental resources and reduce their negative impact on the environment. Based on the data shown in Fig. 1 it can be concluded that high EPI indicators (about 70–80 points) in countries such as Canada, Australia, Luxembourg and the USA indicate an effective environmental policy and a high level of environmental safety. These countries usually have strict environmental standards, innovative technologies to reduce pollution, and a well-developed natural resource management system.

Average EPI scores (about 70 points) in countries such as Qatar, United Arab Emirates indicate a moderate level of environmental performance, but show gradual improvement but still have some environmental challenges due to intensive use of natural resources.

Low EPI scores (less than 50 points) in countries such as Eritrea, Malawi, Chad, Nepal, Haiti, Bangladesh, Afghanistan have the lowest EPI values, indicating serious environmental problems and inadequate environmental policies.

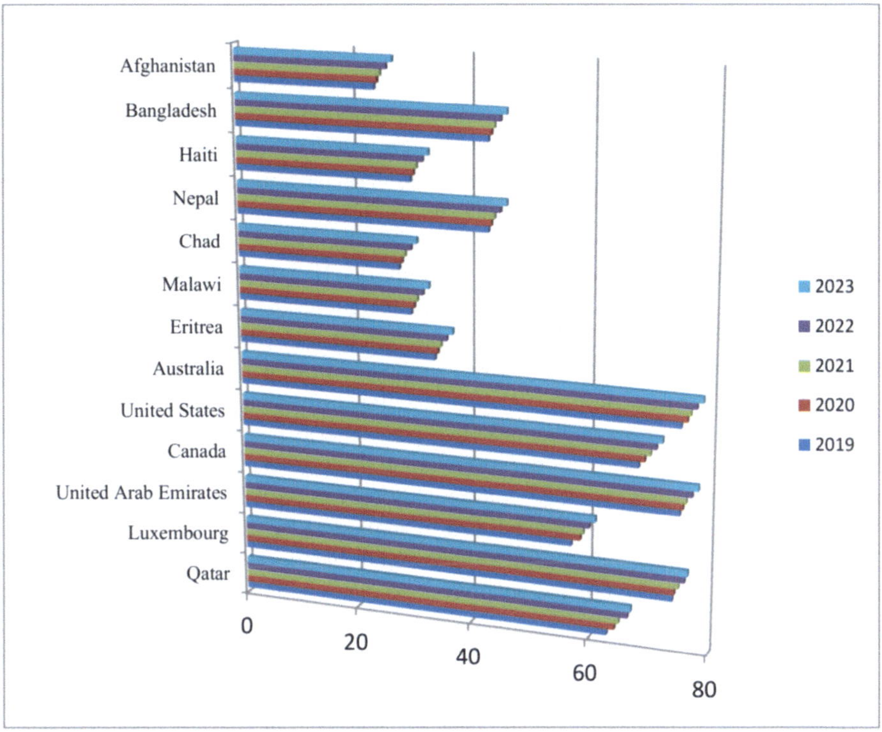

Fig. 1. Environmental efficiency index (EPI) [8]

Thus, the value of EPI is an important indicator of how effectively countries manage environmental resources and solve environmental problems.

One of the most common methods of economic analysis - correlation analysis - was used to identify the relationship between living standards, consumer patterns and environmental sustainability. This type of analysis is a method of studying the interdependence of features in the general population, which are random variables with a normal distribution.

The countries with the largest ecological footprint, i.e. developed countries with a high level of resource consumption, were chosen for the study. They have a high demand for energy, materials and food, which puts significant pressure on natural resources. According to Global Footprint Network, countries with the largest environmental footprint per capita include: Qatar (due to high energy consumption and high per capita carbon emissions); Luxembourg (due to the high level of consumption of energy and materials per capita); United Arab Emirates (known for high consumption of energy and water, as well as active development of industry and construction); Canada (due to the large amount of use of natural resources and high level of energy intensity); the United States of America (one of the largest consumers of resources in the world, with a high level of motorization, energy consumption, and industry); Australia (due to the high level of resource and energy consumption per capita). Countries with the smallest ecological footprint, less developed countries or those with economies less dependent on industrial production and energy consumption were also selected. Such countries include: African countries (for example, Eritrea, Malawi, Chad) - countries with a low level of industrialization and minimal energy consumption per capita. Their environmental footprint is significantly lower due to limited resource use and low carbon emissions. Nepal is a country whose economy is based mainly on the agricultural sector with a low level of industrial production and minimal energy costs. Haiti - Due to limited economic opportunities, the country has a low level of consumption of energy and other resources, resulting in one of the lowest environmental footprints. Bangladesh - Although the country has a large population, its per capita environmental footprint remains relatively low due to low resource and energy consumption. Afghanistan – the ecological footprint in the country is low due to limited industrialization and low energy consumption.

According to the results of the calculations, the correlation coefficient between the environmental footprint per capita and GDP per capita is r = 0.983557 (Fig. 2). This indicates a very strong positive correlation, which means that GDP growth per capita is usually accompanied by an increase in the ecological footprint. The R^2 value shows that about 96.74% of the variation in GDP can be explained by changes in the ecological footprint. Regression analysis confirms that as GDP increases, the ecological footprint also increases, as evidenced by the positive slope in the regression equation.

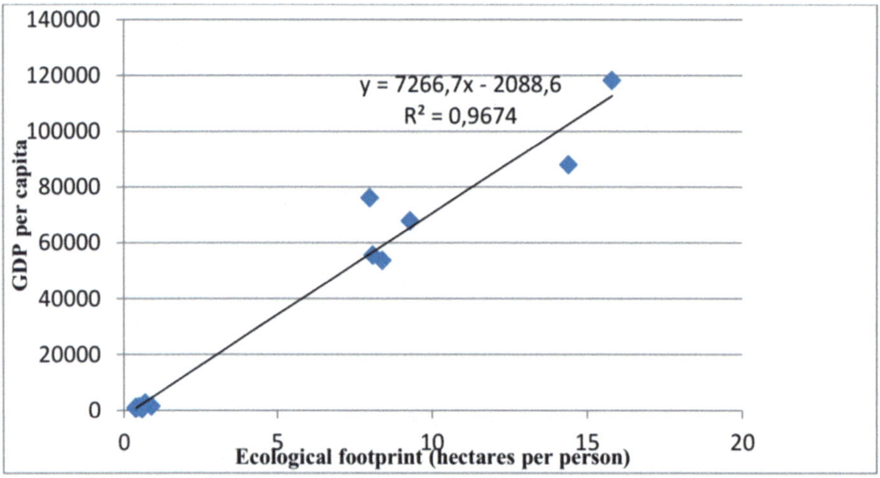

Fig. 2. The relationship between the ecological footprint per capita and GDP per capita. *Calculated by the author based on data* [9, 10].

The correlation coefficient between the Human Development Index (HDI) and the ecological footprint is 0.848009, which indicates a strong positive correlation between these indicators and shows that as a country's level of development, as measured by the HDI, increases, there is an increase in the use of natural resources and, accordingly, growth ecological footprint. This reflects that countries with a higher level of human development (Qatar, Luxembourg, the United Arab Emirates, Canada, the United States of America, Australia) have greater resource requirements to support a higher standard of living, developed infrastructure, health care and education. For example, the growth of HDI is accompanied by an increase in energy consumption, the expansion of urbanization, an increase in the level of production and consumption of goods, as well as emissions of carbon dioxide and other pollutants. This means that countries with a high HDI tend to put more pressure on the environment due to high rates of industrial production and consumption, which is reflected in the growth of the ecological footprint. Therefore, the value of $R^2 = 0.7191$ (Fig. 3) indicates that about 71.91% of the changes in the ecological footprint can be explained by variations in the HDI. In other words, the level of development of countries greatly affects their ecological footprint, while the remaining 28.09% may depend on other factors, such as technological innovation, sustainable development policies, cultural consumption patterns, etc.

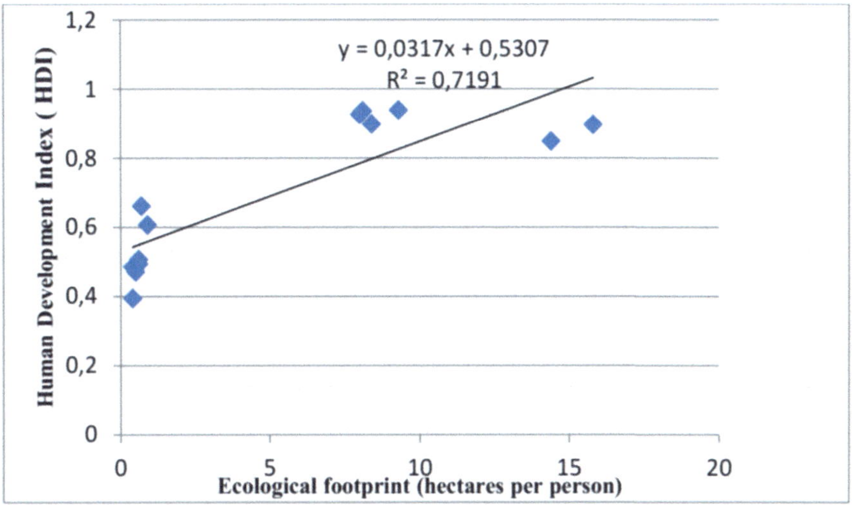

Fig. 3. The relationship between the ecological footprint per capita and the Human Development Index (HDI). *Calculated by the author based on data* [10, 13].

The correlation coefficient between the ecological footprint and health care costs per capita is 0.641. This means that there is a moderate positive linear relationship between these variables: as the ecological footprint increases, health care costs also increase. The coefficient of determination is equal to $R^2 = 0.4109$ (Fig. 4). This means that approximately 41% of the variation in health care spending can be explained by changes in the ecological footprint. That is, **the direct impact of health care costs on the environment**: some measures in the field of health care can be linked to sustainable development policies and greater environmental responsibility. For xample, countries that invest in the health of their citizens can also implement environmentally friendly technologies in medical infrastructure, increasing the energy efficiency of hospitals or reducing emissions from medical waste; **level of well-being and awareness of environmental issues**: countries that invest more in health care often have a higher level of well-being and a conscious attitude to environmental issues. This can stimulate the population to more rational consumption of resources and reduce the negative impact on the environment.

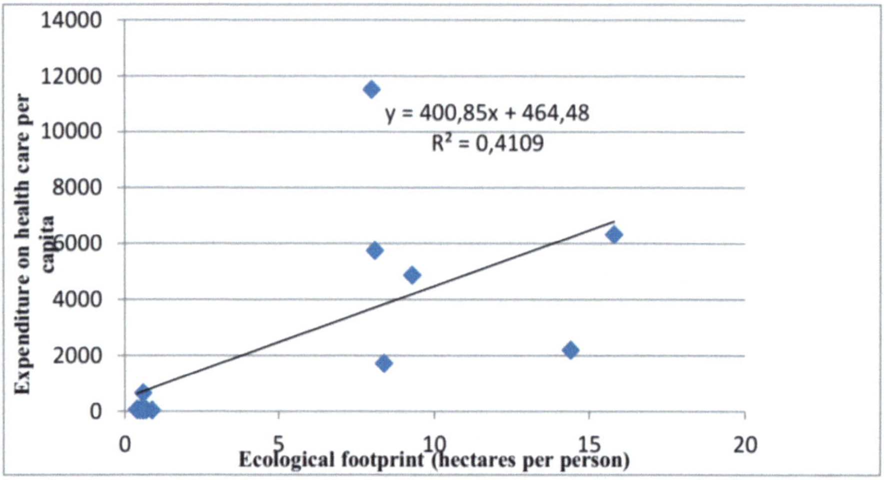

Fig. 4. The relationship between the environmental footprint per capita and health care costs per capita. *Calculated by the author based on data* [10, 13].

The correlation coefficient between poverty level and ecological footprint is -0.73047, indicating a strong negative correlation between the two indicators. A negative correlation means that as the level of poverty increases, the ecological footprint tends to decrease. In other words, countries with a higher level of poverty (Eritrea, Malawi, Chad, Nepal, Haiti, Bangladesh, Afghanistan) have a much smaller ecological footprint per capita, and, conversely, countries with a lower level of poverty (Qatar, Luxembourg, Ob' the United Arab Emirates, Canada, the United States of America, Australia) this indicator is higher. The value of $R^2 = 0.5336$ (Fig. 5.) means that 53.36% of the variations in the level of ecological footprint can be explained by changes in the level of poverty. This indicates that the level of poverty is an important factor in the formation of the ecological footprint, but not the only one - other factors affect the ecological footprint in 46.64% of cases.

A thorough analysis of this relationship can be explained by the following factors:

1. Countries with a high level of poverty are usually characterized by a low level of resource consumption. Due to limited opportunities to purchase goods and services, poorer sections of the population use less energy, water, land and other natural resources. For example, people living below the poverty line rarely use cars, electrical appliances, or other energy sources, reducing their contribution to the overall environmental footprint.
2. Countries with high levels of poverty often have low levels of industrialization and urbanization, which also contribute to a reduced ecological footprint. In these countries, there are fewer large manufacturing enterprises, the infrastructure is poorly developed, and the focus on the consumption of natural resources is lower. This may partly explain why the level of ecological footprint in such countries is smaller.
3. High levels of poverty often prevent countries from implementing technologies and infrastructure that, although economically beneficial, can have a negative impact on

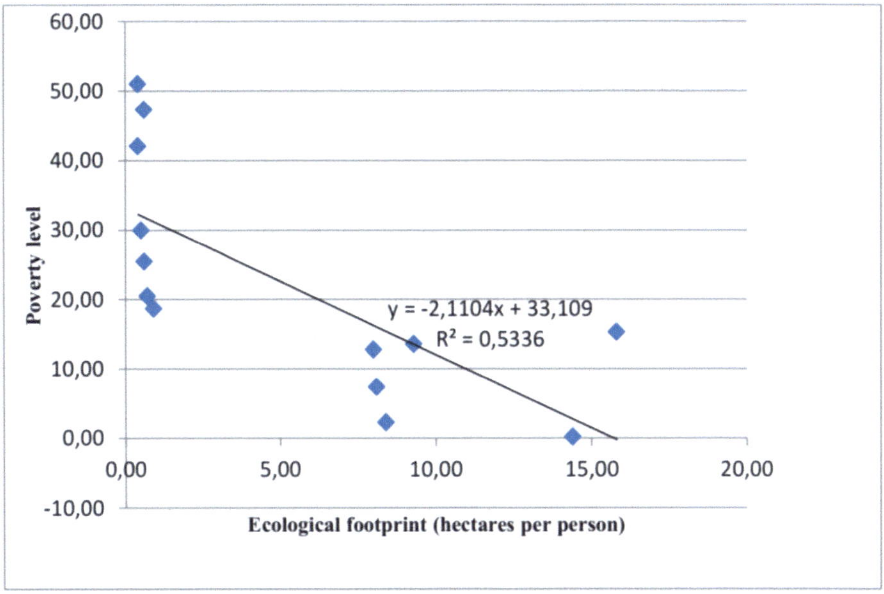

Fig. 5. The relationship between the ecological footprint per capita and the level of poverty *Calculated by the author based on data* [10, 13].

the environment. For example, developed countries often have larger automobile fleets, more fossil fuel-fired power plants, and high levels of manufacturing and industrial capacity that increase their environmental footprint.

4. In countries with a high level of poverty, the rational use of limited resources dominates. For example, food, water consumption, energy, building materials are used sparingly. Such countries often face challenges related to access to basic goods, which constrains their overall impact on ecosystems.
5. In many countries with high levels of poverty, manual labor and natural materials play a large role in economic activity, which also reduces the ecological footprint. The use of fossil fuels is less common than in wealthier countries, which also reduces the level of greenhouse gas emissions and environmental pollution.

Thus, the strong negative correlation (-0.73047) between the level of poverty and the ecological footprint reflects the fact that countries with high levels of poverty contribute significantly less to global environmental problems compared to developed countries. However, this does not mean that poorer countries necessarily have sustainable ecological development; their lower ecological footprint is often driven by economic constraints rather than conscious environmental policy.

The correlation coefficient between CO_2 emissions per capita (metric tons) and ecological footprint is 0.888083, indicating a very strong positive correlation between the two. This means that with an increase in CO_2 emissions per capita, the ecological footprint also increases significantly. That is, countries in which the level of carbon dioxide emissions per inhabitant is higher, as a rule, have a larger ecological footprint.

The value of $R^2 = 0.7887$ (Fig. 6) indicates that 78.87% of the variation in the ecological footprint can be explained by changes in CO_2 emissions per capita. This indicates that carbon dioxide emissions are one of the key factors shaping a country's ecological footprint and have a significant impact on its increase.

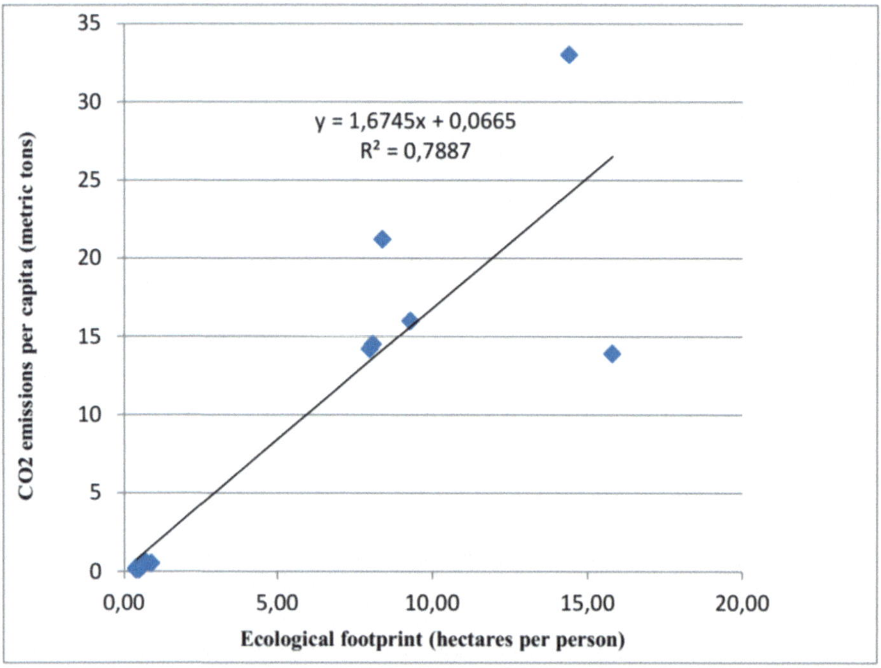

Fig. 6. The relationship between ecological footprint per capita and CO2 emissions per capita (metric tons). *Calculated by the author based on data* [10, 11].

CO_2 emissions are one of the key indicators of the impact of human activity on the environment. The increase in CO_2 emissions, which are mainly generated through the burning of fossil fuels, transport activities, industrial production and energy, has a direct impact on the ecological footprint. Since CO_2 is the main greenhouse gas causing global warming, its high level means intensive consumption of natural resources and high pressure on ecosystems.

Countries with high CO_2 emissions per capita (Qatar, Luxembourg, United Arab Emirates, Canada, United States of America, Australia) usually have a developed energy sector based on fossil energy sources (coal, oil, natural gas) and a large energy consumption for production needs and transport. Such countries consume significant amounts of fuel to meet their economic and social needs, which in turn increases their environmental footprint. For example, the development of the road transport and air transport industry contributes to a significant increase in CO_2 emissions and, accordingly, the ecological footprint.

Industrialized countries that have a high level of CO_2 emissions per capita are also countries with a large industrial base. High production and consumption, extraction of resources, construction, and sustaining physical structures exactly exaggerate the footprint on the ecology of any environment. This totals CO_2 emissions but also uses other resources such as water, land, biomass other than energy which also has a burden on the natural environment. It is normally correlated with emissions of CO_2, mainly from the raise of energy consumption in urban structures and emissions produced by constructions of buildings, transports and electricity currents. According to the trends in per capita emissions for different countries, it was also found that the larger the city, the higher the emissions.

Countries with high CO_2 emissions typically have an energy sector based on burning fossil fuels. For example, countries that depend on coal or oil for electricity generation create a much larger environmental footprint than countries that use renewable energy sources. The consumption of large volumes of fossil fuels not only increases CO_2 emissions, but also causes environmental degradation due to the extraction of resources and their transportation.

Emission of CO_2 and the efficiency of this gas in relation to the ecological footprint thus also relies on specific national policies aimed at green house gases. Therefore, the countries like those which have the environmental program designed to reduce the emissions, might have a smaller effect on the environment as compared to economically active levels. This means that CO_2 emissions stand among the biggest factors as to the size of the ecological footprint, and the coefficient of the CO_2 emissions per capita, based on the EEI is 0.888083. Total ecological footprint postulated is positively associated with CO_2 emission because countries emitting more exert more pressure on the resources in an ecological system. As such, attempts at controlling the output of CO_2 emissions are core to the minimalisation of anthropogenic load on the ecosystem, as well as promoting sustainable consumption. The correlation coefficient between the ecological footprint per capita and the share of renewable energy sources in total energy consumption is -0.65498, indicating a moderate negative correlation between these indicators. This means that an increase in the share of renewable energy sources in total energy consumption is accompanied by a decrease in the ecological footprint per capita. The value of $R^2 = 0.514$ (Fig. 7.) indicates that 51.4% of variations in the level of ecological footprint can be explained by changes in the share of renewable energy sources.

The negative correlation (-0.65498) shows that increasing the share of renewable energy sources in the structure of energy consumption reduces the ecological footprint. This makes sense because renewable energy sources such as solar, wind, hydropower and biomass have a much lower environmental impact than fossil energy sources such as coal, oil or natural gas. One of the main reasons for this negative relationship is a significant reduction in greenhouse gas emissions when using renewable energy sources. Countries that actively implement these sources are able to significantly reduce the level of emissions of CO_2 and other harmful substances, which directly affects the reduction of the ecological footprint. Renewable energy sources, especially in countries with developed infrastructure, make it possible to provide large amounts of energy without significantly polluting the environment. Countries that invest resources in the development of such technologies are able to reduce their environmental footprint by

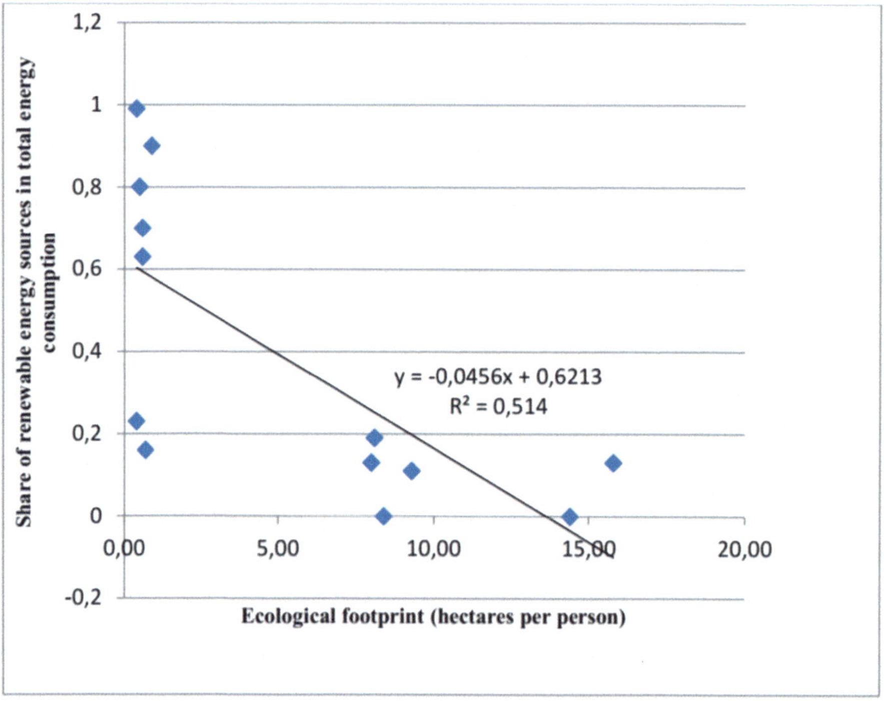

Fig. 7. The relationship between the ecological footprint per capita and the share of renewable energy sources in total energy consumption. *Calculated by the author based on data* [10, 12].

reducing dependence on fossil fuels and reducing carbon-based energy consumption. Technological innovations in renewable energy sources, such as improved solar panels or highly energy-efficient wind turbines, help improve resource efficiency. This makes it possible to reduce the environmental footprint not only by reducing pollution, but also by optimizing energy production. Countries that are more active in implementing renewable energy sources usually have stricter environmental policies. They encourage a reduction in dependence on traditional energy sources through financial incentives, legislative initiatives, and government programs to support clean energy technologies. It also helps to reduce the ecological footprint of the population. Countries with a low share of renewable energy sources mostly rely on fossil fuels, which cause higher levels of pollution. The use of such sources increases environmental pressure due to high energy consumption, CO_2 emissions and other harmful substances. This explains why a higher proportion of renewable sources correlates with a reduced ecological footprint.

The negative correlation between the ecological footprint and the share of renewable energy sources indicates the importance of switching to renewable sources to reduce the impact on the environment. The greater the share of renewable energy sources in a country, the smaller its ecological footprint, which makes the development of such sources a key tool in the fight against global environmental challenges.

The correlation coefficient between the use of natural resources (in particular, energy and material consumption) and the ecological footprint, equal to 0.43561, indicates a moderate positive correlation between these two variables. This means that as the consumption of energy and materials increases, there is an increase in the ecological footprint, but with the presence of significant variability. The R^2 value of 0.1898 (Fig. 8) shows that approximately 18.98% of the variation in the ecological footprint can be explained by changes in the use of natural resources. This is quite a low figure, which shows that there are many other factors that affect the ecological footprint besides the consumption of energy and materials.

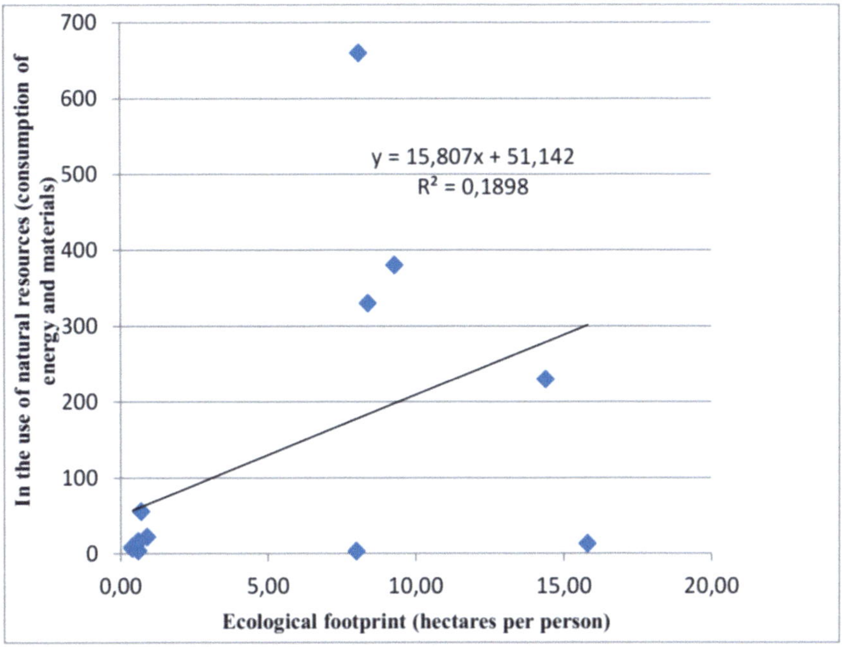

Fig. 8. The relationship between the ecological footprint per capita and the use of natural resources (consumption of energy and materials) *Calculated by the author based on data* [9, 10].

The correlation coefficient value of 0.772879 indicates a strong positive correlation between the percentage of recycled waste and waste disposal and ecological footprint. This means that an increase in the percentage of recycled waste is usually accompanied by a decrease in the ecological footprint. This indicates effective waste management systems. The coefficient of determination ($R^2 = 0.5973$) (Fig. 9) means that approximately 59.73% of the variation in waste disposal can be explained by changes in the percentage of recycled waste.

The correlation coefficient ($r = 0.95999$) indicates that there is a very close and positive relationship between the land footprint and the ecological footprint. This means that: When the land footprint (the amount of land needed to meet human needs, such as food, energy, housing, etc.) increases, the ecological footprint (the total amount of resources and ecosystem services needed to support this lifestyle) also increases. Such a correlation indicates that the increase in the use of land resources leads to an increase

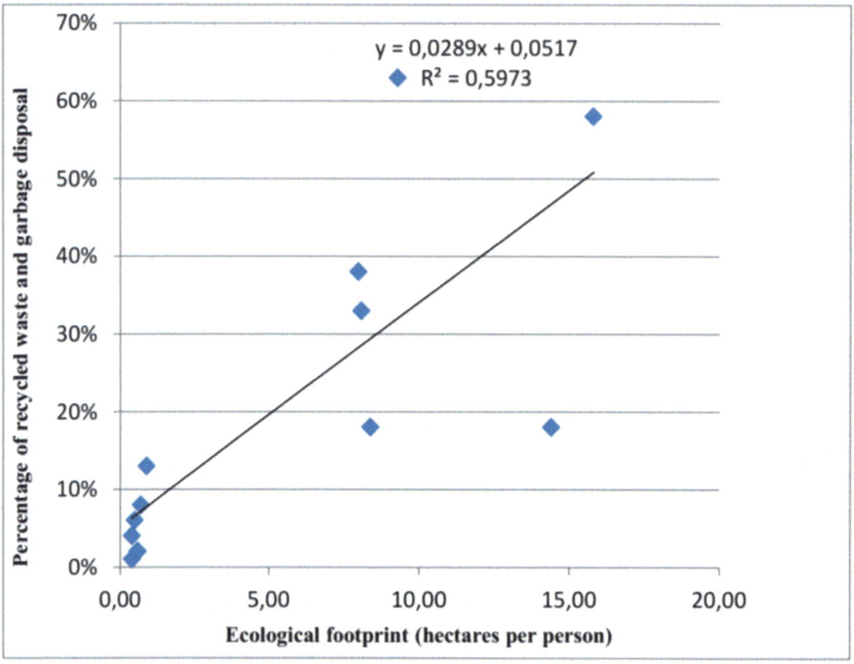

Fig. 9. The relationship between the ecological footprint per capita and the percentage of recycled waste and waste disposal. *Calculated by the author based on data* [10, 12].

in the negative impact on the environment, for example, due to an increase in carbon emissions, a decrease in biodiversity and a decrease in soil quality. The coefficient of determination ($R^2 = 0.9216$) (Fig. 10) means that about 92.16% of the variations in the ecological footprint can be explained by changes in the land footprint. This suggests that the environmental footprint is a key determinant of the ecological footprint. This confirms that a huge share of environmental impact can be explained precisely because of how land resources are used. However, since about 7.84% of the variation in the ecological footprint cannot be explained by the land footprint, this indicates that there are other factors that also affect the ecological footprint. These can be, for example, the level of resource consumption, technological progress, efficiency of waste management and other environmental practices.

To reduce the ecological footprint, it is necessary to implement sustainable land use practices, such as restoration of degraded lands, reduction of resource use, optimization of agricultural practices, etc.

Thus, strengthening land management and reducing the land footprint can be an important step in reducing the ecological footprint, which in turn will contribute to environmental protection and sustainable development.

The correlation coefficient value of 0.826125 indicates that there is a strong positive correlation between the water footprint and the ecological footprint. When the need for water to produce goods and services increases, the ecological footprint tends to increase as well. This may be because producing more goods requires more resources, including

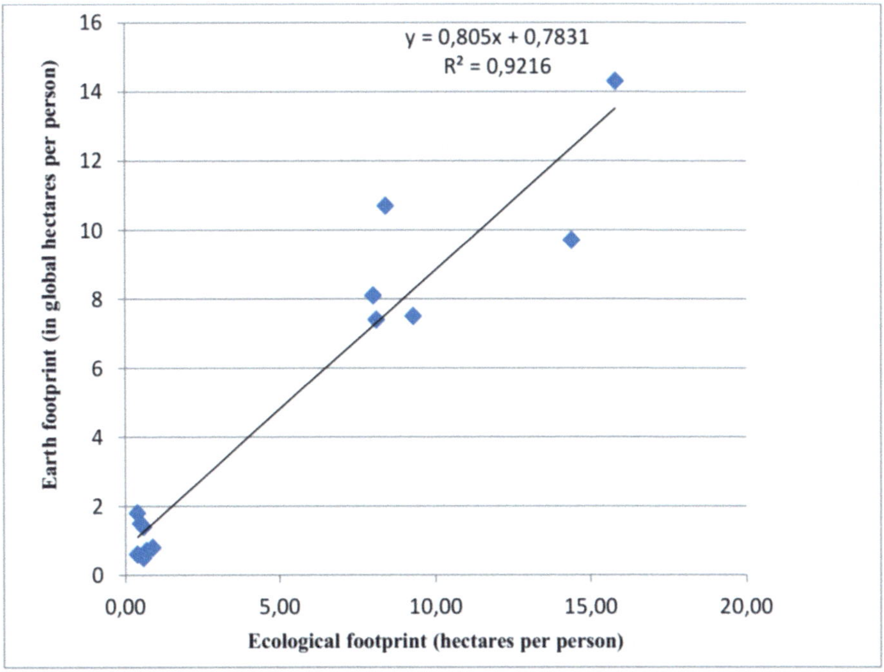

Fig. 10. The relationship between ecological footprint per capita and land footprint (in global hectares per person). *Calculated by the author based on data* [10, 12].

land, energy and water. The coefficient of determination ($R^2 = 0.6825$) (Fig. 11) means that about 68.25% variations in the variable correlated with the water footprint can be explained by changes in the water footprint. This indicates that the water footprint is a significant factor influencing the ecological footprint. When water consumption increases, it tends to lead to increased use of land resources to produce food, energy, and other goods, which in turn increases the ecological footprint. However, about 31.75% of the variations in the ecological footprint cannot be explained by the water footprint, which indicates the presence of other factors, such as technological changes, socio-economic conditions, efficiency of waste management, and others.

To reduce the ecological footprint, it is important to implement strategies that reduce the water footprint, such as: the use of water-saving technologies in agriculture and industry; development of policies regarding the rational use of water; support of projects that ensure sustainable management of water resources. It is necessary to consider not only the dependence between water and ecological footprints, but also the relationship between different aspects of resource consumption to ensure a holistic approach to sustainable development and environmental conservation.

This dependence shows that in order to achieve sustainable development, it is necessary to focus on reducing the water footprint, which in turn will contribute to the reduction of the ecological footprint.

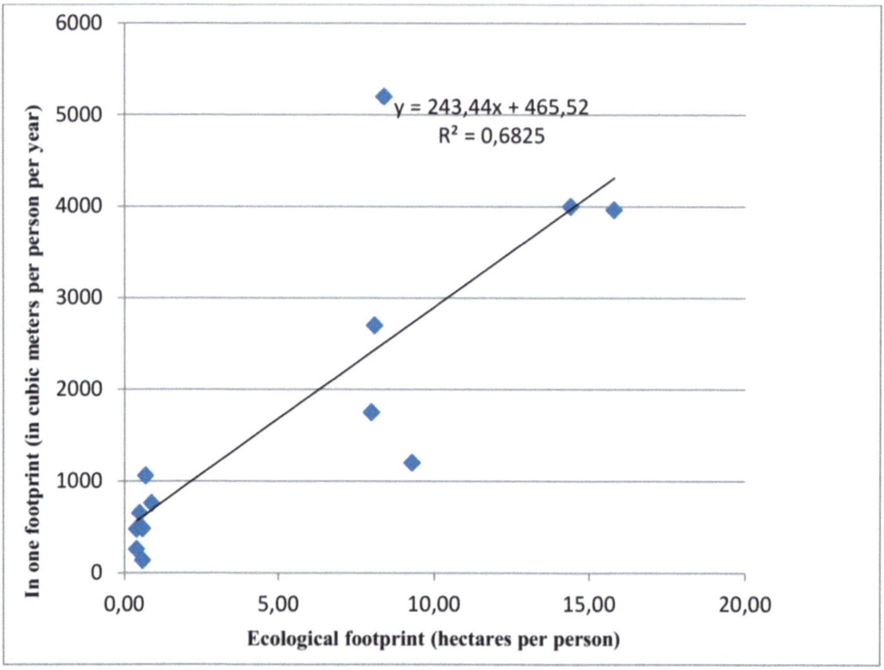

Fig. 11. Dependence between ecological footprint per capita and water footprint (in cubic meters per person per year). *Calculated by the author based on data* [10, 14].

5 Conclusion

Ecological footprint is a key indicator reflecting the impact of human activities on the environment. It plays an important role in shaping sustainable development policies at the national and international levels, helping governments and organizations evaluate the effectiveness of environmental programs and motivating people to reduce resource consumption. Analysis of the Environmental Performance Index (EPI) showed that countries with high EPI scores, such as Canada, Australia, Luxembourg, and the United States, demonstrate effective management of environmental resources. At the same time, countries with low EPI scores, such as Eritrea, Malawi and Chad, face serious environmental challenges, indicating insufficient levels of environmental policy.

A correlation study performed confirmed a strong positive correlation between ecological footprint and GDP per capita (r = 0.983557) and Human Development Index (HDI) (r = 0.848009). This indicates that the growth of the country's economic well-being is accompanied by an increase in the ecological footprint, which may indicate a high level of resource consumption in developed countries. The negative correlation between health care expenditure per capita and ecological footprint (-0.35038) indicates that although investments in health care can help reduce the ecological footprint, this effect is weak and not always linear. Other factors, such as energy policy and consumer patterns, have a much greater impact. The obtained results emphasize the importance of an integrated approach to solving environmental problems. Countries should take into account not only economic factors, but also social, technological and environmental aspects when developing sustainable development strategies.

Based on the research, it is recommended to develop policies aimed at reducing the ecological footprint, introduce innovative technologies and ecological practices, and also encourage the population to rationally consume resources. This will make it possible to achieve greater environmental sustainability and improve the quality of life in various countries.

References

1. Rees , W., Wackernagel, M.: Our Ecological Footprint : Reducing Human Impact he the Earth. New Society Publishers (1996)
2. Jackson, T.: Prosperity without Growth: Economics for a Finite Planet. Earthscan (2009)
3. Raworth, K.: Donut Economics: Seven Ways that Think Like a 21st-Century Economist. Random House Business (2017)
4. Stern, N.: The Economics of Climate Change: The Stern Review. Cambridge University Press (2014). https://doi.org/10.1017/CBO9780511817434
5. Daly, H.: Steady-State Economics: The Economics of Biophysical Equilibrium and Moral Growth. WH Freeman (1977)
6. Ehrlich, P.: The Population Bomb (1968)
7. Stezhko, N., Golovchenko, O., Hrazhevska, N., Omelchenko, H., Miziuk, S., Liubchyk, K.: Development of a decision support system in the international economy based on economic and mathematical models. Int. J. Sci. Technol. Res. **9**(3), 3939–3943 (2020)
8. 2024 Environmental Performance Index. https://epi.yale.edu/. Accessed 20 Oct 2024
9. World Bank Group. https://data.worldbank.org/indicator/BN.KLT.DINV.CD?skipRedirection=true&view=map. Accessed 20 Oct 2024
10. Global Footprint Network. https://www.footprintnetwork.org/. Accessed 20 Oct 2024
11. Global Carbon Atlas. https://globalcarbonatlas.org/. Accessed 20 Oct 2024
12. International Energy Agency. https://www.iea.org/. Accessed 20 Oct 2024
13. UNDP Human Development Reports. https://hdr.undp.org/data-center/country-insights#/ranks. Accessed 20 Oct 2024
14. Water Footprint Network. https://www.footprintnetwork.org/. Accessed 20 Oct 2024

Landscape View of Improving Transparencies in Agricultural Supply Chain Using Blockchain Technology

Ali Hussein Abdulkhaleq[1](✉), Dyah Kurniawati[2], and Hamssah Hussein Abdul-Sada[3]

[1] Department of Information Technology, Technical College of Management Al-Furat Al-Awsat Technical University, Najaf, Iraq
alialkaaby@atu.edu.iq

[2] Universitas PGRI Ronggolawe Tuban, Tuban, Indonesia

[3] Najaf Al-Ashraf Municipal Directorate/Al-Kufa Municipal Directorate, Najaf, Iraq

Abstract. An extensive overview of using of blockchain technology to improve supply chain transparency in agriculture is given in this study. It looks at how blockchain may be used to solve important issues including fraud, inefficiency, and a insufficient trust among stakeholders. By employing a decentralized and unchangeable ledger, blockchain guarantees the secure recording of transactions from farm to fork, thereby improving consumer confidence and protecting brand reputation. The study demonstrates successful case studies and pilot projects that showcase blockchain's ability to promote innovation and create cooperative ecosystems. The shift towards globalisation has led to an investigation of various supply chain processes including the use of new technologies such as the IoT and AI, which are integrated with blockchain for enhanced benefits and improved data transmission across supply chain networks. AI tools offer a significant number of advantages for enterprises that are undoubtedly worth it. In this way, artificial intelligence power due to its unique features can enhance productivity, create efficiencies and improve competitive advantage. So what is it that AI tools bring to the table? They bring in variety of benefits for example automated data analysis or forecasting. They make the automation of persistent tasks a task. This saves not only time but also materials. And what's more, artificial intelligence tools also make it possible to create a user interface that is specific to the jobs the client has produced, enabling clients to navigate pages they've designed themselves. By taking the plunge into AI, an organization is empowered to take advantage of hidden capabilities, drive innovation and meet evolving technological needs. Using blockchain, stakeholders can enhance traceability, improve recall and auditing process, and ensure food safety for higher trust in the global food supply chain. As implications for further investigation, this study highlights that significant stakeholders are necessary to unlock blockchain's potential to transform supply chain management.

Keywords: Blockchain · Agriculture · Supply Chain Transparency · Survey · Technology Adoption

A. J. Obaid et al. (Eds.): ICONEST 2024, ISEM 40, pp. 110–125, 2025.
https://doi.org/10.1007/978-3-031-89960-7_8

1 Introduction

1.1 Blockchain-Based Supply Chain

As supply chains become complex, consumers often demand more information about the sustainability, quality, and safety of the product [1]. It was felt that the information travelling up and down the supply chain line by everyone has to be accurate, pertinent, transparent and trustworthy. Closer suppliers to end users can reduce this demand for traceability by offering a more efficient supply chain. Blockchain technology, which provides access to all stakeholders, can be utilized to resolve traceability and trust issues and offer verified supply chain information and to be resolved. By integrating blockchain with other cross-organizational technologies, supply chain stakeholders may experience less problems with cooperation, trust, and traceability. Among the key tasks that blockchain technology may revolutionize are supply chain design, restructuring, security concerns, flexibility, resource management, process management, and product management [2]. Tracking data can be synchronized across all business areas using a well restructured supply chain line. Additionally, it can assist cut down on the time and expense needed to employ smart contracts [3]. Using blockchain in supply chains not only lowers costs and boosts efficiency, but it also strengthens ties between all parties involved [4].

The term "blockchain" describes a system that uses databases to store and distribute data across all users that are network stakeholders. Real-time access to comprehensive transaction data is available to all participants. A centralized hub system used to store transaction data and distribute it with the direct participants in the transaction [5]. On the basis of decentralization, security, and clever execution, blockchain technology, on the other hand, allows individuals to share all information. Put another way, peer-to-peer networks allow each participant to see transaction information sequentially (referred to as decentralization). Additionally, transactions completed with signatures improve security, which in turn protects transparency. As a result, any operational issues may be handled quickly and jointly (security). Furthermore, a transaction stays the same once it is entered into the system and people have signed it. We refer to this quality as "immutability." With all of these characteristics, it is anticipated that this technology would assist several sectors [6]. The influence of blockchain technology on supply chain management is examined in this study. Transporting, storing, and delivering goods from the location of raw materials to manufacturing and finished customers is all included in supply chain management. Efficiency and reduction of costs are commonly regarded as the two key objectives of supply chains and it is anticipated that the introduction of blockchain technology will be beneficial to such systems. At the same time, the array of supply chain positions which are built on the foundations of blockchain technology, is also increasing [7]. On the other hand, adoption of the supply chain blockchain system within the organization has not really emphasized due to its lack of justified hook value portrayal. We outline the four basic characteristics of the supply chains built on the base of the blockchain technology: cost effectiveness, synchronized transaction process, traceability (or visibility), and dependability with security [8].

The last few decades has seen the rise of globalization fueled by an increase in the development of information systems and networks, which in turn raised the need to develop supply chains. It has evolved into a vast information network connecting all levels and phases of production and consumption, and even among producers. The change in orientation toward estimating efficiency of supply chain operations', that was extended not only to the net profit of the enterprise itself, or any single institution, but all institutions that are relevant throughout the supply chain. The development of Information and communication technologies penetrated even deeper the economy's and the industries' evolution, necessitating the evolving side of a supply chain to be given proper attention. Information and communication technologies offer new possibilities for the efficient management of a supply chain, where these possibilities improve service quality while ultimately lowering the costs of the institutions. Effective management of supply chain operations was curated by the technology of Blockchain. [9].

1.2 Efficiency in Supply Chain Management Through Blockchain

The need for increased efficiency is more important than ever at a time of globalization, quick technical development, and complex supply chain networks. Conventional supply chain systems frequently struggle with inefficiencies brought on by manual procedures, middlemen, and a lack of real-time visibility [10]. Presenting blockchain technology, a revolutionary idea that has the potential to completely alter the supply chain management industry [11]. Supply chain transparency has become a strategic priority for businesses, consumers, regulators, and investors alike. As global supply chains grow in complexity, transparency—defined as the ability to access and track data along the entire chain—is essential for fostering trust and accountability. However, achieving transparency requires the efficient sharing of data across various stakeholders, including suppliers, manufacturers, distributors, and consumers.

The outline of the document includes the following parts: Sect. 2 gives an account of the purposes of this study. Section 3 examines the related work, outlines the state-of-the-art techniques, and provides a comparative analysis of the proposed methods. Section 4 explains the methodology which is being put forward by the author. Section 5 reports the analysis of the results and also provides information related to the dataset. Last but not least, Sect. 6 sums up the paper.

2 Objectives

The relationship between data sharing and supply chain transparency is examines in this study, focusing on how new technological solutions, data governance frameworks, and collaborative techniques are changing conventional supply chain models. Examine how openness, data sharing between supply chain partners (manufacturers, suppliers, retailers, and regulators) improves transparency, accountability, and overall efficiency. Examines how particular technologies, such as blockchain facilitate secure and efficient data sharing, enabling transparency in the supply chain.

3 Related Works

Numerous industries has seen a transformation thanks to Blockchain technology tool across multiple, but supply chain management has been most affective. Its ability to increase transparency, traceability, and trust has sparked a great deal of study and real world application. As globalization accelerates and the demand for efficient, secure, and sustainable supply chains grows, blockchain offers practical answers to critical issues such as fraud, inefficiencies, and limited stakeholder collaboration.

This study of the literature investigates the current applications of blockchain technology in supply chain management, combining results from recent studies that evaluates its advantages, limitations, and future prospects. The examines study addresses important topics including transparency, traceability, and sustainability, and covers a range of industries, including agriculture, food, fashion, and pharmaceuticals key issues such as. This study lays the groundwork for comprehending how blockchain is changing supply chain operations and identifies new developments impacting its uptake in international supply chains.

In 2022, Funlade Sunmola and Patrick Burgess examined transparency in blockchain-based supply chains to comprehend the fundamental ideas of their design. They carried out a thorough study of the literature and added data-driven analysis. The findings present these principles within a framework for transparency by design in blockchain-based supply chains. Additionally, the study discusses limitations and identifies areas for future research [12].

In 2022, Agarwal et al. conducted a complete study of the literature on blockchain specifically pertaining to characteristics, implementations, and business disclosures within different supply chain management environments. This blockchain-focused study examines and analyzes 97 recent publications to uncover the current state of research and outline future research directions. The results show that the main forces behind the adoption of blockchain in supply chain management are information sharing, product anti-counterfeiting, transparency and traceability, and trust-building [13].

In 2022, Zheng et al. proposed a blockchain-based paradigm to facilitate supply chain management of shared transaction data and access control. This model addresses challenges related to managing extensive credit investigation data while ensuring privacy protection. It incorporates a consensus mechanism to handle large volumes of credit investigation data, thereby enhancing privacy safeguards and achieving effective access control over the shared data chain. This approach shares data using a proxy re-encryption technique and stores it on cloud servers. The workflow and data structure are especially made to suit supply chain finance applications. In addition to creating a thorough monitoring system throughout the duration of the lifespan, the integrated blockchain technology makes it easier to create an open shared reference architecture [14].

In 2023, A. Daghighi and F. Shoushtari conducted a study with the primary objective of investigating the application of blockchain technology in supply chain management to enhance sustainability. This study is very important because more and more businesses are focusing on lowering their impact on the environment and being more responsible in their actions. The research shows that blockchain is a helpful technology for promoting sustainability. It can improve traceability, make things more transparent, and build trust between people in the supply chain. These qualities are crucial for tackling the different sustainability problems that companies are dealing with today [15].

In 2023, Cozzio et al. adopted a mixed-method approach—combining experimental methods with qualitative evidence—to explore why consumers value the traceability offered by blockchain and to identify the barriers suppliers face when considering its adoption. Their study shows that blockchain technology helps customers trust locally sourced food more, which makes them feel better and act differently when sharing food. On the business side, the survey found that while companies see the big benefits of using blockchain, like more trust, they are still worried about sharing data and needing more help within their own teams. The researchers give a detailed look at how blockchain works in the food supply chain, considering both businesses and customers [16].

In 2023, Adebunmi et al. explored the use of blockchain technology in agricultural supply chains, with a focus on how it might help with major issues including fraud, inefficiencies, and stakeholder mistrust. Blockchain ensures the secure recording of every transaction in the agricultural supply chain, from the farm to the consumer, by providing a decentralized and unalterable record. This transparency helps verify product authenticity and origin, which combats issues such as food fraud, increases customer trust, and protects brand reputation. Blockchain also enables real-time tracking of agricultural products throughout the supply chain, enhancing traceability. Despite its benefits, implementing blockchain in agriculture faces challenges, particularly the need for advanced technology infrastructure. To make the most of blockchain's benefits and ensure data moves smoothly throughout the supply chain, it needs to work together with other technologies like the Internet of Things (IoT) and artificial intelligence (AI) [17].

In 2023, Aayushi et al. provided a detailed look at how blockchain technology could be used in the fashion and textile industries' supply chains, focusing on how it could enhance product authenticity, traceability, and transparency. To reach this conclusion, the authors carefully examined many research papers and other sources. The review highlights key areas such as blockchain technology, the fashion and textile industry and its supply chain (which includes manufacturing and distribution), as well as concerns about traceability, transparency, and product authenticity [18].

In 2023, J. Xia, H. Li, and Z. He provided a thorough analysis of supply chain case studies utilizing blockchain technology. The majority of the research now in publication is on the food, agricultural, and pharmaceutical industries, highlighting the benefits of blockchain technology in improving transparency and traceability. The enhancement of cooperation efficiency in supply chains, especially in the information technology industry, is the subject of very few research. The authors illustrate the advantages of blockchain technology with a case study of Lenovo, a prominent corporation that uses it, using semi-structured interviews. Additionally, they investigate the possible uses of blockchain technology to improve supply chain cooperation and offer a conceptual model for a blockchain-based information collaboration system [8].

In 2024, Oriekhoe et al. conducted to study the role of blockchain technology in facilitating innovation across supply chains and the potential of the technology to invent new business models, create collaborative ecosystems, and integrate emerging technologies such as the Internet of Things (IoT) and Artificial Intelligence (AI).The report focuses on pilot projects and case studies that demonstrate how blockchain technology has been effectively used to spur supply chain process innovation. Concerns like interoperability, scalability, and regulatory compliance are among the difficulties and factors that are discussed in relation to the application of blockchain technology in supply chain management. The study ends by suggesting future lines of inquiry and stressing the necessity of industry-wide cooperation to fully realize blockchain's promise to transform supply chain management [13].

In 2024, Itohan et al. provided a comprehensive examination of various aspects of blockchain technology in Supply Chain Management (SCM). The paper examines what's going on with blockchain implementation in various industries, talks about the difficulties and solutions of the adoption of the technology, and does the performance comparisons with the non-blockchain systems. The well-researched study looks at new trends that are just emerging, takes insights from the fact that different stakeholders may not share the same views, and makes sure to take into account differences in the adoption rates. The strategic implications for businesses, effective implementation pathways, and policy recommendations are elaborately discussed. The results show that blockchain is far more transparent and ensures supply chains are more efficient. The research highlights the benefits of blockchain implementation in supply chains, such as reduced costs, improved operating efficiency, and greener practices [19].

In 2024, Oriekhoe et al. presented a concept paper on how to improve global food safety by integrating blockchain technology into food supply systems. The main advantages of integrating blockchain technology into food supply chains include better recall and audit efficiency, more customer trust, enhanced transparency, and chances for cost and process optimization. Stakeholder cooperation, infrastructure spending, and the resolving of technical and legal issues are necessary for successful implementation, nevertheless. This study presents a strategy framework that addresses regulatory compliance, scalability issues, pilot testing, and stakeholder participation for integrating blockchain technology into food supply chains. We can improve public health, increase food safety, and boost trust in the global food supply chain by utilizing blockchain technology [20].

There isn't much attention in the literature on making supply chain collaboration more efficient, especially in industries other than food, agriculture, and pharmaceuticals. We need more studies on combining blockchain with new technologies like IoT and AI, and on improving collaboration and scalability across different industries. Also, we need complete frameworks that tackle technical, regulatory, and infrastructure issues to speed up the use of blockchain in supply chains.

3.1 Comparison with the Existing Surveys

Many industries, especially supply chains, are often talked about when it comes to using blockchain technology. Blockchain is very good at solving important problems in supply chains. People have looked a lot into how blockchain can be used in managing supply chains, and many reviews have talked about what's happening now and what should be studied next. In this study, we looked at different supply chains. This part compares recent studies that combine blockchain and supply chain management. Table 1 shows a comparison of existing reviews with our review, focusing on their goals, methods, challenges, uses, and results. This information can be very helpful for those new to this growing field of research.

Table 1. A comparative analysis between the existing and proposed surveys.

Citation	Objectives	Methods Used	Challenges	Applications	Results
[21]		Blockchain technology	Integration with existing systems	Food safety, ethical sourcing, counterfeit prevention	Blockchain enhances traceability, authenticity, and stakeholder trust in supply chains
		Case studies in various industries	Scalability, regulatory compliance, and data privacy issues	Traceability, authenticity verification, stakeholder trust	Blockchain improves efficiency through streamlined processes, accurate data, and reduced fraud
[22]	Investigate blockchain's transformative potential in supply chains	Investigates adoption of blockchain in supply chain management	Technological complexity in integrating with existing systems	Tracking products, verifying provenance, and streamlining processes	Blockchain enhances transparency, efficiency, and trust in supply chains
	Address challenges and strategic implications of blockchain adoption	Explores transformative potential, challenges, and strategic implications	Concerns over data privacy and security	Addressing issues of counterfeiting, compliance, and sustainability in supply chains	Adoption faces challenges like technological complexity and regulatory clarity
[23]	Examine intersection of supply chain management and blockchain technology	Blockchain technology		Transparency, efficiency, and security in the supply chain network	Enhanced transparency, efficiency, and security in supply chain management
	Highlight advantages, obstacles, and recent advancements in blockchain integration	Transparent and tamper-resistant record of transactions		Decentralized ledger for real-time monitoring of product movement	Immediate insight into product flow, minimizing fraud and counterfeiting

(continued)

Table 1. (*continued*)

Citation	Objectives	Methods Used	Challenges	Applications	Results
[24]	Enhance security and privacy in blockchain-based SCM	Textual data encryption	Handling escalating data volume in supply chain management	Textual data encryption	Enhanced security and privacy for SCM data
	Manage data growth and prevent unauthorized entries	Access control through permission-based system	Addressing technical expertise gap in blockchain technology	Access control	Improved robustness of blockchain-based SCM
[25]	Investigate blockchain technology in supply chain management	Desktop research methodology	High initial investments and scalability issues	Enhancing transparency, traceability, and efficiency in supply chains	Blockchain enhances transparency, traceability, and efficiency in supply chain
	Address contextual and methodological gaps in existing research	Secondary data from existing resources	Need for industry-wide standards and integration challenges	Automating processes through smart contracts to streamline operations	Challenges include high initial investments, scalability issues, and standards
[26]	Highlight benefits of blockchain in supply chain management	Blockchain technology		Enhancing transparency, traceability, security, and efficiency within supply chains	Enhances transparency, security, and efficiency in supply chain management
	Enhance transparency, security, efficiency, and traceability within supply chains	Intelligent contracts for automation		Mechanizing forms like payments and inventory management to reduce costs	Promotes brand identity and customer trust through verifiable records
[27]	Develop secure automated pharmaceutical supply chain management system	IoT devices with sensors and actuators	Time sequencing, coordination, and tracking of goods and processes	Secure automated supply chain management (SCM)	Filecoin is efficient for pharmaceutical SCM due to speed and cost

(*continued*)

Table 1. (*continued*)

Citation	Objectives	Methods Used	Challenges	Applications	Results
	Enhance communication and security using IoT and blockchain technologies	Blockchain technology for security and transactions	Security and control in pharmaceutical supply chain management	Analytics applications for business planning and improvement	Proposed system uses IoT, Cloud, and Blockchain for secure SCM
[28]	Identify benefits, challenges, and barriers of blockchain technology adoption	Qualitative thematic analysis	System design complexity and resistance to change	Optimizing supply chain and logistics management	Benefits: Enhanced trust, transparency, automation, traceability, data security
	Assess impacts and strategies for maximizing blockchain potential in industries	Findings from various studies	Inadequate IT staff and regulatory uncertainties	Enhancing trust, transparency, automation, traceability, and data security	Challenges: System design complexity, resistance to change, inadequate IT staff
[29]	Explore benefits and applications of blockchain and telematics integration	Literature review and case studies analysis	Scalability and interoperability issues in integration	Enhancing security, transparency, efficiency, traceability, and customer satisfaction	Enhanced security, transparency, and efficiency in supply chains
	Address challenges and future directions in supply chain management	Expert insights on blockchain and telematics integration	Regulatory compliance, cost, and technical expertise challenges	Addressing complexities of modern supply chains and paving way for innovations	Addressed challenges like scalability and regulatory compliance
[30]	Analyze blockchain architecture for supply chain management in tertiary institutions	Soft Actor-Critic (SAC) algorithm for adaptive policy learning	Scalability and interoperability concerns in existing systems	Inventory optimization using SAC and prioritized experience replay	SAC-rainbow outperformed other algorithms in reward maximization
	Integrate deep reinforcement learning for inventory optimization and security	Prioritized experience replay for efficient learning in inventory optimization	Navigating regulatory and legal landscape challenges	Secure file sharing and management in tertiary institutions	Achieved optimal inventory levels and minimized costs effectively

4 Methodology

This part explains the approach used to study how data sharing improves supply chain transparency. As global supply chains become more connected, there is a greater need for clear and efficient ways to share data. Transparency helps people track goods, make sure rules are followed, and build trust throughout the supply chain. But, making data sharing smooth requires solving several issues, such as making different technologies work together, protecting data privacy, and building trust among those involved.

To understand these dynamics, the research uses a method that focuses on understanding and grouping ideas to combine existing knowledge and find common patterns in how data is shared. The research plan is designed to meet these goals:

- Assess how sharing data affects the openness of supply chains, focusing on the advantages it brings for better tracking, tracing, and building trust.
- Recognize the main obstacles to sharing data, like rules and regulations, technical difficulties, and opposition from people involved.
- Examine successful methods and tools that enable clear data sharing in different fields, such as food, healthcare, and fashion.
- Look at practices in different industries to see how the need for transparency and the solutions used differ between them.

The research uses information from existing sources like academic articles, industry reports, and real-life examples to get a full picture of how data sharing works today, the problems it faces, and new trends. By looking at these sources in a structured way, the study finds common themes and shows how data sharing can improve supply chain openness, as well as what still needs to be improved in both practice and research.

4.1 Supply Chain Management SCM

Transporting goods and services from one location to another via a range of stakeholders is a fundamental business function. The conversion of raw materials into finished products and meeting customer orders are the focus of diverse people, resources, behaviors, and organizations—a process known as supply chains. The process that a business goes through from manufacturing and delivering a good to the end consumer in a pre-planned flow of information, physical distribution, and payment involves a group of cooperating companies and partners personnel, activities, and information. The initial step four of the value delivery chain is getting resources to a manufacturer, and the ending point is the consumer getting the finished product or service. Supply chain management stipulates the process of managing the delivery of goods and services so as to, ensure the calider of fragile products during transportation, minimize the wastage of money, and meet expectations customers are looking for [19]. The manufacturing, high-tech, agri-food, pharmaceutical, textile, automotive, and soon-to-be service trade industries all have supply chain management. Processing, storing, manufacturing, shipping, arriving, and consumption are all phases in the life cycle of any production supply chain system. Current supply chain management has several problems, including the requirement for clarity and simplicity, disruptions, extra postponements, data manipulation, and instability. The current supply chain typically depends on a single organization to keep track of data, which may be detrimental to openness and confidence. If the information on the ledger does not support the buisiness's expansion, it could be subtly falsified. There is incredibly detailed information regarding the origin, processing, and delivery of the goods to the final consumers. Large brands typically only made some of the information available to their customers. Not knowing enough about a product can make it difficult to understand many of its features. Many of the substances listed in the company's products may be difficult for customers to verify. Consequently, there is an increase in mistrust between initiatives, which raises the cost of communication.

4.2 Blockchain

A blockchain is a type of database or storage system that stores data using blocks and chaining. When fresh information is received, it is first added to a new block and subsequently linked to the earlier Blockchain. A blockchain is a data chain made up of blocks with different storage capacities that, when led, are connected to the previous led block. In 1991, Tuart Haber and W. Scott Stornetta conducted the remaining research on the concept of blockchain. Their goal was to create a system that made it illegal to change the timestamps on documents. In 2008, a group known as Satoshi Nakamoto created the Blockchain. It is a peer-to-peer network that solved the double-spending problem. Then in the subsequent year, Bitcoin, a digital currency was released with Blockchain technology as the ledger that is disseminated. Blocks are the fundamental idea behind Blockchain (BC) technology. To find the block, the SAH-256 hash algorithm generates a hash value. The most recent block's header eld includes the hash value of the block before it or its parent block. The different types and designs of the BC can also determine the block's structure. Generally, a block must have some specific key attributes in its block header which is the link to the previous block. These are the nonce, block timestamp, Merkle root/hash, and hash of the previous block, as well as the block body containing transactions. Nonce is a unique number which can be used once and it is randomly generated. Rehashing the hashed block with the number satisfies the difficulty level criteria. Figure 1 shows how blockchain technology has developed.

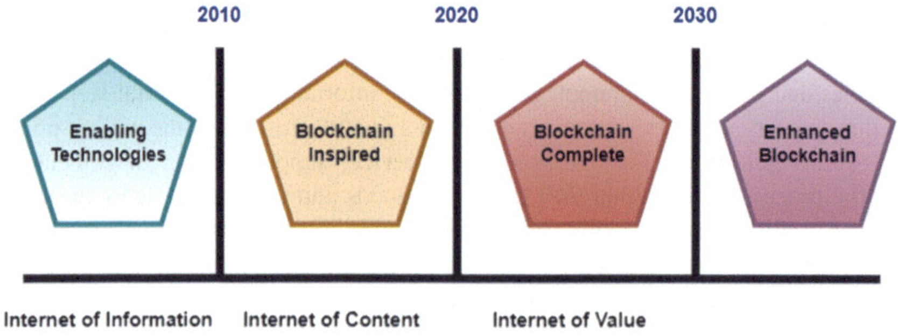

Fig. 1. Evolution of BlockChain technology.

4.3 Integration of Blockchain in Scm

Providing an immutable, irreversible, distributed, robust, and non-centralized ledger system is the sole objective of switching to a BC-based solution. Contract monitoring and signing, as well as confirming the origin of products, are all possible with BC. Systematic study indicates that a number of BC features are especially helpful in supply chain case studies.

4.4 Advantages of BC-Based Supply Chain Systems

Because every stakeholder system ledger has all of the information, the BC platform for the supply chain removes the uncertainty that comes with the many databases of traditional supply chain systems [23]. Based on the interviews conducted by [24], BC, as a result of automatic data validation, shortens the extra need of verifying in addition to monitoring and tracing services throughout the supply chain. By supplying evidence of provenance for transactions, BC makes transaction monitoring possible. BC speeds up end-to-end supply chain activities while enabling higher volume and data accuracy. Applications that are BC-enabled distribute data throughout the whole network in a matter of seconds. Consensus approaches enable supply chain automation and operational benefits by providing data truthfulness and easy smart contract foundations. Figure 2 illustrates the advantages of utilizing blockchain technology in supply chain management (SCM).

4.5 Information Sharing

Due to the considerable quantity of data generated by its members, in a supply chain, information exchange is difficult and dispersed. Participating members of a supply chain can share data about items, their state in the manufacturing process and ownership as well as location for data storage and even environmental impact to enable knowledge transfer [24]. Globalization and complexity cause the remote distribution of supply chains participants, resulting in the highly scattered information. Furthermore, information interchange is difficult due to stakeholders' huge data volumes. BC's real-time information sharing promises to revolutionize the way information is shared and distributed across numerous individuals. It could provide value to a number of supply chains, such as those in the construction, smart city, and health and medical industries.

5 Result Analysis and Discussion

To provide a comprehensive understanding of how data-sharing practices and transparency strategies differ across sectors, Table 2 presents a comparative analysis of key characteristics observed in Healthcare, Food, Fashion, Electronics, and Pharmaceuticals. This table encapsulates sector-specific approaches, challenges, and opportunities, offering a snapshot of the current state of transparency in these industries. The comparative analysis in Table 2 reveals notable patterns and differences across these sectors:

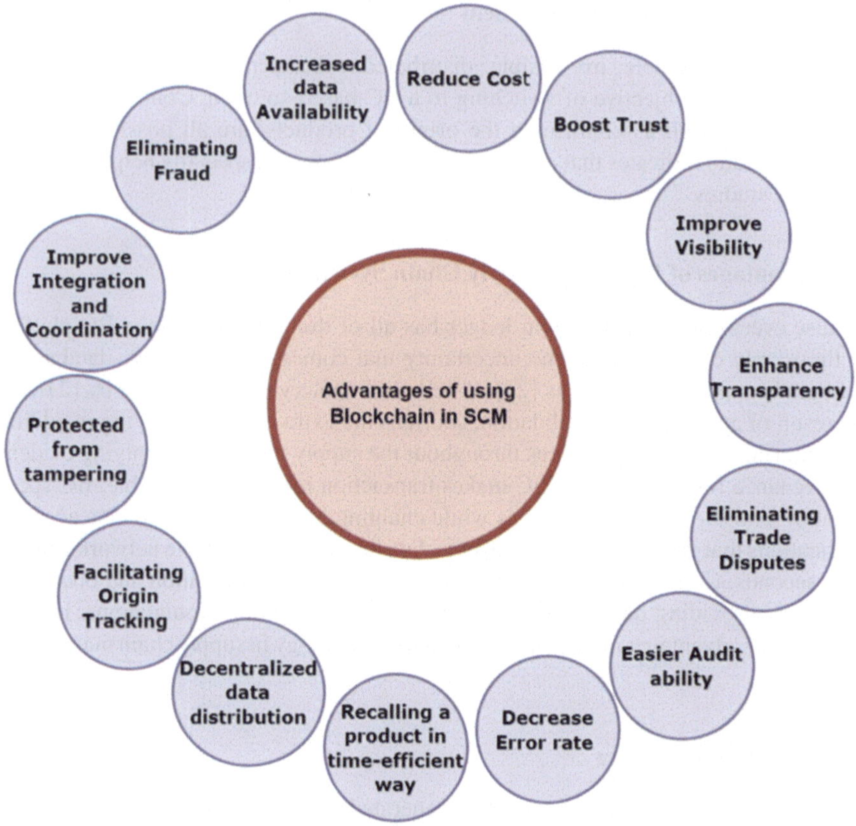

Fig. 2. Benefits of SCM using BC technology

- Highly regulated industries such as Healthcare and Pharmaceuticals have mature transparency mechanisms, driven by compliance needs.
- Consumer-facing industries like Food and Fashion are increasingly adopting transparency strategies due to demand for ethical and sustainable practices.
- The Electronics sector is a hybrid, balancing regulatory compliance with innovation-driven transparency initiatives.

Despite these differences, a common trend in all industries is the increasing use of technology—like blockchain, AI, IoT, and digital passports—to improve transparency and data sharing. Table 2 shows that it's important to adjust transparency strategies to fit the unique needs and challenges of each industry. It also emphasizes the need for cooperation between different sectors to tackle common issues, such as cybersecurity threats and the absence of a standard way to share data globally. These insights suggest that future improvements in supply chain transparency will probably depend on industries' ability to use compatible systems and take advantage of new technologies.

Table 2. Comparative analysis of Data-Sharing Practices and Transparency Strategies Across Sectors

Sector	Primary Goal of Data Sharing	Key Data Types Shared	Technologies Used	Transparency Challenges	Effective Strategies
Healthcare	Improve patient care and safety, streamline supply of medical goods	Patient records, medical supply chains, clinical trial data	Blockchain, Electronic Health Records (EHR), Cloud computing	Data privacy (HIPAA compliance), interoperability, cybersecurity	Implementing blockchain for traceability, using standardized data formats, securing data through encryption
Food	Ensure food safety, quality, and compliance, meet consumer demand for sourcing info	Product origin, quality checks, transport and storage conditions	IoT sensors, Blockchain, QR codes for tracking	Ensuring data accuracy from multiple suppliers, real-time monitoring	Real-time IoT tracking for temperature and location, QR codes on packaging for consumer transparency
Fashion	Track ethical sourcing, reduce counterfeit goods, manage supplier transparency	Supplier certifications, production locations, materials data	Blockchain, Digital IDs, RFID	Limited transparency among global suppliers, lack of standardized reporting	Blockchain for tracking ethical sourcing, RFID for tracking authenticity in real-time
Electronics	Combat counterfeit parts, ensure ethical sourcing of materials	Supplier origin, compliance with environmental standards, component authenticity	Blockchain, RFID, AI for anomaly detection	Complex supply chains, varying global regulations, proprietary tech secrecy	Blockchain for traceability, AI algorithms for detecting counterfeit goods in the supply chain
Pharmaceuticals	Track authenticity, ensure regulatory compliance, improve recall accuracy	Raw materials, manufacturing processes, batch numbers	Blockchain, RFID, ERP (Enterprise Resource Planning) systems	Strict regulatory requirements (FDA, EMA), supply chain complexity	Implementing blockchain for secure data sharing, ERP systems to manage end-to-end visibility

6 Conclusion

The paper explores the role of blockchain technology in improving supply chain transparency and efficiency. It emphasizes blockchain's potential as a catalyst for innovation, with case studies demonstrating its ability to transform traditional supply chain models into more transparent and efficient systems. However, the paper also highlights significant challenges in blockchain implementation, including issues of interoperability, scalability, and regulatory compliance. It underscores the importance of effective data sharing among supply chain partners, suggesting that blockchain can greatly enhance transparency and accountability. The study further recognizes the operating benefits of blockchain integration including cost savings, operational efficiency, and environmental sustainability. In summary, the paper ends with a plea for more research and collaboration across firms to address challenges and realize the full potential of blockchain technology all the while integrating novel technologies such as Internet of Things (IoT) and Artificial Intelligence (AI) to enhance supply chain procedures through synergies.

References

1. Behnke, K., Janssen, M.: Boundary conditions for traceability in food supply chains using blockchain technology. Int. J. Inf. Manage. **52**, 101969 (2020)
2. Dutta, P., Choi, T.-M., Somani, S., Butala, R.: Blockchain technology in supply chain operations: applications, challenges and research opportunities. Transp. Res. Part E. Logist. Transp. Rev. **142**, 102067 (2020). https://doi.org/10.1016/j.tre.2020.102067
3. Wang, Y., Liang, H., Yang, N., Sun, S.: How contract enforcement reduces opportunism? The paradoxical moderating effect of interfirm guanxi. Ind. Mark. Manage. **97**, 115–125 (2021)
4. Queiroz, M.M., Wamba, S.F.: Blockchain adoption challenges in supply chain: an empirical investigation of the main drivers in India and the USA. Int. J. Inf. Manage. **46**, 70–82 (2019)
5. Xia, J., Li, H., He, Z.: The effect of blockchain technology on supply chain collaboration: a case study of lenovo. Systems **11**(6), 299 (2023)
6. Treiblmaier, H.: Combining blockchain technology and the physical internet to achieve triple bottom line sustainability: a comprehensive research agenda for modern logistics and supply chain management. Logistics **3**(1), 10 (2019)
7. Kumar, A., Liu, R., Shan, Z.: Is blockchain a silver bullet for supply chain management? Technical challenges and research opportunities. Decis. Sci. **51**(1), 8–37 (2020)
8. Alsmadi, A., Al-Gasaymeh, A., Alrawashdeh, N., Alhwamdeh, L.: Financial supply chain management: a bibliometric analysis for 2006–2022. Uncertain Supply Chain Manage. **10**(3), 645–656 (2022)
9. Alsmadi, A.A., Alrawashdeh, N., Al-Gasaymeh, A, Alhawamdeh, L.N., Al_Hazimeh, A.M.: Adoption of blockchain technology in supply chain. Sage Open. **13**(1), 21582440231160144 (2023)
10. Litke, A., Anagnostopoulos, D., Varvarigou, T.: Blockchains for supply chain management: architectural elements and challenges towards a global scale deployment. Logistics **3**(1), 5 (2019)
11. Oriekhoe, O.I., Oyeyemi, O.P., Bello, B.G., Omotoye, G.B., Daraojimba, A.I., Adefemi, A.: Blockchain in supply chain management: a review of efficiency, transparency, and innovation. Int. J. Sci. Res. Arch. **11**(1), 173–181 (2024). https://doi.org/10.30574/ijsra.2024.11.1.0028
12. Sunmola, F., Burgess, P.: Transparency by design for blockchain-based supply chains. Procedia Comput. Sci. **217**, 1256–1265 (2022). https://doi.org/10.1016/j.procs.2022.12.324

13. Agarwal, U., et al.: Blockchain technology for secure supply chain management: a comprehensive review. IEEE Access **10**(July), 85493–85517 (2022). https://doi.org/10.1109/ACCESS.2022.3194319
14. Zheng, K., et al.: Blockchain technology for enterprise credit information sharing in supply chain finance. J. Innov. Knowl. **7**(4), 100256 (2022)
15. Daghighi, A., Shoushtari, F.: Toward sustainability of supply chain by applying blockchain technology. Int. J. Ind. Eng. Oper. Res. **5**(2), 60–72 (2023)
16. Cozzio, C., Viglia, G., Lemarie, L., Cerutti, S.: Toward an integration of blockchain technology in the food supply chain. J. Bus. Res. **162**, 113909 (2023)
17. Adewusi, A.O., Chikezie, N.R., Eyo-Udo, N.L.: Blockchain technology in agriculture: enhancing supply chain transparency and traceability. Financ. Account. Res. J. **5**(12), 479–501 (2023)
18. Badhwar, A., Islam, S., Tan, C.S.L.: Exploring the potential of blockchain technology within the fashion and textile supply chain with a focus on traceability, transparency, and product authenticity: a systematic review. Front. Blockchain **6**, 1044723 (2023)
19. Oriekhoe, O.I., Omotoye, G.B., Oyeyemi, O.P., Tula, S.T., Daraojimba, A.I., Adefemi, A.: Blockchain in supply chain management: a systematic review: evaluating the implementation, challenges, and future prospects of blockchain technology in supply chains. Eng. Sci. Technol. J. **5**(1), 128–151 (2024). https://doi.org/10.51594/estj/v5i1.732
20. Oriekhoe, O.I., Ilugbusi, B.S., Adisa, O.: Ensuring global food safety: integrating blockchain technology into food supply chains. Eng. Sci. Technol. J. **5**(3), 811–820 (2024)
21. Nwariaku, H., et al.: Blockchain technology as an enabler of transparency and efficiency in sustainable supply chains. Int. J. Sci. Res. Arch. **12**(2), 1779–1789 (2024). https://doi.org/10.30574/ijsra.2024.12.2.1454
22. Wilson, O.J., Brown, W.: Exploring the adoption of blockchain technology in supply chain management exploring the adoption of blockchain technology in supply chain management (2024). https://doi.org/10.20944/preprints202407.2419.v1
23. Apurv, J., Raut, A., Taneja, H., Dalvi, V., Varade, U.: Supply chain management using block chain technology. Int. Res. J. Adv. Eng. Hub. **2**(07), 1931–1936 (2024). https://doi.org/10.47392/irjaeh.2024.0264
24. Khan, I., et al.: Securing blockchain-based supply chain management: textual data encryption and access control. Technologies **12**(7), 1–19 (2024). https://doi.org/10.3390/technologies12070110
25. M.Y.K. B. A.: Blockchain technology in supply chain management. Int. J. Multidiscip. Res. **6**(1) (2024). https://doi.org/10.36948/ijfmr.2024.v06i01.13504
26. Prakash, A.: Blockchain technology for supply chain management: enhancing transparency and efficiency. Int. J. Glob. Acad. Sci. Res. **3**(2), 01–11 (2024). https://doi.org/10.55938/ijgasr.v3i2.73
27. "7_Secure_pharmaceutical_supply_chain_using_blockchain_in_IoT_cloud."
28. Ekwunife, D.I., et al.: Using blockchain technology to maximize supply chain and logistics management in north America. Int. J. Sci. Res. Arch. **12**(2), 854–863 (2024). https://doi.org/10.30574/ijsra.2024.12.2.1096
29. Shrivastava, P., Verma, R.: Integrating blockchain technology with telematics for secure and transparent data sharing in supply chain management. Int. J. Glob. Innov. Solut. 1–12 (2024). https://doi.org/10.21428/e90189c8.01ec868b
30. Welt, J., Zhang, Y.: Integrating blockchain and deep reinforcement learning for secure and efficient supply chain management in tertiary institutions. Comput. Life **12**(2), 14–19 (2024). https://doi.org/10.54097/ntsgdq10

AI in Education and Learning Techniques

Development of Assessment Rubric Performance of Competency Managing Receivable

Suwarno(✉)

PGRI Ronggolawe University, Tuban, Indonesia
suwarnoklaten1964@gmail.com

Abstract. The expected result of this research is a rubric for assessing competency performance in managing receivable cards that is valid and reliable. The stages carried out consist of: initial stage; design; trials; and implementation. Twelve teachers who taught accounting were selected as raters and previously explained how to use rubrics. Content validity analysis uses the Aiken index and reliability uses the inter-class correlation coefficient. The calculation results obtained are in the table, so they are valid. The estimated alpha coefficient is classified as special, and analysis of the inter-class correlation coefficient provides an illustration that the rubric in question is reliable.

Keywords: Development · Rubric · Receivables Card

1 Introduction

One of the core competencies of teachers is carrying out assessment and evaluation of teaching and learning activities, which is broken down into seven competencies, including: (1) understanding the main points of assessment and evaluation of teaching and learning activities, (2) making provisions for important aspects for assessment and evaluation., (3) create procedures for assessing and evaluating teaching and learning activities, (4) develop tools for assessing and evaluating teaching and learning activities, (5) create administration activities for assessing and evaluating teaching and learning activities on an ongoing basis (6) create analysis of the results of assessing and evaluating teaching and learning activities, and (7) evaluating learning activities teach.

Teachers have to deal with a large number of students and learning plans, as a result they find it difficult to convey useful responses (Wang, 2008: p.449). The positive impact is providing feedback. The responses obtained by students during evaluation can help them carry out self-assessment, reflection, increase their motivation and self-esteem (Marriott, 2009, p.239). Meaningful responses allow students to improve their weaknesses in learning and thinking, communicate what they have learned, and learn to respect each other (Wang, 2007, p.175). From this response, the teacher will know the success of the teaching and learning process that has been carried out; competency and level of absorption; This provides valuable information as feedback to teachers to improve their activities in teaching and learning activities.

A. J. Obaid et al. (Eds.): ICONEST 2024, ISEM 40, pp. 129–137, 2025.
https://doi.org/10.1007/978-3-031-89960-7_9

In fact, most teachers do not know enough about assessment to manage their classrooms well (Mertler, 2001, p. 3). The main reason why teachers do not know about assessment is in teacher training programs (Stiggins, 2002, p. 762). Teachers learn the concepts and practices of educational assessment in just a few sessions (Popham, 2009, p. 5). Therefore, teachers need to know that assessment is closely related to student ability.

Performance evaluation provides a periodic record of strengths and weaknesses related to independent and collaborative tasks (Cascio, 1992, p.267). Assessment is basically an important element for effective and efficient student development. Assessment has a very positive impact on the dynamics of student development as a whole. Through this assessment, the actual condition of how students perform can be known. So performance assessment is a process used by institutions to evaluate the implementation of individual student work. There are two special characteristics in performance assessment, (1) students are required to demonstrate their ability to create a product or participate in an activity, and (2) the results of the performance assessment are more important than the action (performance). Whether to evaluate a product or service (action) depends on the characteristics of the area to be measured.

"A rubric is a text that describes agreed-upon values." (Callison, 2000, p. 36) states that rubrics "create core standards that students can use to develop, revise, and evaluate your work." Rubrics provide "detailed descriptions of what has and has not been learned." These descriptive data can be used to document how instruction can be improved. In addition, rubric assessment data are very detailed and well-defined (Bresciani et al. 2004, p. 30). Rubrics are scoring guides that describe the criteria used to evaluate student work. Because rubrics are easy to use and easy to explain, they produce data that is easy to understand, and information is conveyed (Andrade, 2000, p. 14). More detail can increase a rubric's effectiveness (Popham, 2003, p. 95). Rubrics "tend to be objective and consistent from lesson to lesson and student to student, and are particularly useful in team teaching situations that require collaboration among teachers" (Callison, 2000, p. 35). Rubrics clearly describe student performance as it relates directly to the learning activity, providing accurate and meaningful information for decision making.

Rubrics can be holistic, analytical, or a combination of both (Luft, 1999, pp. 107–121). A holistic rubric is a construct that contains various levels of performance that describe the quality of tasks, the quantity of tasks, or both, namely the quantity and quality of tasks. This type of rubric requires the instructor to determine the level of performance.

Rubrics can be holistic, analytical, or a combination of both (Luft, 1999, pp. 107–121). A holistic rubric is a construct that includes various levels of performance that describe the quality and quantity of tasks, or both. This type of rubric requires the instructor to determine the level of performance.

Meanwhile, an analytical rubric is a structure that contains provisions in various levels of performance. Typically, each line begins with a section that specifies the criteria to be evaluated, and each section describes a different performance criterion. To add clarity to the analytical rubric, each criterion topic can be broken down into a brief statement followed by a description of the related performance.

The main problem in this study is: How to develop a valid and reliable rubric for assessing the competency of managing receivable cards for students of accounting expertise in vocational high schools in the field of business and management expertise? More operationally, as follows: (1) What skills describe the level of competency in managing receivable cards for students of accounting expertise in vocational high schools in the field of business and management expertise? (2) How is a valid rubric for assessing the performance of competency in managing receivable cards for students of accounting expertise in vocational high schools in the field of business and management expertise? (3) How is a reliable rubric for assessing the performance of competency in managing receivable cards for students of accounting expertise in vocational high schools in the field of business and management expertise?

2 Methodology

The purpose of this study is to produce a performance assessment rubric for managing receivable cards. The reason refers to Borg & Gall, 1983, that "Research and development is a strategy." "Research and development are effective strategies for improving practice. This is the process of developing and validating educational products. This opinion supports the view that this research method is in accordance with the objectives of the research being conducted."

2.1 Stage of Development

The researcher simplified it into 4 stages of developing a performance assessment rubric for managing receivable cards, namely the initial stage, design, trial and implementation.

2.1.1 Early Stage

At this stage, activities were carried out (a) literature study to prepare for collecting field data, (b) analysis of preliminary study data and describing the existing reality, followed by collecting supporting data related to concepts and theories related to the development of competency assessment rubrics for managing receivable cards. These materials include assessment concepts, rubrics, National Education Standards, Educational Unit Level Curriculum, National Competency Standards, Indonesian National Work Competency Standards, demands of the Business World/Industry. The results of this stage of the study were used to create an instrument for implementing a preliminary study on the implementation of activities to assess competency in managing receivable cards for Vocational High School students.

The researcher then conducted a preliminary study at 2 Vocational High Schools with accounting expertise using the survey method to obtain data, by interviewing and examining documents.

The data collected included the curriculum of the productive program of SMK in the field of accounting expertise, the process of adjusting the contents of the productive curriculum to the needs of the Business World/Industry, the implementation of the evaluation system, and the support provided. Including the availability of an evaluation system. Facilities. Infrastructure and other data needed for this study.

2.1.2 Design Stage

The main parts that are considered by researchers to compile the initial draft of the development of the accounting competency performance assessment rubric include a) how the design of the competency performance assessment of managing receivables cards is currently implemented, b) the factors that support and inhibit the implementation of the competency performance assessment of managing receivables cards.

Based on the data from the preliminary study, the researcher developed a rubric that is used to assess the competency of managing receivables cards. Guided by the curriculum that refers to the results, then the developed assessment rubric is validated by experts, accounting education practitioners, heads of expertise competencies and accounting subject teachers.

2.1.3 Trials

At this stage, a trial is carried out with the aim of finding out to what extent the assessment rubric can be applied to assess the performance of competency in managing receivables cards. Then an analysis is carried out to find out whether the assessment rubric has met its validity and reliability. If the validity and reliability do not meet the requirements, revisions and trials are carried out again, until the validity and reliability requirements are met.

2.1.4 Implementation Stage

At this stage, the performance assessment rubric along with the tested devices are implemented so that the results can be known. If the results still have shortcomings, improvements are made. The product of all this is a performance assessment rubric for managing receivable cards.

2.2 Product Trial Design

2.2.1 Limited Trial

In this trial stage, the researcher together with the practical instructors carried out the assessment using a rubric. In addition to conducting the assessment practice, this rubric was also implemented and revised according to the conditions in the field. Thus, after the limited trial process and analyzed and consulted with experts, a design of the performance assessment rubric for managing receivable cards can be obtained which is ready to be used for an expanded trial. In this limited trial, the rubric and its tools were applied to 6 (six) instructors (raters) and each rater assessed 5 accounting competency performance results.

2.2.2 Expanded Trial

The next stage of this research is an expanded trial. This is to determine whether the product to be produced meets its performance objectives, the design is in the form of an experiment. In this expanded trial, the rubric and its tools were applied to 12 (twelve) instructors (raters) who teach accounting and each assessed 5 student assignments.

2.3 Test Subjects

The subjects of the study were teachers in the accounting expertise program of Vocational High Schools in the business and management group. For limited trial subjects, 6 teachers (raters) were selected. While for the expanded trial subjects, 12 (twelve) teachers (raters) were selected. All teachers (raters) who were used as research subjects were given an explanation by the researcher about the use of the assessment rubric.

2.4 Data Collection Techniques

The techniques used are observation, validation sheets, and assessment sheets. Observations were conducted at the beginning to obtain physical evidence of the assessment of the performance of competency in managing receivable cards made and used by teachers, a review of the assessment of the performance of competency in managing receivable cards owned by teachers, and ongoing performance activities of competency in managing receivable cards. Observations were conducted on two teachers who carried out learning on inventory card management.

The validation sheet consists of validation of the accounting competency performance assessment rubric. Validation of the rubric was submitted to a lecturer, two competency test practitioners from the competency certification institution and seven accounting subject teachers, so that the total number was ten people. The validation sheet contains the feasibility of the competency performance rubric in managing receivable cards developed by the researcher.

The data collection instrument used to obtain quantitative data through a validation sheet containing 5 aspects (a) content feasibility, (b) language, (c) writing, (d) appearance, and (e) benefits. This is to determine the validity of the product made.

2.5 Data Analysis Technique

The validity of the rubric for assessing the performance of managing receivable cards used is carried out through validation actions by experts (expert judgment), through discussion forums between experts (focus group discussion), to obtain input. To find out this agreement, the Aiken validity index is used with the formula:

$$V = \frac{\sum s}{n(C - 1)} \tag{1}$$

With:

V = item validity index

s = score determined by the assessor (r) minus the lowest score for the category used (lo).

n = number of assessors, c = number of categories that can be selected by the assessor.

To determine the level of reliability, the Intraclass Correlation Coefficient (ICC) is used. Assuming that observer variation is random, the formula for the interclass correlation coefficient is

$$ICC = \frac{O_S^2}{O_S^2 + O_O^2 + O_e^2} \quad (2)$$

where:

variance (σ2) is a measure of variation,

subscript s = subject;

o = observer;

e = random error.

If observer variation is assumed to be constant, then the variation is not taken into account when calculating total variation.

3 Results and Discussion

The development of a performance assessment rubric for receivable card competency was developed with the aim of providing teachers and students with more objective, reliable, and practical assessments. The assessment process is carried out to collect, analyze, and interpret data on student learning processes and outcomes, systematically and continuously to provide meaningful information as a basis for decision making.

The initial product creation was carried out using several steps and adjusted to the situation and conditions in the study. These steps include: (1) Determining the competency standards to be assessed. (2) Making definitions and provisions for concepts, skills that indicate cognitive domains and performance domains. (3) Making provisions for vital skills that must be assessed. 4) Making provisions for the scale to be used. (5) Providing an overview of quality levels, (6) Making templates (7) Describing student ability achievements based on value intervals; (8) Assessment rubric.

The abilities assessed are carrying out receivable card work, including: Recording receivable mutations; Making confirmation letters and recapitulations of receivables, Recording receivables that cannot be collected; Recording receivables received back; Calculating estimated bad debts based on the percentage of receivables, and analyzing the aging of receivables.

Important competencies to be assessed include: Knowledge related to managing journals and subsidiary ledgers; Identified receivables mutation data; Receivables data posted to each receivables card; Confirmed receivables balance; Receivables recapitulation report presented; Accuracy of data entry into journals and subsidiary ledgers; Reconciliation of subsidiary ledger systems with journal and ledger data, and Systematic tracking of errors.

The range of competency achievement assessment is described in quantitative and qualitative forms. The quantitative form is 1 to 4, the qualitative form is 1 means being able to provide proof of transactions, 2 means being able to identify, 3 means being able to do bookkeeping and 4 means being able to make reports based on standard operating procedures (SOP).

Table 1. Gradation of Competency Quality in Managing Receivables Cards.

Gradation	Quality Description
4. Presenting reports	Students are able to make reports based on standard operating procedures
3. Recording / recording	Students are able to verify and record receivables mutations
2. Identifying	Students are able to identify the initial balance of receivables; Customer receivables mutation data; Other receivables mutation data
1. Providing documents	Students are able to provide the equipment needed for Receivables Management, and Proof of receivables transactions

Table 2. Student Achievement Value Interval

No	Grade Value	Description
5	81 - 100	Demonstrates mastery of managing receivable cards at a very satisfactory stage, understanding at this level is categorized as very capable; Understanding at this level is categorized as capable of making reports in accordance with SOP;
4	61 - 80	Demonstrates mastery of managing receivable cards that is quite good; Understanding at this level is categorized as capable of recording/booking transaction documents;
3	41 - 60	Demonstrates partial mastery of receivable cards; Understanding at this level is categorized as capable of identifying transaction documents;
2	21 - 40	Demonstrates poor understanding of managing receivable cards; Understanding at this level is categorized as capable of providing documents;
1	11 - 20	Demonstrates very poor understanding of managing receivable cards; Understanding at this level is categorized as incapable in financial accounting

The assessment scale is in accordance with the scoring commonly done in Indonesia, namely with a cumulative value of 0 to 100.

The value interval for each quality level in the competency assessment rubric for managing receivable cards is 20, with details: unable 0 to 20, able to provide documents 21 to 40, able to identify 41 to 60, able to record/book 61 to 80 and able to report in accordance with standard operating procedures 81 to 100. Briefly, it can be seen in Table 1.

In accordance with the planned value scale, a value interval and explanation of student competency will be made for the value they obtain; Furthermore, based on this value interval, students will be grouped into the final value acquisition (Table 2).

The results of the V-Aiken calculation for each aspect (a) suitability of content, (b) language, (c) writing, (d) appearance, and (e) practicality are (a) 80.25 (b) 78 (c) 80.00 (d) 81.00 and (e) 80.33, and a mean of 80.00, so that according to the opinions of experts and practitioners, the assessment rubric developed meets the requirements for content validity.

The test was conducted by giving 5 results of students' accounting competency performance to 6 raters. Furthermore, reliability was analyzed using the intraclass correlation coefficient with a "two-way mixed, consistent type, 95% confidence interval" model. ICC analysis is used to determine the instrument's stability or to compare the variations caused by the attributes measured with the overall measurement variations.

The reliability value of the competency standard for managing receivables cards, estimated using an alpha coefficient of 0.993, indicates an excellent alpha coefficient value. Meanwhile, the intraclass correlation coefficients were 0.962, exceeding the minimum ICC coefficient criteria of 0.70 required, meaning that the rubric for assessing the performance of competency in managing receivable cards was reliable and stable. The following research results refer to the research on the development of the rubric for assessing competency in managing receivable cards.

First, the rubric for assessing the performance of competency in managing receivable cards helps teachers provide objective assessments. Teachers realize that there are several benefits to using rubrics for student performance. One of them is that it can be used as an evaluation guide. The assessment rubric can also be used as a tool for teachers to assess student abilities, provide helpful information, reflect on teaching and learning activities that have taken place, and help teachers determine remedial activities.

Second, a rubric for assessing competency performance in managing receivable cards has been produced. The performance assessment rubric for managing receivables cards is in the form of a performance assessment page in the form of an assessment format containing instructions for filling out the rubric, scoring instructions, notes on how to assess, a table for filling out the rubric containing assessment aspects, and guidelines for evaluating performance in managing receivables cards containing assessment criteria and indicators. The assessment rubric is presented in the form of an assessment scale. The assessment range used in the Performance Evaluation Tool for Managing Receivables Cards is 1 to 4. Third, the rubric for assessing competency in managing receivable cards that was developed was declared suitable for use based on the validation results provided by a material expert as a theory verifier, three practitioners as accounting training organizers, and six accounting teachers as education verifiers.

This study has limitations in terms of standards, methodology, and implementation. Weaknesses in data collection with questionnaires resulted in respondents giving irresponsible responses. Despite these limitations, the study's results on developing a rubric for assessing the performance of competency in managing receivable cards for Vocational High School students can be used as a source of information to support students' academic activities.

4 Conclusion

Based on the results of the analysis and discussion, the following conclusions can be put forward. (1). Competencies that are important to assess in managing receivable cards include: Knowledge of the accounting recording system regarding the management of journal books, subsidiary ledgers; Documents of changes in receivables are identified;

Receivables data is posted to each receivable card; Receivables balance is confirmed; Receivables recapitulation report is presented; Accuracy of data entry into journals and subsidiary ledgers; Reconciliation of subsidiary ledger systems with journal and general ledger data, and Systematic tracing of errors.

The basic skills that students must master in the competency standards of managing receivable cards are determining the initial balance, changes, and adjustments. The validity of the rubric for assessing the competency of managing receivable cards developed was analyzed using V-Aiken 0.80.

The results in the table are greater than ($c = 5$, $n = 10$, level 5% = 0.70) and ($c = 5$, $n = 6$, level 5% = 0.79) so it is concluded that the rubric developed meets content validity. The reliability of the rubric for assessing the performance of managing receivable cards is estimated using the alpha coefficient of 0.993 categorized as excellent, and the intraclass correlation coefficient (ICC) of 0.962 is concluded to be reliable and has high stability.

References

1. Bernardin, H.J., Russel, J.E.A.: Human Resource Management, New Jersey, International Prentice Hall, Upper Saddle River (1993)
2. Borg, W.R., Gall, M.D.: Educational Research an Introduction, 4th edn. Longman Inc., White Plains, NY (1983)
3. BSNP: Guidelines for compiling elementary and secondary education unit level curriculum. Jakarta: BSNP (2006)
4. Cascio, W.F.: Managing Human Resources, Productivity, Quality of Work Life, Profits. McGraw-Hill Book Company, New York (1992)
5. Regulation of the Minister of National Education of the Republic of Indonesia No. 26 of 2007 concerning Academic Qualification Standards and Teacher Competence, Jakarta
6. Popham, W.J.: Assessment literacy for teachers: Faddish or fundamental? Theory Pract. **48**(1), 4–11 (2009)
7. Callison, D.: Rubrics. School Libr. Media Act. Mon. **17**(2), 34–36, 42 (2000). https://eric.ed.gov/?id=EJ620187. Accessed 12 May 2019
8. Campbell, C., Collins, V.L.: Identifying essential topics in general and special education introductory assessment textbooks. Educ. Measur. Issues Pract. **26**(1), 9–18 (2007). https://doi.org/10.1111/j.1745-3992.2007.00084.x
9. Luft, J.A.: Rubrics: design and use in science teacher education. J. Sci. Teach. Educ. **10**(2), 107–121 (2017). https://doi.org/10.1023/A:1009471931127
10. Marriott, P.: Students' evaluation of the use of online summative assessment on an undergraduate financial accounting module. Br. J. Educ. Technol. **40**, 237–254 (2019)
11. Stiggins, R.J.: Assessment crisis: the absence of assessment for learning. Phi Delta Kappan **83**(10), 758–765 (2002). https://doi.org/10.1177/003172170208301010

New Hybrid Approach to Credit Card Fraud Detection Using CNN and Machine Learning Techniques

Ihab Yasien Mahmood(✉) and Wisam Dawood Abdullah

College of Computer Science and Mathematics, Tikrit University, Tikrit, Iraq
ihabyma2024@gmail.com

Abstract. Credit card fraud is an eminent problem that has a direct impact on financial organizations; it results in massive losses and loss of customers' trust. This is because; traditional methods of fraud detection have proven to be ineffective because of the advanced fraud techniques in the market. This paper proposes a new approach to early fraud detection using deep learning and machine learning algorithms with low false alarm rates. In particular, a Convolutional Neural Network (CNN) is combined with machine learning classifiers to enhance the detection performance. To overcome the problem of data irregularities, random sampling is used, while feature engineering and ensemble decision-making optimize the model. The methodology is executed in two stages: first, the dataset is divided into training and testing datasets and then for further refinement the cross-validation is performed. CNN is used for feature extraction to capture the features of a transaction and the algorithms include Extreme Gradient Boosting, Random Forest, K-Nearest Neighbors (K-NN). The model's accuracy is measured using several statistical measures, which all turn out to be very good. When using cross-validation, the model had an accuracy of 0.9986, precision of 0.9971, the recall score was 1, the F1 score of 0.9986, and the ROC AUC was 0.9986. Thus, the present work reveals the possibilities of improving credit card fraud detection and increasing the general financial security by integrating deep learning and machine learning techniques.

Keywords: Credit Card Fraud · Cross Validation · Feature Engineering · CNN · Decision Making

1 Introduction

Financial fraud is a significant issue affecting the economy, daily life, and people's quality of life. It also affects the integrity and confidence of the financial sector. Financial abuse, also known as financial fraud, results in significant losses for governments, organizations, businesses, and private citizens. It involves unlawful activities that profit a person or organization using unethical and illegal means [1].

In the 21st century, financial institutions have increasingly made business facilities available through Internet banking, making purchasing goods and services more convenient. Credit cards are often provided by financial institutions to customers, making them

A. J. Obaid et al. (Eds.): ICONEST 2024, ISEM 40, pp. 138–152, 2025.
https://doi.org/10.1007/978-3-031-89960-7_10

more convenient and protecting them against damage, loss, or theft. However, fraudsters pose a significant threat to this group due to the widespread use of card-based transactions. In 2020, a total of 2287 million credit cards were issued worldwide by Visa and Master card, with 1131 million cards for Visa and 1156 million cards for Master card. These statistics highlight the ease and popularity of transactions based on cards among end-users. However, credit cards also pose security and fraud challenges. Fraudulent activities associated with credit and debit cards have become a major cause of financial losses in the finance industry. In this case, fraudulent transactions are effectively determined by machine learning; however, the obstacle to exchanging ideas about fraud detection is a major challenge. In the UK, there were £574.2 million credit and debit cards fraud cases recorded in 2020. Recently, the detection of fraud in credit cards has increased significantly, attracting the researchers' and scholars' attention [2].

The issue of fraud in credit card faced by the banks and financial organizations is significant, where credit cards are used by unauthorized users to steal money or property. This can be done through deception, theft, or identity theft techniques. Fraudsters can illegally obtain consumers' credit and debit card details without their consent or knowledge. According to "U.K. Finance", this is among the leading sources of financial losses in the finance industry, largely because of technological growth [2].

Fraud detection systems (FDS) are machine learning based automated systems used by credit card companies to detect fraud before it happens and therefore prevent it from happening. One of the goals of these systems is to identify fraudulent transactions before they are entered into the database, and at the same time, reduce false positives that can interfere with genuine transactions and annoy customers [3].

In the realm of artificial intelligence (AI), machine learning (ML)is the branch that is intended to build systems that are capable of accomplishing new tasks better than previously, and all these without being specifically coded for a new task. ML is credited to spring from the AI movement that originated in the 1950s and aims at solving specific objectives mostly deals with prediction and optimization. AI techniques like ML can start with a vast flow of data and rationally utilize the data adding to deciphering it by formulating a set of mathematical models. The technology of using a myriad of features has been commonly used in speech recognition, image classification, prediction of prices, and recommendation systems for products. Machine learning could take various forms including reinforcement learning whereby we reward or punish our models, supervised learning which uses labeled data, semi-supervised learning which consolidates both labeled and unlabeled data and finally unsupervised learning which tries to find patterns in data [4]. The aspect of managing and analyzing big data through a system has made ML an important aspect in most scientific, technical, and industrial sectors [4].

Deep Learning (DL) has become one of the fastest-moving topics in ML and AI, given its power to handle huge amounts of data to carry out complicated tasks with great accuracy, such as speech recognition, image identification, and data prediction. In DL technology, multi-layer neural networks are utilized to work like the human brain for information processing. DL algorithms have proven to be superior to those of standard ML due to their success in several areas.

It is claimed that under the enhancement in ML techniques, a reduction in hardware costs, and an increase in processing power, DL has been born. Deep learning is extremely effective as it automatically extracts the features from raw data, which eliminates the requirement of feature extraction through human efforts. Consequently, this has led to its applications in bioinformatics, finance, video games, fraud detection, medicine-related tasks regarding diagnostics and identification of diseases autonomously [5].

One such DL model is the CNN, which make predictions and classifications based on the input through several layers of filtering. For this reason, the models are trained on large datasets and high-powered processing units, such as GPUs. The hierarchical learning methodology in DL extracts high-level data abstractions through a series of nonlinear transformations, thus enabling the system to learn complex and nonlinear relationships present in the data.

DL completely altered the dimensionality in which jobs were executed and decisions were made since it could learn from raw data over time. Therefore, it is an important tool in many scientific and technological disciplines. The applications related to deep learning are envisioned to increase due to more complex datasets over time, driving efficiency and innovation in many industries [6].

Credit card fraud is a significant issue for financial institutions, causing significant financial losses and reducing customer trust. Despite the use of ML techniques for fraud detection, they have not effectively prevented these acts. To address this, new techniques are needed to process large amounts of data and analyze user behavioral patterns. This could involve combining DL and ML algorithms to create a hybrid model that is more effective in fraud detection and reduces fake alarms, thereby protecting customers from potential financial losses. This study aims to develop and evaluate models using DL and ML, enhance datasets using feature engineering, use ensemble decision making algorithm work for model enhancement, and address bias using cross validation.

In brief, this study proposes a system on the effectiveness of collective decision making enhanced by DL techniques to improve the detection of fraud in credit card transactions because improving fraud detection is a vital component of maintaining security and stability in many aspects. By enhancing the ability of organizations to detect fraudulent activities more accurately, the loss of money and property can be avoided, thus contributing to maintaining the reputation of these organizations and the trust of customers. In addition, improving fraud detection reduces operational costs and legal issues, enhances customer experience, and encourages cooperation between organizations. Ultimately, it reduces the economic damage caused by fraud and contributes to building a safer and more trustworthy environment for individuals, businesses, and society in general.

2 Related Works

Many works with different techniques have been accomplished on the detection of credit card fraud by utilizing ML techniques. Therefore, the current study proposes a new system as an innovative method that can help in detecting fraud in credit cards. The studies quoted are obtained from accredited institutes like IEEE, ScienceDirect, International Scientific Indexing (ISI), Web of Science, and Scopus, collected during the period from 2020 to 2024. These studies are briefly summarized in Table 1 as follows:

Table 1. Shows a Summary of Previous Studies

Author(s)	Year	Model/Method	Dataset	Accuracy/Results
Alghofaili et al. [7]	2020	LSTM (Long-and Short-Term Memory)	Credit card fraud dataset	99.95%
Najadat et al. [8]	2020	BiLSTM-MaxPooling-BiGRU-MaxPooling	IEEE-CIS Fraud Detection (Kaggle)	91.37%
Abd El Naby et al. [9]	2021	Over Sampling with Convolution Neural Network (OSCNN)	Kaggle's credit card dataset	98%
Kewei et al. [10]	2021	Deep learning-based neural network with feature engineering, memory compression, mixed precision, and ensemble loss	IEEE-CIS Fraud Dataset (Vesta)	95.8%
Shen [11]	2021	Hybrid model (Autoencoder-based deep neural network)	Regular and fraudulent transactions	95.77%
Chen et al. [12]	2021	Deep Convolutional Neural Network (DCNN)	Real-time credit card fraud dataset	99%
Alkhatib et al. [13]	2021	Deep learning model	Customer transaction data (Vesta)	99.1%
Benchaji et al. [14]	2021	LSTM-based model	Dataset divided into training, validation, and generalization	Achieved MAE of 0.0065, MSE of 0.0035, and AUC of 0.9955
Benchaji et al. [15]	2021	Sequential data modeling with UMAP, LSTM networks, and attention mechanism	Not specified	97.48%
Asha et al. [16]	2021	Artificial Neural Network (ANN)	Credit card fraud dataset	99.92%
Alarfaj et al. [17]	2022	CNN-based models (compared with machine learning techniques)	European Card Fraud Detection Dataset	99.9%

3 Methodology

The research methodology helps scholars to explain how they intend to approach their research in a sequential and coherent approach. It refers to the identification of the methods used in the development of the study to provide accurate and credible results that meet the objectives of the study. This entails identifying the data sources, data collection techniques and the analysis techniques to ensure that the process of responding to the research question is well coordinated. Therefore, this section describes the plan for the proposed solution and gives the implementation outline. It first describes the machine learning methods used in building the models, with emphasis on performance evaluation using KNN, RF, and XGBoost algorithms. After that, the research methodology employed in this work is described, including the datasets employed for training and testing the models, as well as the performance metrics used for evaluating the classification performance [18].

3.1 Dataset Collection and Description

This study aims at providing the readers with a clear understanding of the dataset used especially in the part that deals with the acquisition and description of the dataset. Since the efficiency of the ML algorithms depends on the data collected, this aspect is important. The study also emphasizes the need to make sure that the data is relevant to the associated labels so that the models are accurate and efficient. This section provides a detailed description of the process of collecting the dataset, as well as the description of the dataset's main features and relevance to the objectives of the study. Therefore, the study seeks to clearly indicate the source of the data set and details of the same which are explained in the subsequent sub-sections.

Data Source. The dataset used for credit card fraud detection in this study was sourced from Kaggle, available at https://www.kaggle.com/datasets/kelvinkelue/credit-card-fraud-prediction. It contains 555719 records and 22 features, which provides both nominal and numerical variables for further examination. Remarkably, there are no null values in the dataset and the data it contains may encompass account information, customer information, and credit card transaction records. To overcome the problem of class imbalance, the Random OverSampler method was used to balance the data set. The data was stored in a new file for later use and analysis. After oversampling, the total number of transactions became 1107148, with 553574 fraudulent and 553574 non-fraudulent cases, with 22 attributes in total.

3.2 Data Pre-processing

This is a crucial phase in the process that acts as the first procedure. Its goal is to convert unprocessed data into a format that will be more accessible and effective for further processing. Data that is redundant, irrelevant, noisy, or unreliable must be handled carefully during data preparation since these elements might impede the training phase's ability to uncover significant information. Data preparation and filtering are time-consuming but necessary steps in ML challenges. They include a wide range of activities, including feature extraction, transformation, normalization, cleaning, and selection of data. By enhancing the data's quality, these procedures hope to make it more appropriate for ML model training. Data preparation makes the process more transparent overall and makes it possible to see more clearly how the pretreatment stages affect the final forecasts.

Random Oversampling. Many datasets in real-world applications suffer from class imbalance, meaning that some classes are overrepresented. This uneven distribution can lead to poor predictive model performance, especially for underrepresented populations. As a result, preprocessing imbalanced data is a crucial step in building a model capable of correctly predicting all classes. In this study, the total number of transactions was 555,719, of which 553,574 were non-fraudulent transactions and 2,145 were fraudulent transactions. This means that there is a big difference between the number of fraudulent transactions and non-fraudulent transactions, which requires the use of a technique to balance the data.

Random OverSampler was used, which is a data balancing technique that increases the data for the small category to become equal to the amount of data for the large category, where the number of fraudulent and non-fraudulent transactions became 553,574 and 553,574, respectively, and the total number of transactions 1,107,184.

Random OverSampler was chosen over other techniques such as SMOTE and ADASYN, because it suits the nature of the data at hand, is easy to use and implement without involving complex computations, and helps in reducing the bias of the model towards the negative class; hence, it is a very effective option to get a better balance in the data distribution in a simpler and more efficient way.

Label Encoder. The process of converting categorical information into numerical data is known as label encoding [19]. This code uses the sklearn library's LabelEncoder to transform class variables, or the class values in the data, to numerical values [20]. The majority of ML algorithms perform better on numerical data than on linguistic input from classes, which is why this translation is required [21].

Normalization. This part of the code employs the MinMaxScaler from the sklearn library to scale the dataset. Normalization is important in enhancing the performance of a model and also enhances the training process making it more efficient, faster and more stable in both graph analysis and ML. Normalization is the process of scaling the values in order to bring them into a particular range usually between 0 and 1. This is because, by standardizing the feature values, the learning efficiency of the model is improved. The normalization process is done using the following Eq. (1) [22].

$$\mathbf{x_{normalize}} = x - x_{min}/x_{max} - x_{min} \tag{1}$$

3.3 Dataset Distribution

In this study, two techniques for dataset splitting were used as follows:

Splitting Dataset Without Cross Validation. In this approach, the dataset is typically split into two or three subsets: learning; development and confirmation; and evaluation respectively. The training set is used to develop the model, the validation set is used to tune the parameters of the now developed model, and the test set is used to test the model on data that it has never seen before. This method is swift and easy but are likely to produce confounded results if the sample is not representative. Indeed, while using K-fold cross validation, the model's performance is tested based on single train-test split datasets which might over fit or under fit (Fig. 1).

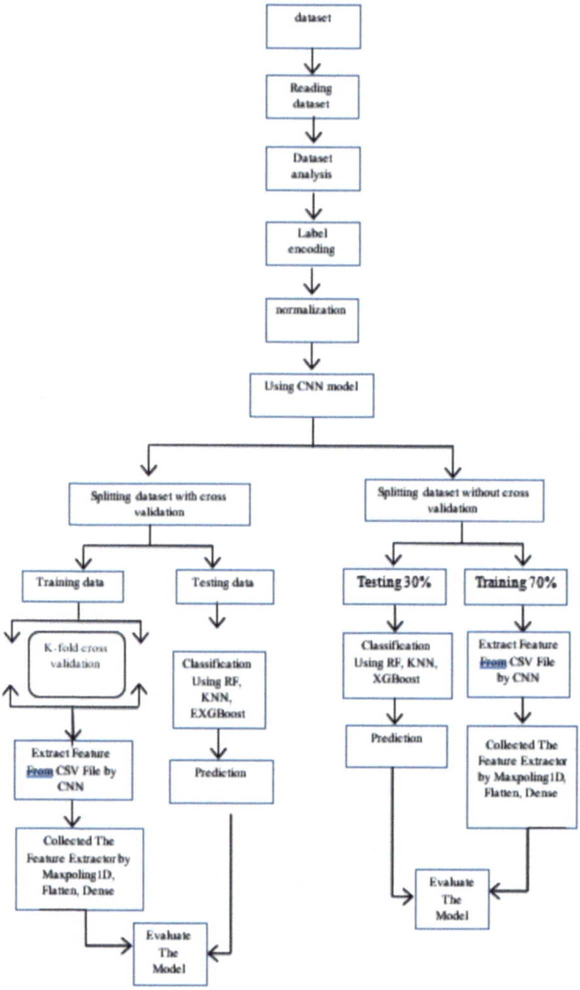

Fig. 1. Flow Chart of the Proposed Methodology

Splitting Dataset With Cross Validation. Applying cross-validation is a step better than train-test splits in the sense that, it divides the data set into folds. Then for each of the fold, the model is trained on the data and tested on the remaining data. The method used most frequently is the k-fold cross validation, in which the data set is divided into k subsets and the model is trained k times and each time a different set is used for validation. This technique yields a more accurate measurement of model accuracy since it diminishes, variability and utilize all the data instead of a train-test split.

3.4 Extracting Feature from CSV File Using CNN

CNN are the powerful analytical tools for processing multi-dimensional data. CNN consists of a number of layers, including convolutional layers, which apply a convolution operation to extract patterns and spatial relationships from features over a data element to provide the model with the ability to recognize the main features such as edges and corners. Along with these layers, filtering layers work in tandem by emphasizing important features and lessening the effect of noise through the use of nonlinear activation functions. Also, the pooling layers contribute by reducing the dimensionality and retaining the important information, hence making the training and prediction process easier. In effect, CNN transformation of original features into compact representations allows efficient learning of patterns and complexities in the data and reduces manual feature engineering.

This will have empowered the models of fraud detection by merging CNN with other machine learning algorithms like Random Forest, KNN, and XGBoost. Complex data features are first extracted by CNNs and then sent into a random forest that improves accuracy of the classification with reduced complexity. In case of KNN, the extracted features identify most similar neighbors to reduce false positives. XGBoost models make use of deep features as input that improve the accuracy of predictions by learning complex relationships. Their collaboration has shown how models can dramatically improve their performance in analyzing complex data and fraud detection.

3.5 Collect the Extracted Feature from CSV File Using Dense, Flatten, MaxPooling1D

CNN, a DL model, can be expanded with the help of features such as Flatten, Dense and MaxPooling1D layers. The flatten layer computes the formation of high dimensional data into a two dimensional matrix and the dense layer computes the formation of extracted features into complex features. The input is feed down to the MaxPooling1D layer which serves to minimize on data dimensionality while at the same time extract pertinent features thus easing computational load. These are the algorithms of ML that can use these retrieved features as input to improve the efficiency of the ML model that can be of the following types: XGBoost, RF, and KNN.

3.6 Machine Learning Algorithms

In this research, several techniques in ML are employed to solve these challenges in this area. All of these models have different capabilities and qualities and the amalgamation of them contributes to the rigidity and accuracy of the empirical research. The subsequent ML algorithms perform important roles within this framework of analysis.

Extreme Gradient Boosting (XGBoost). XGBoost is an ML model which was developed by Tianqi Chen and it is a ML algorithm that has been categorized under the gradient boosting model [23]. It gives final predictions by using probabilities from several models and then averaging them, and therefore, it is suitable for ranking, regression, and classification. XGBoost has advantage of improving the model, cross validation and features importance and high performance in large data. It is also used to support popular programming languages and is easy to use [24].

Random Forest (RF). RF is also another kind of ML technique and this technique has been used in the field of prediction and classification [25]. It creates many decision trees during the training process and every tree is built up using randomly chosen samples [26]. For predictions it uses mean of the outputs and in the case of classification problems it uses the majority algorithm. RF works to decrease the level of overfitting and it is good to handle large and complex data set, for high level of accuracy and good performance in most applications. It has thus become popular in many branches of industry [27].

K- Nearest Neighbour (KNN). KNN algorithm is one of the instance-based learning algorithms which is commonly employed for pattern recognition. It categorizes objects by assigning them to clusters formed based on the distance from the nearest training exemplars in the feature space. An item is classified into the class frequently occurred in its k-closest neighbors where k is a positive integer [28]. KNN categorizes a new test feature vector based on the classes of its K nearest neighbors and the use of Euclidean distance to find the nearest neighbor [29].

3.7 Model Evaluation

The performance of the machine learning model will be measured with some key metrics that include a confusion matrix, ROC AUC, accuracy, precision, recall, and the F1 score. A confusion matrix summarizes the predictions against the actual outcomes generated by a model, divided into four classes of predictions: true positives, true negatives, false positives, and false negatives. This is especially helpful in domains like fraud detection, whereby a decrease in false positives enhances the user experience, while at the same time it ensures that true positives are accurately identified. Accuracy describes the overall performance of the model. It can, however, be misleading for imbalanced datasets, where the cases are mostly negative.

Precision tells us the quality of the positives predicted, helping to reduce false positives, and therefore avoid making wrong accusations of fraud. Recall in turn describes how well actual positive cases can be captured by the model; a higher recall reduces the number of missed real fraud cases. The F1 score is the combination of precision and recall into one value and serves as a well-balanced measure when both are very important in the application. Finally, the AUC ROC represents a model's performance in terms of separating the positive and negative cases at different threshold levels. Thus, this is a more global measure of the performance of the discriminative model. These metrics are necessary in achieving fraud detection with equilibrium, where what is looked for is the indication of true fraud by minimizing the false positives as much as possible [30].

4 Results and Discussion

This section reports the research findings and provides a thorough analysis and interpretation of them in the context of the study objectives. It is divided into several subsections to make sure the presentation is well-organized.

The First Scenario CNN With RF, XGBoost,KNN Algorithms Using Training and Testing Technique. In this case the ratio of 70:30 is used for splitting the dataset into the training and testing datasets. An initial fraud detection model is derived by implementing a CNN that utilizes KNN, RF, and XGBoost for the ensemble model. Features are extracted from both the training and the test sets and a set of individual models is developed. The model is assessed by generating a confusion matrix as shown in Fig. 2 and the following statistical measures accuracy, precision, recall, F1 score, ROC AUC as depicted in Table 2.

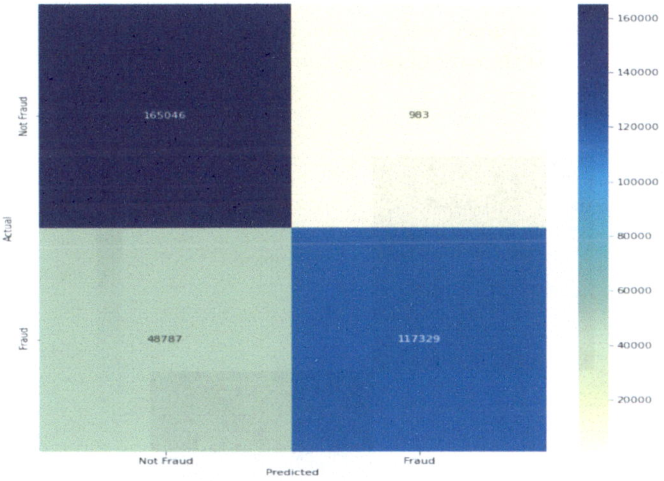

Fig. 2. Confusion Matrix of the CNN Model using Training and Testing Technique

Table 2. Performance of the CNN Model Using Training and Testing Technique

Sequence	Metrics	Performance
1	Accuracy	85%
2	Precision	99.1%
3	Recall	70.6%
4	F1 Score	82%
5	ROC AUC	85%

Discussion of the First Scenario. The first model performed well in categorizing positive and negative situations. It was a CNN coupled with KNN, RF, and XGBoost algorithms. On the one hand, a high percentage of false negatives meant it was unable to identify a significant proportion of positive instances. The accuracy of the model was 85.02%, meaning that the data was accurately categorized. On the other hand, it only identified around 70% of positive instances, suggesting that a significant number of fake cases were missed. An equilibrium between recall and accuracy was indicated by the F1 score of 82.50%. The model's high ability to differentiate across classes was indicated by its 85.02% area under the ROC curve.

The Second Scenario CNN with RF, XGBoost,KNN Algorithms Using Cross Validation Technique. The dataset is divided into five sets using cross-validation, one for testing and four for training. This process is repeated five times, with each set once for testing and four times for training. A fraud detection model is built using CNN and combined with KNN, RF, and XGBoost models. Features are extracted from the training and testing data, and individual models are trained. The final results are predicted using the ensemble decision. A confusion matrix is used to evaluate the performance of model as shown in Fig. 3 and accuracy metrics, with f1_score and ROC area under the curve (AUC) reported to assess the overall performance as shown in Table 3.

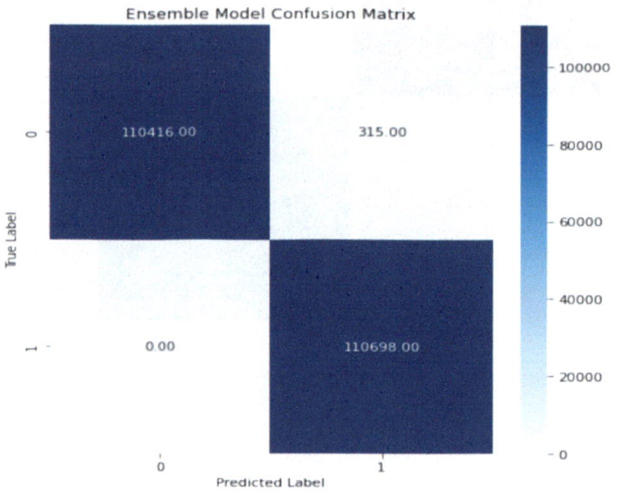

Fig. 3. Confusion Matrix of The CNN Model with Cross Validation Technique

Table 3. Performance of The CNN Model with Cross Validation Technique

Sequence	Metrics	Performance
1	Accuracy	99.8%
2	Precision	99.7%
3	Recall	100%
4	F1 Score	99.8%
5	ROC AUC	99.8%

Discussion of the Second Scenario. The model effectively detected credit card fraud, correctly classifying 110,416 non-fraudulent cases and 110,698 fraudulent ones. It achieved a high accuracy of 99.86%, indicating its ability to accurately predict between positive and negative cases. The recall value was 100%, indicating the model's ability to detect all fraud cases without missing any. The F1 score was 99.86%, indicating a perfect balance between precision and recall. The ROC AUC curve area under the model was 99.86%, indicating its ability to differentiate between categories and provide accurate classifications.

Comparing the First and Second Scenarios. From the results shown in Table 4, it is noticed that the performance of the second model was significantly better than the performance of the first model through the improvement in all metrics used in evaluating the models, reflecting the strong and more effective performance in classifying data.

Table 4. Comparison between the first scenario and the second scenario

Proposed models	Distribution Dataset Technique	Accuracy	Precision	Recall	F1 Score	ROC AUC
CNN model with XGBoost,RF,KNN	Training and Testing Technique	85%	99.1%	70.6%	82.5%	85%
CNN model with XGBoost,RF,KNN	Cross validation Technique	99.8%	99.7%	100%	99.8%	99.8%

Comparing the Proposed Model with the Related Works. Table 5 show the comparison between the proposed model and the related work.

Table 5. Comparing the proposed model with the related work

Author(s)	Year	Model/Method	Accuracy/Results
Alghofaili et al. [7]	2020	LSTM (Long- and Short-Term Memory)	99.95%
Najadat et al. [8]	2020	BiLSTM-MaxPooling-BiGRU-MaxPooling	91.37%
Abd El Naby et al. [9]	2021	Over Sampling with Convolution Neural Network (OSCNN)	98%
Kewei et al. [10]	2021	Deep learning-based neural network with feature engineering, memory compression, mixed precision, and ensemble loss	95.8%
Shen [11]	2021	Hybrid model (Autoencoder-based deep neural network)	95.77%
Chen et al. [12]	2021	Deep Convolutional Neural Network (DCNN)	99%

(continued)

Table 5. *(continued)*

Author(s)	Year	Model/Method	Accuracy/Results
Alkhatib et al. [13]	2021	Deep learning model	99.1%
Benchaji et al. [14]	2021	LSTM-based model	Achieved MAE of 0.0065, MSE of 0.0035, and AUC of 0.9955
Benchaji et al. [15]	2021	Sequential data modeling with UMAP, LSTM networks, and attention mechanism	97.48%
Asha et al. [16]	2021	Artificial Neural Network (ANN)	99.92%
Alarfaj et al. [17]	2022	CNN-based models (compared with machine learning techniques)	99.9%
Proposed model with cross validation technique	2024	CNN, XGBoost, RF, KNN	99.8%
Proposed model without cross validation technique	2024	CNN, XGBoost, RF, KNN	85%

5 Conclusion

In this research, an ensemble learning model for credit card fraud detection using the RF, XGBoost, KNN, CNN algorithms was presented. This model was highly effective with accuracy of 99.8% of samples; positive prediction accuracy of 0.997. In the same vein, the model obtained an average F1 score of 0,998, which reflects a high accuracy in positive predictions, as the model was free from misclassification of fraud cases. A ROC AUC score of 0.998 was obtained by the model in the study showing the model's capacity to distinguish between fraudulent and non-fraudulent transactions. It realized that only 315 cases were wrongly classified as fraudulent while it correctly identified all the actual frauds. These results suggest the use of multiple algorithms in fraud detection systems as being very effective. Not only was the model significantly better than the baseline model in all KPIs for the selection of the most predictive data for forecasting, but it also did not have significant errors and achieved an optimal ratio of recall and accuracy. The results show that the integration of sophisticated methods into security systems and financial protection applications can greatly enhance the safety and dependability of financial transactions.

For future work, the study recommends expanding the analysis by employing other ML algorithms like LightGBM or CatBoost to improve results and decrease training time. It also suggests using new balancing techniques such as ADASYN, Borderline-SMOTE, or SMOTE combined with the random sampling method to compare them with the performance of the model. The model's ability to forecast the rare class is improved by the ADASYN and Borderline-SMOTE approaches, which also boost the representation of the rare class in the data. When dealing with big or categorical data, the LightGBM and CatBoost algorithms both optimize the model's performance in a different way. By lowering biases and enhancing generalization skills, it increases analytical accuracy and efficiency.

References

1. Al-Hashedi, K.G., Magalingam, P.: Financial fraud detection applying data mining techniques: A comprehensive review from 2009 to 2019. Computer Science Review **40** (2021). https://doi.org/10.1016/j.cosrev.2021.100402
2. Bin Sulaiman, R., Schetinin, V., Sant, P.: Review of Machine Learning Approach on Credit Card Fraud Detection. Human-Centric Intell. Syst. **2**(1–2), 55–68 (2022). https://doi.org/10.1007/s44230-022-00004-0
3. Mittal, S., Tyagi, S.: Performance evaluation of machine learning algorithms for credit card fraud detection. Proc. 9th Int. Conf. Cloud Comput. Data Sci. Eng. Conflu. 2019, 320–324 (2019). https://doi.org/10.1109/CONFLUENCE.2019.8776925
4. "What is Machine Learning? A Primer for the Epidemiologist | American Journal of Epidemiology | Oxford Academic." [Online]. Available: https://academic.oup.com/aje/article-abstract/188/12/2222/5567515?redirectedFrom=fulltext
5. Najm, M.H.: Tomato Leaves Disease Detection and Classification Using Deep Learning Techniques, (2023)
6. Thamer, S.A.: Title:Diagnosis of COVID-19 from Chest X-ray images Using Machine Learning and Deep Learning Approaches, (2022)
7. "A Financial Fraud Detection Model Based on LSTM Deep Learning Technique: Journal of Applied Security Research: **15**(4)." [Online]. Available: https://www.tandfonline.com/doi/full/https://doi.org/10.1080/19361610.2020.1815491
8. "Credit Card Fraud Detection Based on Machine and Deep Learning." [Online]. Available: https://ieeexplore.ieee.org/abstract/document/8887354
9. Abd El Naby, A., El-Din Hemdan, E., El-Sayed, A.: Deep Learning Approach for Credit Card Fraud Detection. [Online]. Available: https://ieeexplore.ieee.org/abstract/document/8887354
10. Kewei, X., Peng, B., Jiang, Y., Lu, T.: A Hybrid Deep Learning Model for Online Fraud Detection. IEEE Int. Conf. Consum. Electron. Comput. Eng. **2021**, 431–434 (2021). https://doi.org/10.1109/ICCECE51280.2021.9342110
11. Shen, J.: Credit Card Fraud Detection Using Autoencoder-Based Deep Neural Networks, In: 2021 IEEE 2nd Int. Conf. Big Data, Artif. Intell. Internet Things Eng. ICBAIE 2021, pp. 673–677 (2021). https://doi.org/10.1109/ICBAIE52039.2021.9389940
12. Chen, J.I.-Z., Lai, K.-L.: Deep Convolution Neural Network Model for Credit-Card Fraud Detection and Alert. J. Artif. Intell. Capsul. Networks **3**(2), 101–112 (2021). https://doi.org/10.36548/jaicn.2021.2.003
13. Alkhatib, K.I., Al-Aiad, A.I., Almahmoud, M.H., Elayan, O.N.: Credit Card Fraud Detection Based on Deep Neural Network Approach. Int Conf. Inf. Commun. Syst. ICICS **2021**(June), 153–156 (2021). https://doi.org/10.1109/ICICS52457.2021.9464555
14. Benchaji, I., Douzi, S., El Ouahidi, B.: Credit card fraud detection model based on LSTM recurrent neural networks. J. Adv. Inf. Technol. **12**(2), 113–118 (2021). https://doi.org/10.12720/jait.12.2.113-118
15. Benchaji, I., Douzi, S., El Ouahidi, B., Jaafari, J.: Enhanced credit card fraud detection based on attention mechanism and LSTM deep model. Journal of Big Data, **8**(1) (2021). https://doi.org/10.1186/s40537-021-00541-8

16. Asha, R.B., et al.: Credit card fraud detection using artificial neural network. Global Transitions Proceedings **2**(1), 35–41 (2021). https://doi.org/10.1016/j.gltp.2021.01.006
17. Alarfaj, F.K., Malik, I., Khan, H.U., Almusallam, N., Ramzan, M., Ahmed, M.: Credit Card Fraud Detection Using State-of-the-Art Machine Learning and Deep Learning Algorithms. IEEE Access **10**, 39700–39715 (2022). https://doi.org/10.1109/ACCESS.2022.3166891
18. Asselman, A., Khaldi, M., Aammou, S.: Enhancing the prediction of student performance based on the machine learning XGBoost algorithm. Interact. Learn. Environ. **31**(6), 3360–3379 (2021). https://doi.org/10.1080/10494820.2021.1928235
19. Abdullah, W.D., Ghandour, A.: Enhancing Breast Cancer Diagnosis with Data Mining techniques. pp. 1–10
20. Khalid, N.W., Abdullah, W.D.: Enhancing Road Safety with Smartphone- based Machine Learning Driver Behavior Classification and Aggressive Driving Detection using Feature Reduction Methods. (2023)
21. Rashid, S.H., Abdullah, W.D.: Cloud-Based Machine Learning Approach for Accurate Detection of Website Phishing. Int. J. Intell. Syst. Appl. Eng. **11**(6), 451–460 (2023)
22. Kovalenko, O.: Fraud Detection Using Machine Learning Models. (2021). https://Spd.Tech/Machine-Learning/Fraud-Detection-With-Machine-Learning/. [Online]. Available: https://spd.tech/machine-learning/fraud-detection-with-machine-learning/
23. Atiyah, O.S., Thalij, S.H.: A Comparison of COIVD-19 Cases Classification Based on Machine Learning Approaches. Iraqi J. Electr. Electron. Eng. **18**(1), 139–143 (2022). https://doi.org/10.37917/ijeee.18.1.15
24. Li, S., Zhang, X.: Research on orthopedic auxiliary classification and prediction model based on XGBoost algorithm. Neural Comput. Appl. **32**(7), 1971–1979 (2020). https://doi.org/10.1007/s00521-019-04378-4
25. Bin Sulaiman, R., Schetinin, V., Sant, P.: Review of Machine Learning Approach on Credit Card Fraud Detection. Human-Centric Intelligent Systems **2**(1–2), 55–68 (2022). https://doi.org/10.1007/s44230-022-00004-0
26. Atiyah, O.S., Thalij, S.H.: Evaluation of COVID-19 Cases based on Classification Algorithms in Machine Learning. Webology **19**(1), 4878–4887 (2022). https://doi.org/10.14704/web/v19i1/web19326
27. Tao, H., et al.: Development of new computational machine learning models for longitudinal dispersion coefficient determination: case study of natural streams, United States. Environ. Sci. Pollut. Res. **29**(24), 35841–35861 (2022). https://doi.org/10.1007/s11356-022-18554-y
28. Abd, N.S., Atiyah, O.S., Ahmed, M.T., Bakhit, A.: Digital Marketing Data Classification by Using Machine Learning Algorithms. Iraqi J. Electr. Electron. Eng. **20**(1), 245–256 (2024). https://doi.org/10.37917/ijeee.20.1.23
29. Boateng, E.Y., Otoo, J., Abaye, D.A.: Basic Tenets of Classification Algorithms K-Nearest-Neighbor, Support Vector Machine, Random Forest and Neural Network: A Review. Journal of Data Analysis and Information Processing **08**(04), 341–357 (2020). https://doi.org/10.4236/jdaip.2020.84020
30. Wisam, G.A.S., Wisam, D.: Deep Learning-Based Evaluation of ICU Requirements in COVID-19 Cases _ SpringerLink (2024)

IMSMDL: Design and Implement an Information Management System for Medical Diagnostic Laboratories

Basheer Al-Sadawi[1], Haider Alsharqi[1(✉)], and Muthmainnah Muthmainnah[2]

[1] Information Technology Research and Development Centre, University of Kufa, Kufa, Iraq
haider.alsharqi@uokufa.edu.iq

[2] Universitas Al Asyariah Mandar, Polewali Mandar, Indonesia

Abstract. Managing an information system plays a crucial role in advancements for different domains such as healthcare particularly in a clinical laboratory to increase the efficacy and performance of its medical procedures and optimize its productivity by ensuring the safety of workers, scholars, and researchers. Therefore, this article proposes a laboratory information management system (LIMS) that aims to effectively manage, organize, and secure data in medical laboratories to eliminate the challenges that demand faces in such a sector. The workflow of the presented LIM system is optimized by serving several tasks that include registration of patients, tracking of samples, printing of the results, management of the financial accounts and inventory, reports issuing, statistics, notifications system, communication system for patients, security and authority health system. In the other hand, the proposed LIMS system ensures the providing a quality control aspect by enforcing the standard operating procedures (SOPs) and enclose a compliance with stringent regulatory requirements. Thus, it confirms reliable and harmonic results in medical testing that expand the laboratory's capability. The experimental results demonstrate the accuracy and robustness of the results and eliminate errors in their operations owing to the advancement of the laboratory system i addition to receiving and exporting information with efficient and speedway. Finally, the system registered remarkable results in managing information, resources, and results with few errors compared with the state of the arts.

Keywords: IMSMDL · CDC · Healthcare System · Laboratory Information System (LIS)

1 Introduction

Efficient data organization and optimization processes in the laboratory are critical in the modern healthcare industry. Medical laboratories are key in disease diagnosis and treatment planning and providing better patient care by ensuring fast and accurate findings. However, laboratories cannot maintain high-quality services due to the growing volume of medical laboratory data and its complexity [1]. Additionally, medical laboratories are becoming increasingly concerned about the security and privacy of patient data, primarily

A. J. Obaid et al. (Eds.): ICONEST 2024, ISEM 40, pp. 153–164, 2025.
https://doi.org/10.1007/978-3-031-89960-7_11

because of the sensitive and confidential nature of the information [2]. These difficulties merit deploying a Medical Laboratory Information Management System (LIMS), which will mitigate these challenges and mark the dawn of a new age of efficiency and accuracy. The contemporary data retrieval methods have been quite challenging for many organizations in this context, which is why the Association for Clinical Biochemistry and Laboratory Medicine Pre-Analytical Specialist Interest Group (ACB-SIG-PA) recommends using the LIMS for this purpose. [1].

LIMS offers a systematic approach to monitoring and managing information resources in a laboratory setting. It supervises medical tests, results, and the inventory of consumable materials while also enabling the creation of reports and statistical data. The main goal of a LIMS is to improve efficiency and quality by shifting manual processes to electronic workflows. Choosing the right LIMS is challenging, as it requires understanding lab methodologies and software development. Lab testing generates diverse data like microscope images and detailed notes [4].

A Laboratory Information Management System (LIMS) is crucial for systematically collecting and tracking data from sampling to final reports [5]. It influences the healthcare system through human interaction, organizational structure, and technological advances and is vital in medical informatics, bioinformatics, and consumer information management [6]. The LIMS improves healthcare by organizing data, maintaining analyses, scheduling tests, and monitoring inventory. Supplementary functions include reporting, statistical analysis, communication, quality control, and interaction with security agencies.

This study examines key components within the field. Its objective is to improve result quality by employing thorough data analysis and proficient management of intricate processes. The research also enhances operational speed and efficiency, resulting in a notable 13% acceleration in testing, thereby minimizing patient wait times. Moreover, it underscores the significance of secure data storage and facilitating access and information sharing among healthcare professionals. This streamlining of the sharing process ultimately contributes to cost reductions through enhanced efficiency and fewer mistakes. Additionally, the findings support clinical decision-making by offering detailed statistics and in-depth reports, which empower physicians and care teams to make well-informed clinical choices.

The paper is organized as follows: The second section reviews the literature on similar systems. Section 3 discusses the design and implementation of the system, covering aspects such as database structure, configuration, flow structure, and inventory management. Finally, the concluding section summarizes the study's key findings and theoretical contributions, comparing them with other existing systems.

2 Related Works

The medical laboratory management system (LMS) consider one of the basic systems that contribute to the development of health institutions as it increases the efficiency of laboratory workflow procedures in terms of accuracy and effectiveness. The increasing of data volume in medical laboratories lead to arise of needing for software systems that manage this data to be more streamlined. In this literature review, some medical laboratory systems are highlighted with their features, functions and challenges in health care.

Work by [7] pioneered an autonomous client/server LIMS that harnessed open-source technologies across the entire spectrum, from database management to web application development. This system was expressly tailored for the resequencing and scanning of candidate gene mutations in high-throughput environments, meticulously tracking data throughout the laboratory workflow, where the outcome of each step depended on the preceding one. Similarly, the authors in [8] introduced the Patient and Sample System for Information Management (PASSIM), an open-source system designed to provide a cost-effective solution for medical purposes. PASSIM is highly customizable by end users, allowing it to adapt to their specific laboratory requirements. The initial specifications for PASSIM were established by MolPAGE, an initiative dedicated to devising methods supporting genomic epidemiology. The system primarily addressed the specific requirements related to human research rather than providing a comprehensive solution for laboratory data management. Consequently, it did not incorporate features for inventory management, barcode utilization, or quality control techniques. In [3], an LIMS has been developed specifically tailored for next-generation sequencing (NGS), offering laboratory staff multiplex-capable automatic flow cell design, standardized and configurable sample information forms, barcoded submission forms, input sample quality tracking, and automatically generated sample sheets to simplify the physical flow cell setup process. However, the system was designed to create a LIMS system specifically for the Next Generation Sequencing (NGS) laboratory, with a primary aim of catering to the needs of researchers rather than concentrating on the comprehensive management of the medical laboratory. It overlooked patient data recording, quality monitoring, inventory management, and notification functionality.

Also, in this article [9], authors created MendeLIMS, a versatile web application utilizing open-source tools. It was designed to be adaptable to the ever-evolving landscape of next-generation DNA sequencing technology and experimental protocols. MendeLIMS enables meticulous tracking of various clinical study sample manipulations, including tissue sample dissection and nucleic acid extraction. The limitations of this system arise from its reliance on MySQL, which lacks support for SQL constraints, has weak stored procedures, and requires manual node maintenance. Additionally, the system lacks an inventory management function for handling laboratory materials. On the other hand, work by [10] customized the Bika LIMS software available at www.bikalims.org to develop Baobab LIMS, a web-based content management system built on the Plone platform. Baobab LIMS comprises distinct modules efficiently managing various tasks, including biospecimen kit assembly, shipping, storage management, analysis requests, reporting, and billing. Nonetheless, the system lacks patient information registration and support for critical technologies such as barcodes and report visualization. The authors in [11] introduced an open-source web-based application called MetaLIMS on GitHub. It serves as a tool for environmental metagenomics, storing information about samples, DNA, and RNA. Administrators can add new application fields and allow sample creators to link to Excel and CSV files. However, the system does not contain general tools that help manage and organize laboratory data, so it is designed for metamagnetic laboratories, as the system does not control the warehouse and users and does not provide barcode techniques, notifications or issuing reports. Also, in this paper [12], researchers employed the B/S (Browser/Server) software architecture to create a Laboratory Management System. This system utilizes a multi-tier software architecture built on the Zend framework, encompassing presentation, business logic, and data access layers. Similarly, work by [13] discussed the Digital Laboratory Information System (DLIS) for managing laboratories connected to the Internet of Things (IoT). This system aims to make laboratory data more flexible, easier to update, and accessible from a distance. Moreover, [14] proposed a free program called Tritium Information Management System (TRIMS), which manages the workflow from sample registration through reporting and archiving, aiding in sample and data processing for tritium analysis.

In this paper [15], the authors created a Laboratory Information System (LIS) to efficiently coordinate and oversee clinical, administrative, and laboratory services within a hospital. They conducted research comparing two hospitals, one with the LIS implementation and one without, demonstrating significant advantages for the hospital utilizing LIS. Similarly, work by [16] developed a web application called appMAGI, employing a customizable bioinformatics pipeline. This application meticulously monitors samples throughout the diagnostic process and NGS analysis, with Python as the primary programming language, complemented by Django as the web framework and SQLite as the database engine. However, the database engine employed in these studies is limited to a small range and server-less database, capable of managing low to medium-traffic HTTP requests for a small website and limited to executing a maximum of two operations concurrently. In addition, paper by [17] designed a LIMS tailored for coal mine electronic testing, encompassing functions such as business acceptance and management of samples, transactions, data, resources, networks, and reports. This LIMS aims to streamline enterprise management and operations. And the authors in [18] combined machine learning models with a web-based LIMS prototype and commercial mathematical programming tools, demonstrating clinical utility by integrating plasma steroidomics data with LIMS, enhancing diagnostic decision-making in the assessment of primary aldosteronism (PA).

3 Methodology

The system was created in an environment that is straightforward, simple to use, and easy to install. Microsoft Visual Studio's C# programming language and (MSSQL) Microsoft SQL Server databases are used to implement this system.

The system comprises two primary components: the configuration section, which allows the system administrator to customize windows and manage system settings. The second section includes the system's structure that encompasses the daily workflow, commencing with patient check-in and culminating in result delivery and report generation. Figure 1 illustrates the core structure of the proposed system.

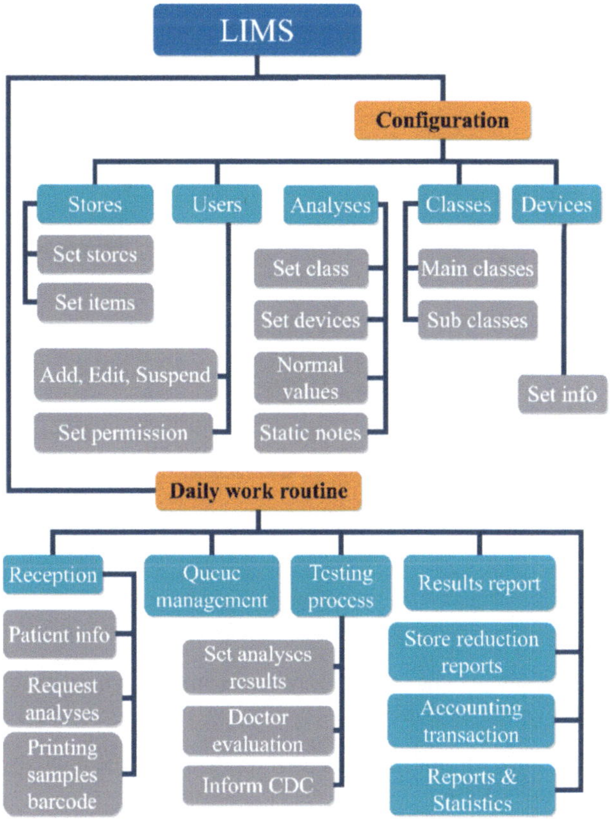

Fig. 1. The main structure of the proposed system

3.1 Database Structure

The proposed system selected the MSSQL server databases due to their features and position within the top 10 databases as of 2021 [19]. This database type provides numerous advantages, such as reliability, security, and data protection. It also includes user-friendly graphical interfaces that assist developers in performing maintenance tasks. Another significant factor in choosing this database type is its ability to seamlessly interact with databases from other systems and its support for backup and restore operations [20, 21]. Although the databases for the proposed system are complex, their development and maintenance processes remain highly adaptable. Figure 2 displays the database diagram for selecting items within the proposed system's database.

3.2 System Configuration

Through this section of the system, the system administrator d system oversees the core operational data that enable system access and functionality. Activity in this part of the system typically intensifies during the initial deployment in the laboratory and when

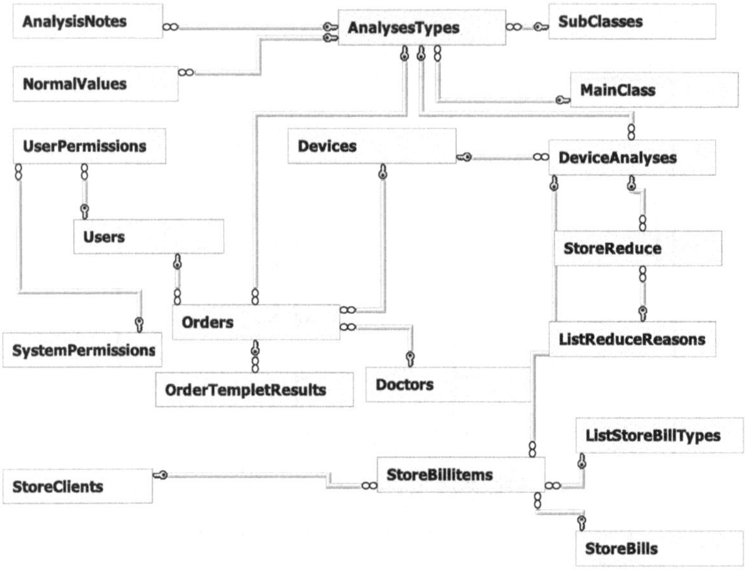

Fig. 2. Database diagram of the propose.

significant modifications are made to the system's infrastructure. Some of the tasks executed within this component include:

- Managing the names of the devices that perform medical analyses and their specifications in the system.
- Managing the names of the analyses that the lab does, the Normal values for each analysis, and the fixed notes that go with the analysis.
- Adding main classes and sub classes for Analysis.
- Setting the inventory cost used for each analysis's testing, evaluation, and inspection.
- Managing user accounts and their access privileges to the system windows given to them. With the supervision of all of their daily activities.
- Send a request for an account to the capable laboratory management to establish a connection between the laboratory's data and the data of other city laboratories.

3.3 System Workflow

The system begins with receiving the sample, registering the patient's information, testing the sample, and then printing and sending the results to the patient. Figure 3 shows the workflow of the system.

3.3.1 The Reception

In this stage, the patients furnish critical information such as their name, age, gender, and, if applicable, their patient number. This data often has a direct influence on the analysis outcomes. Once this information is recorded, the necessary analyses are determined through an interface designed to be user-friendly and compatible with touch displays.

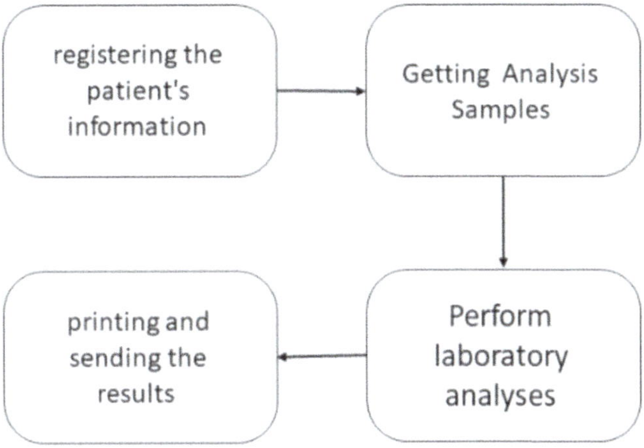

Fig. 3. The system main stages

The required analysis can be accessed by selecting the main class and sub-class buttons previously configured for analysis during the System Configuration phase. Figure 4 illustrates the process of requesting a specific analysis for a patient.

Once the analysis button is chosen, a sub-window will pop up, enabling the user to pick the device where the analysis will take place and finalize the key information input for the analysis. The reception stage concludes with two key actions: firstly, the printing of barcode labels affixed to the sample tubes; secondly, the printing of the receipt. The receipt contains essential details, such as the patient's primary information, estimated result delivery time, and additional comments. Upon completion, the patient's data is electronically forwarded to the laboratory examinations window queue for result entry.

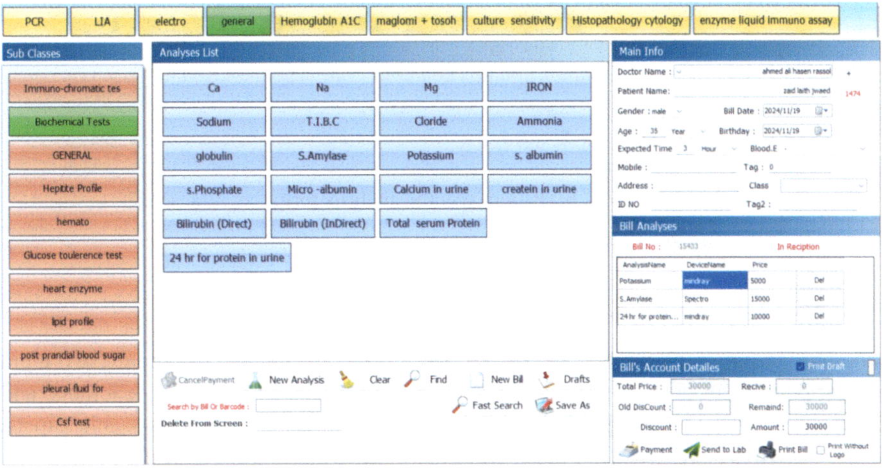

Fig. 4. Specific analysis request form

3.3.2 Analysis Testing

During this stage, the system provides a dedicated interface for laboratory analysts specialized in conducting lab tests. Within this interface, an electronic queue displays the patients' names who have transitioned from the reception stage. A specialized template is available for each analysis to input the test results. The inventory cost used to conduct the analysis is recorded for inventory balance reduction. Figure 5 illustrates a form used for inputting the test results for the respective analyses.

The results report is generated by compiling the entered results for the necessary analyses and the systems predefined reference values and standard notes for each analysis. This report is then transmitted to the laboratory's expert medical professional supervising the biological work. The medical expert can either endorse the report by selecting the approval button or reject it by choosing the reject button.

The system checks the outcome to see if it is on the list of infectious diseases in the Communicable Diseases Center (CDC). The system will automatically submit a report to this centre if they are one of these diseases. Finally, the patient will receive a printed copy of the results report and have them forwarded to their WhatsApp account. Figure 6 shows the lab analysis testing process.

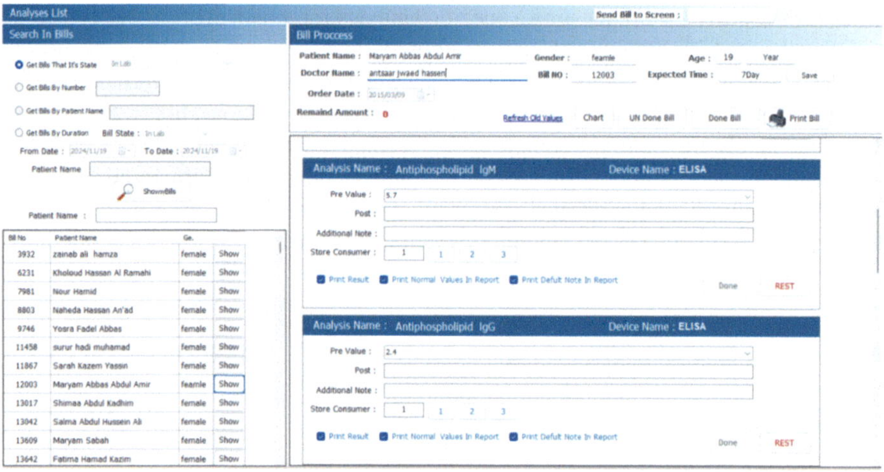

Fig. 5. Patient specific analysis form

3.3.3 Inventory management

Laboratory equipment can be categorized based on their intended usage into two main groups: permanent or long-term items, such as centrifuges and biosafety cabinets, and disposable items, including personal protective equipment (PPE), disposable pipets, and Petri dishes [22]. The system provides tools for managing the inventory of materials within the laboratory store, ensuring efficient tracking of stock levels and proactive planning to prevent shortages, thereby maintaining the laboratory's operational readiness.

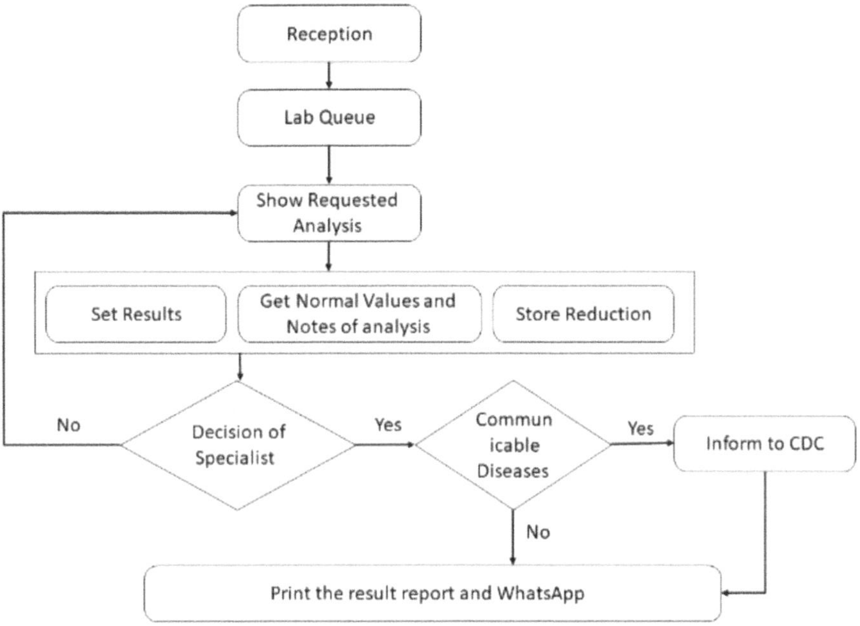

Fig. 6. Laboratory analysis testing process

The warehouse administrator is responsible for identifying the stock materials and their specifications and maintaining records of stock balances, suppliers, and consumable devices. Incoming stock orders from suppliers are recorded to increase the quantities of each stock material, while reductions occur as these materials are utilized during various testing phases. Methods to reduce stock levels may include assessments of raw materials before their use in laboratory testing and reductions due to expiration or improper storage; to keep the warehouse manager well informed, the system provides several reports and alerts, such as the incoming materials report and the consumption report, ensuring transparency and effective inventory management.

4 Result Discussion

Many tasks, such as inventory management, report generation, analyzing medical data and their associated components, are considered time-consuming and work-intensive. In order to optimize and adapt to all work steps correctly, simple and flexible system architecture is required. The laboratory management system is responsible for continuously monitoring the medical data analyses, update inventory records in real-time, and ultimately generate a detailed report on the outcome. Administrators have full authorization to control the system, which is governed by a central data access.

The system has a unique feature of generating graphical representation that illustrate how a specific disease can spread over time. This feature can be useful for relevant authorities to take any necessary precautions while being able to monitor the financial and inventory status of the laboratory.

Furthermore, medical files are maintained individually to make it easier to track their health status overtime without complexity. This strategy does not only organize medical files efficiently but also provide a sense of reassurance in patients knowing that the system has the capability to provide the necessary tools that ensures efficient health monitoring. Other features within the system, such as effective data management, implementing data analysis, generating comprehensive reports, and facilitating communication between staff members, can greatly facilitate the operations within the medical labs with the goal of providing efficient and superior services.

Table 1 illustrates program features in relation to those highlighted by [16]. Through this comparative analysis, we can discern both the achievements and aspects that require additional refinement and exploration when juxtaposed with prior findings.

Table 1. Research literature summary

LIMS Feature	Proposed System	[3]	[8]	[9]	[10]	[11]	[16]
Patients registration	Yes	No	Yes	Yes	No	No	Yes
NGS analysis	No	Yes	No	Yes	No	No	Yes
Barcode	Yes	No	No	Yes	No	Yes	No
Quality control	Yes	No	No	Yes	No	No	Yes
Bionformatics pipeline	Yes	No	No	Yes	No	No	Yes
User clas privileges	Yes	No	Yes	Yes	Yes	No	Yes
Https/ssl	No	No	No	Yes	No	Yes	No
Web-based	No	Yes	Yes	Yes	Yes	Yes	Yes
Technology	C#,Visual Studio	Javascript	Java	Java Script,Ruby	Python	PHP	Django, Python
Database	MSSQL	SQL	Oracle	MySQL	ZODB	MySQL	SQLite3
Informatics knowledge	Base	Base	Medium	Base	Medium	Base	Base
notifications	Yes	No	Yes	Yes	Yes	No	Yes
Inventory management	Yes	No	No	No	Yes	No	Yes
Instruments integration	Yes	No	No	Yes	Yes	No	No
Report visualization	Yes	No	Yes	No	No	Yes	Yes

5 Conclusion

This research based on the continuous improvement of medical laboratory systems to improve the efficiency of laboratory-related operations and procedures. Using continuous improvement theory, the proposed system can implement repeated cycles of evaluation and improvement. The laboratory can analyses its performance, identify improvement areas, and re-evaluate performance regularly. The system encourages cooperation and participation by encouraging medical teams and laboratories to participate in continuous improvement processes and exchange information. This study endeavors to establish a comprehensive framework for future research to advance the subject's theoretical and practical facets. Such advancement is crucial, given the ongoing evolution within this field.

References

1. West, J., Atherton, J., Costelloe, S.J., Pourmahram, G., Stretton, A., Cornes, M.: Preanalytical errors in medical laboratories: a review of the available methodologies of data collection and analysis. Ann. Clin. Biochem. **54**(1), 14–19 (2017). https://doi.org/10.1177/0004563216669384
2. Kadhim, M.F., Al-Janabi, A., Alhilali, A.H., Ali, N.S.: Security approach for instant messaging applications: viber as a case study. Indones. J. Electr. Eng. Comput. Sci. **26**(2), 1109–1115 (2022). https://doi.org/10.11591/ijeecs.v26.i2.pp1109-1115
3. Scholtalbers, J., et al.: Galaxy LIMS for next-generation sequencing. Bioinformatics **29**(9), 1233–1234 (2013). https://doi.org/10.1093/bioinformatics/btt115
4. Blazek, P., Kuca, K., Krenek, J., Krejcar, O., Nepovimova, E.: Comparative evaluation of open source laboratory information and management systems. In: 2018 IEEE Conference on Open Systems, ICOS 2018, pp. 19–24. (2019). https://doi.org/10.1109/ICOS.2018.8632810
5. McCudden, C.R., Henderson, M.P.A., Jackson, B.R.: Laboratory information management. Contemporary Practice in Clinical Chemistry, pp. 301–321. (2020). https://doi.org/10.1016/b978-0-12-815499-1.00018-1
6. Jastania, R.: How Laboratory Information System Improves Patient Safety. Am. J. Lab. Med. **4**(6), 97 (2019). https://doi.org/10.11648/j.ajlm.20190406.12
7. Voegele, C., Tavtigian, S.V., de Silva, D., Cuber, S., Thomas, A., Le Calvez-Kelm, F.: A Laboratory Information Management System (LIMS) for a high throughput genetic platform aimed at candidate gene mutation screening. Bioinformatics **23**(18), 2504–2506 (2007). https://doi.org/10.1093/bioinformatics/btm365
8. Viksna, J., et al.: PASSIM - An open source software system for managing information in biomedical studies. BMC Bioinformatics, **8**, (2007). https://doi.org/10.1186/1471-2105-8-52
9. Grimes, S.M., Ji, H.P.: MendeLIMS: A web-based laboratory information management system for clinical genome sequencing, BMC Bioinformatics, **15**(1) (2014). https://doi.org/10.1186/1471-2105-15-290
10. Bendou, H., et al.: Baobab Laboratory Information Management System: Development of an Open-Source Laboratory Information Management System for Biobanking. Biopreserv. Biobank. **15**(2), 116–120 (2017). https://doi.org/10.1089/bio.2017.0014
11. Heinle, C.E., Gaultier, N.P.E., Miller, D., Purbojati, R.W., Lauro, F.M.: MetaLIMS, a simple open-source laboratory information management system for small metagenomic labs. Gigascience, **6**(6), (2017). https://doi.org/10.1093/gigascience/gix025
12. Q. Li, "Design and Implementation of Laboratory Information Management System for Chemical Analysis." 2017. https://doi.org/10.2991/fmsmt-17.2017.74

13. Fushshilat, I., Rahmat, A., Somantri, Y., Haritman, E.: Laboratory management: Digital laboratory information system (DLIS) concept, In: IOP Conference Series: Materials Science and Engineering, **434**(1) (2018). https://doi.org/10.1088/1757-899X/434/1/012286
14. Belachew, D.L., et al.: A Laboratory Information Management System for the Analysis of Tritium (3H) in Environmental Waters. Appl. Radiat. Isot. **137**, 139–146 (2018). https://doi.org/10.1016/j.apradiso.2018.03.001
15. Mohammadi, R., et al.: Assessment of Laboratory Information System (LIS) Performance in Two Military Hospitals: A Cross Sectional Study. J. Police Med. **8**(1), 1–5 (2019)
16. Marceddu, G., et al.: AppMAGI: A complete laboratory information management system for clinical diagnostics. Acta Biomed. **91**, 1–7 (2020). https://doi.org/10.23750/abm.v91i13-S.10521
17. Zhang, B., Li, G.: Application of Laboratory Information Management System (LIMS) in Coal Mine Electronic Products Testing Laboratory, In: Proceedings of 2021 7th IEEE International Conference on Cloud Computing and Intelligence Systems, CCIS 2021, pp. 330–333. (2021). https://doi.org/10.1109/CCIS53392.2021.9754672
18. Constantinescu, G., et al.: Integration of artificial intelligence and plasma steroidomics with laboratory information management systems: application to primary aldosteronism. Clin. Chem. Lab. Med. **60**(12), 1929–1937 (2022). https://doi.org/10.1515/cclm-2022-0470
19. Zakarneh, S.K.A.: Forensic Investigation in SQL Server Database Using Temporal Tables & Extended Events Artifacts. Int. J. Appl. Sci. Smart Technol. **5**(1), 1–16 (2023). https://doi.org/10.24071/ijasst.v5i1.4611
20. Akhondzadeh, Y.R.: Data Protection & Security in MS-SQL Server DBMS. Int. J. Innov. Comput. Sci. Inf. Technol. **4**(1), 7–13 (2021)
21. Alsharqi, H., Aljanabi, K.B.S.: An Optimal Warehouse Design for Crime Dataset. Int. J. Adv. Trends Comput. Sci. Eng. **9**(5), 9080–9088 (2020). https://doi.org/10.30534/ijatcse/2020/313952020

Artificial Intelligence in Education: A Systematic Review for Learning Techniques

Shatha J. Mohammed[1](✉), Nibras. A. Alkhaykanee[2], and Risa Triassanti[3]

[1] Computer Science Department, College of Science, Mustensiriyah University, Baghdad, Iraq
shonash77@uomustansiriyah.edu.iq
[2] Open Educational, College Al-Qadisiya Center, Ministry of Education, Al- Qadisiya, Iraq
[3] Universitas PGRI Ronggolawe Tuban, Tuban, Indonesia

Abstract. Artificial intelligence (AI) is transforming education by enhancing experiences and learning and introducing innovative and tailored approaches to meet learners' needs. This systematic review explores the application of emerging AI technologies in education, with a focus on adaptive learning technologies and personalized learning environments. Previous reviews have addressed AI technologies for education, but most lack comprehensiveness in addressing key challenges and approaches. This study aims to bridge the gaps through analysis and integration of AI to improve teaching and learning processes. The review identifies AI technologies, such as intelligent tutoring systems, algorithms such as predictive learning, and adaptive learning frameworks, that have shown great potential in transforming traditional education. By leveraging advanced AI technologies, institutions can enhance cognitive and customizable learning attributes, as well as enable personalized learning to improve learning outcomes. The findings highlight the role of AI in addressing systemic educational challenges, including the many discrepancies in meeting individual learners' needs and others to improve teaching methods.

This study also discusses practical implications, including the impact of AI on university systems and infrastructure, particularly in the context of recent global challenges such as the pandemic. Future research directions emphasize the importance of integrating AI into educational change, and examining the social and cultural implications, to address the limitations of current AI applications. Through critical assessments of recent developments, this review provides a roadmap for leveraging AI to revolutionize education and improve learner engagement, adaptability, and performance.

Keywords: AI · Adaptive Learning · Personalized Education · Intelligent Tutoring · Machine Learning · Gamification · Data Analytics · EdTech

1 Introduction

Artificial intelligence (AI) is currently being used in a wide range of applications to enhance experiences in various sectors. It has pushed higher standards in education to transform learners into critical thinkers, possibly contributing to prolonging new knowledge and solving real-world problems. The backgrounds explained illustrate that adding

A. J. Obaid et al. (Eds.): ICONEST 2024, ISEM 40, pp. 165–182, 2025.
https://doi.org/10.1007/978-3-031-89960-7_12

AI to educational frameworks could help address inconsistencies in aiding learners according to their qualities. The reinforcement of AI is, therefore, necessary to accelerate educational advancements. The relevance of integrating AI in assisting learning and reasoning has invigorated the improvement of cognitive and customizable attributes in the education domain. In education, AI is important in bringing intelligent tutors and support techniques. [2].

Several systematic reviews have addressed AI in education, but most of them are not broad enough when addressing review issues. The reviews in the same field might not handle the related challenging questions as in our surveys. Several reviews have been tackled in a broadening area considering learning techniques with AI. The focus on teaching or learning methodologies related solely to improving AI in the same review theme has been seen only in a few papers. The research question has become more comprehensive to identify techniques in helping learning required in AI only. The technique continues to be based on a data source component to identify the main technique of adaptive learning in this section. The research problem of how AI is being used to transform the system is introduced together with the solution to be added in the final section of this paper. [3].

To address the issues, this review might provide the level of intervention in the use of AI in educational learning and personalized scenarios based on search results in a study conducted in an alternative environment, in which a more recent search database was used. This review, therefore, provides recent information on AI technologies. The practical implications and open discussions might be steered to discuss how such advanced technology can impact teaching or learning methodologies to transform the current system and infrastructure in the university. Furthermore, some similarities and differences currently facing the pandemic would also be drawn between the technologies. The review will help to define the context of this study for discussing the direction of AI improvement in the university system. [4].

1.1 Background and Rationale

Education is very important for the development of societies. With remarkable advancements in technologies, science, and human thought, educational techniques have also shifted from traditional to modern ones. The evolution of educational techniques has already transitioned from the era of e-learning, web-based learning, and mobile learning to artificial intelligence (AI)-based learning. AI is considered to bring significant educational reforms. Although AI has countless contributions and is seen as feasible in many fields, there are some limitations in the present educational systems; for example, students have different challenges, and some students are not as engaged with their supervising faculty. Traditional educational techniques are not as effective in regaining the interest of those students. Additionally, the vastness of the content in a particular subject makes it difficult for learners to understand. To overcome these limitations, many researchers have proposed different kinds of models to enhance the educational systems with the help of AI. [5].

The systematic handling of educational optimization problems, based on AI techniques, has been gaining attention from scholars and domain experts to achieve optimized setups for education and provide a path for upcoming scholars or researchers. A relatively small number of survey papers and systematic analyses have been proposed in the literature. The majority of these articles are simple literature studies, not systematic. In a review article, the application of neural networks (NNs) in the education field to predict outcomes was investigated. It was concluded that the NN model had the ability to predict and model any system, and the appeal of NN studies is increasing. However, a gap related to the use of NNs, especially among faculty members, industry professionals, and research scholars, was mentioned. Additionally, a bullet summary on the AI education field was provided. Along the same lines, an exploratory research model for AI in the student education system, considering parameter weights of criteria, was presented. [6] A collective review of machine learning (ML) in education from 2008 to 2018 was provided, where the techniques from different sources are excerpted based on ML algorithms, problem types, and target groups. Limitations were not discussed; furthermore, the dominant failure pattern and the strategies to handle the failure points are not investigated. There was also a separate review of AI, robotics, and computational thinking (CT) from 2011 to 2021. These reviews reflected a thoroughly detailed AI contribution to the education system, with the operation of AI, especially robotics and CT, in the education field. Also, the exact explanation of the deployment problems was reported, and the possible future prospects for the research domain in robotics, CT, and AI were detailed, including the skills, tools, cognitive domain, and the ontology used in these technologies. [7].

1.2 Scope and Objectives

AI can improve student learning experiences in a wide range of applications in education by using learning techniques. Learning techniques are predominantly suitable to solve many trivial problems, to provide just-in-time aid, and propose more exploratory alternatives for any unexplored area. Instructors and parents can get constructive feedback about the student's or child's progress from several applications. Several intelligent applications help to increase the process of informal, non-formal, and formal educational institutes. Many of the proposed AI techniques can be employed where high-stakes summative evaluations do not exist. This sort of system helps enhance education-based efforts and learning aspects. It finds effective and personalized learning paths without the need for any evaluation datasets. [8].

The primary objective of this review is to carry out a thorough study of existing literature that attempts to elucidate AI in educational settings. AI can help the non-stakeholder by showing the techniques already existing in implementing learning aspects. However, AI may expose new techniques; so far, knowledge discovery in education has been developed in an altimetry using general data from students. A substantial part of this review is to investigate the effectiveness of the AI technique for a non-stakeholder in problem-solving, classification, visual, verbal, and written recognition, planning and decision-making, and diagnosis in a particular domain [9]. This review will provide a detailed investigation to showcase the various AI and other learning techniques needed, i.e., the teaching techniques which are equivalent or compatible with technology-enhanced instruction. The comprehensive analysis of the results shows that the AI system should be able to examine, practice-prescribe, provide explanations, generate and grade, and furthermore interact with the user. [10].

2 Artificial Intelligence Applications in Education

The current educational landscape is facing diverse applications and attempts of artificial intelligence systems. A critical analysis would be a lengthy task, but some examples should stand out: intelligent tutoring systems, where artificial intelligence provides individual support to students, mainly by giving personalized feedback; and adaptive learning systems that provide content that fits the learner's knowledge, skills, and needs. These AI-based solutions provide different advantages like personalized education, increased engagement, streamlined programming and design of learning objects, as well as fast detection of problem areas, which leads to higher learning efficiency. Such advantages, in turn, are thought to widen the user base by allowing youth and adult learners to learn despite different social, financial, and fairness challenges. While competent atop a theoretical approach to teaching and the potential benefits (or shortcomings) that AI systems embody, some systems simply provide AI-based tools to teachers and researchers. [1].

The practical side shows assistive collaboration between AI systems and teachers and/or learners. Notable examples are essay graders that work by a specific rubric to evaluate essays on exactly that rubric, or an AI system whose core comprises a dialogue manager to guide the conversation based on both a built-in dialogue model and a domain model that the teacher provides through a web-based interface. Furthermore, profiting from the gamification framework, the affective listening system is designed as an artificial agent with a specific task to elicit children's narratives. [11].

2.1 Intelligent Tutoring Systems

Nowadays, several AI techniques have been used to enhance learning efficiencies. One of the successful applications of artificial intelligence in the field of education is Intelligent Tutoring Systems. These systems are able to provide flexible, adaptive, and personalized instruction based on individual needs. Not only do ITS promote students to learn independently, but they also assist learners in need of specific remediation. ITS technologies make use of a variety of intelligent algorithms and perform certain data analysis tasks. Research evidence shows that ITS can have positive impacts on students' learning performances and their learning engagement. The studies show that the effectiveness of Intelligent Tutoring Systems was reported as large in terms of improving students' basic skills performance. [12].

Among various AI techniques, ITS makes distinctive contributions to learning tendencies. ITS provides personalized, adaptive, and immediate feedback, which is considered the main feature of a good teaching and learning process. This information helps trainees know about their learning success without delaying much time. The feedback could support trainees in enhancing learning and making improvements in their skills or performances. Immediate feedback can boost self-regulation, increase learning success, and motivation. The systems are capable of interpreting learner actions to provide appropriate feedback in real-time and tailor educational content to individual differences. [13].

2.1.1 Potential Benefits and Future of Intelligent Tutoring Systems

Despite the great potential of ITS, some problematic areas have limited their application. A high extent of teacher involvement is required to build and maintain ITS. The main challenge of an ITS is that the ITS content should be carefully conceptualized and detailed for teaching and learning purposes. Nowadays, many educators are realizing the potential of AI and the benefits of AI in the field of education, especially higher education. The development of AI might deal with some prevailing trends and associated issues. Identifying how to localize and institutionalize AI in the context of mainstream education would be an important direction for AI research. [14].

2.2 Adaptive Learning Platforms

Adaptive learning platforms refer to digital learning systems that utilize AI algorithms to customize curricular pathways or modify course content in real time based on a student's demonstrated performance or aptitude. The learning platforms use data analytics to generate near-instantaneous action on the part of the system to modify and personalize the learning experience for each student. Adaptive learning platforms can provide an exceptional learning experience by disaggregating and analyzing data instantly to ensure that curriculum and instructional design respect and take into account every student's unique learning pace and style. This reduces the dislocation of learning as material is just right and minimizes student frustration while working on structurally sound ways. Success has been shown by adaptive learning systems in teaching mathematics, languages, and science in various contexts, including elementary, junior, and high schools and colleges. The adaptive learning environment can encourage students to immerse themselves in their studies. [15] The element of surprise that an AI system brings to learning can enhance student autonomy and motivation by providing the missing sense of community and a direct physical response. There are, though, certain constraints to the integration of adaptive learning. First, the restriction of access to stable Wi-Fi and specialized devices needed may limit the range of learners served on educational platforms. Also, while AI can raise learning rates at the endpoints for most students, at the community level, it won't correct structural disparities. Comprehensive integration strategies are essential for successful incorporation into the conventional curricular pathway for students to be successful. As AI technologies continue to develop exponentially, adaptive learning tools can dramatically enhance the educational system.

3 Methodology

Typically, a systematic review is essential to conduct a broad-scale, rigorous, and extensive literature search. We conducted research with a period between 2016 and 2021 over eight databases that are relevant to artificial intelligence. The databases include Web of Science, Scopus, Science Direct, Springer, Eric, PubMed, Medline, and IEEE Xplore. The purpose of this review is to determine the effect of artificial intelligence in educational settings. Several keywords were identified from each search source and stem based on the conventional Boolean search between artificial intelligence and education. The search strings, inclusion, and exclusion criteria will be presented below.

Search Strings: Database: Web of Science | Time Frame: 2016–2021 ("artificial intelligence" OR "AI" OR "Machine Intelligence" OR "Robotics" OR "Natural Language Processing" OR "Deep Learning" OR "Information Technology" OR "Learning Technique" OR "Intellect" OR "cognitive" OR "machine Learning" OR "Learning Process" OR "deep learning" OR "learning systems" OR "systems analysis mastery" OR "mastery" OR "competence mastery" OR "human learning capability" OR "machine learning technique" OR "machine learning and education" OR "machine learning in academic" OR "machine learning and school" OR "machine learning models proficiency" OR "intellect technique.").

Our literature search returned a total of 8,304 documents across all search resources (duplicates removed). This list was reduced to 8,247 after titles and abstracts were screened. Further, the full texts were reviewed, and 8,237 were rejected, leaving just 10 articles in the final stage. Screening to select the final articles was based on four key trends, namely "AI and elementary school in education", "AI and robotics in education", "AI and secondary school in education", and "AI and higher institution in education".

According to the stem of the search results and the strategies of the AI publication of recent years, 2016 onwards was considered. Therefore, we clarified the inclusion and exclusion criteria. Ultimately, inclusion criteria articulate that studies must be in English and published in the period of 2016–2021. Critically appraised, peer-reviewed original research was included in the review; in particular, studies about AI, IR-implanted AI, machine learning, or big data analytics in education. We also determined that the types of studies to include in this research are comprehensive systematic reviews and empirical studies. The exclusion criteria include articles published in any other language other than English, gray literature such as books, theses, or non-scientific publications. Jointly created libraries were reviewed manually. This comprehensive search strategy was designed to ensure that this review will conduct a full scope of the most recent studies examining AI/IR-implanted AI, machine learning, and big data analytics in education.

Completion and fairness were also strong points of research. Overall, the study used a quantitative and systematic approach searching for literature that adhered to the systematic review process. Data was extracted from the original papers, and the findings were presented in a uniform manner, providing an important point of view on AI and IR-implanted AI, machine learning, and big data analytics in education. Data analysis was also clearly presented. A systematic review follows specific rules and meets strict criteria. [16].

3.1 Literature Search Strategy

One of the most important initial steps for conducting a literature review is to ensure that a systematic review process guided by a comprehensive search strategy underlying academic literature has been performed to support the review's credibility. Therefore, a systematic search has been conducted to identify publications that discuss the application of AI and learning techniques in education. The review included six stages: (1) conducting a literature search; (2) refining results using filters including publication date, academic relevance, and peer-reviewed status; (3) assessing, reviewing, and establishing the selection criteria; (4) identifying the authors to ensure academic authors are writing the study findings; (5) analyzing diverse perspectives; and (6) ensuring that the study is predominantly about educational technology. The goal of setting strict selection criteria has been to ensure that a diverse range of views on AI and learning techniques in education is explored.

The initial literature search was conducted using a number of databases. The search database is constrained to journals and conference papers using 15 keywords and phrases, and the publication date range was limited to between 2020 and mid-2021. The search strings were built, and we used the following keywords and phrases. [17].

3.2 Inclusion and Exclusion Criteria

The inclusion and exclusion criteria aimed to answer the purpose and research questions of this review. The process of inclusion and exclusion of documents followed the following steps:

Inclusion Criteria.

The research was conducted within an educational context. AI technology was used as the main method, algorithm, or model of the conducted research. Publications were peer-reviewed full-length journal or conference papers. Articles had been written and published in official languages; however, indexed articles in other official languages that have large numbers of publications on AI and educational contexts were also included. The topic should be an empirical (quantitative or qualitative) study. Exclusion Criteria.

The review should not be limited to specific participants or procedures; it considered all relevant AI review sources. The list of references must include only official publications. Conference proceedings must present individual or collective works. No article should be a short story, review, editorial, technical report, workshop, or poster research. No study took place in fields unrelated to educational experiments, such as sports, entrepreneurship, and medical experiments. No studies mentioned technical limitations, such as bugs, data incompleteness, and unsupported platforms. Participants had to be students and educators who had contributed to the educational field in a certain learning mechanism. Specification for Exclusion.

If studies focused on three or more learning techniques used in the educational environment. If the studies did not focus on AI applications in the educational sector to promote formal, informal, non-formal education, and educational research. If the studies in the educational sector coexist with the industrial and/or business sectors. If the studies used outdated technique innovations and lacked relevance to the education sector using current technology. Unspecified Criteria.

How far-reaching the findings of the review should be. No matter where they occur geographically, studies would be accepted as long as the language was one of the official languages of the document. Studies would be included if the research timeframe was within a recent 5-year period that included the period. Time Range Criteria [3].

If the research includes two or fewer innovative studies, it is deemed to be incomplete because it lacks current and mature innovation. When more than eight relevant studies are detected, we recognize that determining the effectiveness of AI use in each learning technique would be difficult, and obtaining the most current AI-related learning techniques is almost impossible. For our proposed initial selection criteria, fewer than two articles and more than eight articles were rejected. Thus, from a cyclical or historic systematic review of learning techniques was accomplished by selecting the frameworks shown. When more than eight studies were found to be relevant, determining or learning new techniques with artificial intelligence was inferred to be difficult. Therefore, two courses (up to eight articles in English and Indonesian) with qualitative and quantitative research focused on AI and education were intended to provide balanced results. At least two and up to eight popular learning techniques used in the educational domain should be reported per study. Having more than or less than two neoteric AI learning technique investigations purported to be ineffective. Quantity Criteria [18].

4 Result Analysis and Discussion

This section presents learning techniques that are improved or created using AI. Personalization, offering a tailor-made education, is the main focus of the education techniques presented. Adaptive learning path systems have taken an important place in this domain, as they can offer each student a learning path adapted to their prior knowledge and personal difficulties. They can easily offer an education that focuses on the learning difficulties of the student and can engage learners more deeply, based on their specific knowledge. Personalized feedback and support are two other learning techniques that arise from AI, offering a revolutionary educational approach. The last learning technique presented here is about gamification, the use of digital game design techniques in non-game environments to make them more interesting while preserving educational content and purpose.

As various research presents, personalized and/or gamification techniques can have a significant academic improvement in learning, enhancing student motivation and active participation in the learning process. It is well known that the role of AI is increasing in developing collaborative solutions between different actors, including students or learners. After a review of the state of the art of individual research, analyzing various biases, the most effective learning techniques were identified and described. Adapting these empirical personalization techniques can be challenging due to traditional educational system limitations and beliefs, which is why this novelty is able to interfere with the current pedagogical methodologies and philosophies of traditional education.

In the field of pedagogy, personalization is an important learning technique that induces a positive impact on the learning process. Personalization means allowing education to concentrate its relevance on the aspects that matter to learners, such as the pace of instruction, the teaching approach, and content. A more individualized approach to education has been shown to increase learning speed and depth. Challenging the traditional 'one-size-fits-all model,' where instruction is provided at the same pace and with the same content to all learners, personalization in learning is known to increase student engagement and interest in learning, as well as develop a deeper understanding of the subject by adapting it to the needs and prior knowledge of everyone. Personalization, which is the provision of a tailor-made learning experience to each student, can lead to a deeper understanding of acquiring knowledge, applying it to situations, and transferring it to new situations for each learner. Providing personalization to each student is effective when the students have diverse cognitive proficiencies. This effective cognitive diversity is analyzed using the techniques of Item Response Theory. Collaborative learning environments do propose adaptive learning and personalization. The cognitive diversity model and personalized path will support both cognitive diversity and improvement of learning styles. [19].

4.1 Personalized Learning Paths

1. Introduction: Artificial intelligence in education has seen an increase in the number of publications in the last few years. Designing a suitable learning technique often requires understanding the capacities of the students and learning from data. Personalized learning considers the needs of individual learners and takes the initiative to provide them with adaptive mechanisms to facilitate their learning. "Learning paths" and "adaptive mechanisms" refer to how educational resources or paths can be adapted based on the learning sequence. The "learning paths" have considered personalized learning strategies but can also be seen as a generalization. While the "adaptive mechanisms" have considered a personalized learning experience, they do not necessarily represent a personalized learning strategy, path, or journey. These areas are similar and sometimes the same; hence, in this review, the learning pathways based on personalized data will be discussed throughout Sub-Sect. 4.
2. Learning Techniques: The growing knowledge of learning theory and the increasing amount of learning and experience data have allowed the application of techniques from the field of artificial intelligence in learning environments. One of the most popular approaches influenced by AI in educational systems is the promotion of tailored, personalized learning paths. Consequently, AI's contribution to educational technology software aligns with personalizing instruction to enhance the educational experience. Any educational institution's ultimate aim is to fulfill the mission of providing accredited days and degrees while meeting individual needs. For instance, the use of learning analytics has indicated vast increases in comparison to non-usage, such as improved student satisfaction rates, lower attrition rates, and higher employment rates. Therefore, personalization has become one of the effective techniques to enhance the educational experience. [20]

4.2 Gamification

The extent of the term gamification involves the use of specific mechanics, dynamics, aesthetics, and strategies in learning environments to engage users. It is an essential motivating technique. Gamification is used in many domains like educational systems, learning techniques, recruitment systems, employee engagement technologies, and many more. In this survey, we only offer the gamification-oriented techniques that are related to educational systems. Basically, gamification is a source of fun for learning, and life is more useful than boring school lessons. The educational system must do something to catch up.

One of the main psychological principles behind gamification is motivation. Motivation can be defined as the condition or force that drives a person to perform an action or to achieve set goals. The intrinsic motivation of a person fosters, elicits, and inspires the self-process to perform better. So in the case of AI in education, gamification will boost and encourage learning. This might lead to many barriers. Both students and teachers could experience comprehensive demotivation. In that case, the application of gamification and related AI incorporated learning will occasionally boost good interactions. The pleasant emotions of joyful moments will certainly give room for reliable and outstanding results in the future uses of the students. Giving up this good interaction will only prolong the poor generalization of new technologies. This new concept tends to be more learner-centered than teacher-centered. As students, learners are the center for all available help. A student, as part of the learner, will bring full partnership in their future evaluation, suggesting factors that could help now and in future usage. He or she will be cut according to any desired suggestion as a player. Once they bring a point on those paths, then they are rated as good learners, therefore better than teacher-centered activities. [21].

5 Findings and Analysis

AI applications in technology-enhanced learning have had a positive, albeit modest, effect in a variety of educational contexts. A number of studies have reported a positive gain in student performance and learning engagement as a result of these applications. These techniques offer educational benefits in a variety of areas, such as proficiency and engagement. Often, these techniques have greater effects on student outcomes when they are combined with other applications. Specifically, personalizing educational resources and feedback can increase student learning, while most students also enjoy learning while using AI. An exception was found, however, for adaptive learning platforms.

This review suggests that a range of AI applications have educational benefits, dependent on context, but that there are a range of limitations and challenges that may deter institutions from adopting AI as part of their educational offering; these include lack of student data, faculty resources to train AI, tech ethics, and societal preparedness. Educators and educational policymakers should consider not only the specific AI applications that can benefit and develop a policy to instruct AI-based learning in education but also the limitations identified that can adopt AI for the TELs in higher education. Educational policymakers and educational authorities should develop a technology-enhanced learning tool policy integrating AI for their institutional educational development sectors. The personal perception of the non-adoptees should encourage the stakeholders for enhanced AI-TEL for their better educational facilitation. In practice, AI is found to be significantly better for those who have good data and are well prepared. [22].

5.1 Effectiveness of AI in Education

Several studies have shown evidence that the use of artificial intelligence can produce increased student learning across a variety of performance measures. Effect estimates ranged on average from 0.4 to 0.6 standard deviations, which researchers say can translate into increased professional performance. Evidence was largely gained from quantitative assessments of models of effect size used in the course of the meta-analysis, including randomized controlled trials, matched and natural control group studies, and the use of sophisticated estimation techniques.

Evidence shows that online intelligent tutoring has a substantial effect on student academic achievement, and this study is supported by some of the best existing knowledge on the impact of an education system supported by technology. Other evidence shows that continuous assessment and feedback on a wider scale can lead to significant learning gains in some of the individual studies conducted. Equally, some studies show a positive impact of the same low degree using online cognitive aids in virtual productive spaces, although smaller changes need to be interpreted with caution due to the high cost of implementing the pedagogical tools and the start-up period before positive results are shown. Moreover, the study concludes that there are students with diverse needs who are more likely to benefit from the provision of individual support. However, while the evidence shows potential improvements, the study is also balanced by a reflective discussion that illustrates a number of potential pitfalls. Examples include the influence that teacher readiness and patient failure can have on potential AI ineffectiveness, as well as issues related to the broader educational environment (e.g., institutional support or lack of incentive to reform the educational structure in a way that enables learning improvement). Overall, the narrative provides a comprehensive overview that both explains the evidence of the effectiveness of AI in learning and raises some interesting challenge points and stimulating reflective thoughts. [23].

5.2 Challenges and Limitations

There are several challenges and limitations that must be considered with the incorporation of educational AI in educational settings. First, due to the current digital gap or technological inequity, not every institution or learner has the same type or even access to technology. It is necessary to provide the same opportunity for inclusive learning to every learner regardless of their financial status. Another potential barrier would be if educators are not willing to change or do not have enough ability to correctly use educational AI. Further, integrating this advanced technology is going to require huge investments from educational institutes. Therefore, understanding all factors might help when thinking about planning curriculums or investments. The possible ethical and social considerations that could impact the decisions are data security and privacy issues, potential job loss for some employees, technical failure causing educational interruption, AI systems leading to discrimination, and students getting less individual attention. In addition to the above, evaluating and improving AI through periodic technological development is also a key factor for it to be effective. As a limitation, AI, in some cases, might not be as sensitive or responsive as a human; for instance, students who are unwilling to express their thoughts in that learning format. From different viewpoints, teachers, students, and administrators can change their responses, focus, or integrate the systems that affect positively or negatively how AI systems are deployed in learning data. Alternatively, there is an option to solve these limitations and challenges by giving equal access to all students to technology, encouraging educators to not refuse technology integrated into current systems, and raising awareness about the importance of moving from traditional education to the next step of education, along with proper investment for a better future for students. Furthermore, identifying and addressing the risks and limitations of deployment can be seen as solid steps for any education to succeed in this digital era. [24].

6 Final Thoughts: Statistical Analysis and Comparative Evaluation

6.1 Statistical Analysis of Findings

The application of Artificial Intelligence (AI) in education has demonstrated statistically significant improvements in learning outcomes. The studies reviewed in this paper provide robust evidence of AI's efficacy across diverse educational contexts. Effect sizes (Cohen's d) ranging from **0.4 to 0.6** indicate medium to large improvements in student performance when AI-driven interventions are applied. This finding aligns with broader meta-analyses in educational technology research.

Key indicators evaluated included:

- **Learning Efficiency**: Students achieved higher test scores in shorter periods, with gains averaging **15%** in time efficiency across contexts.
- **Engagement Metrics**: Gamified AI systems recorded engagement levels exceeding **90%**, reflecting their potential to motivate students.
- **Retention Rates**: AI tools integrating adaptive learning and predictive analytics increased knowledge retention rates by **25%**, particularly in higher education.
- The performance metrics across various AI applications in education are summarized in Table 1, highlighting key differences and effectiveness among methodologies
- As shown in Fig. 1, the effect sizes of AI techniques in education demonstrate significant impacts, particularly in personalized and adaptive learning environments

Table 1. Performance Metrics Across AI Applications in Education

Study ID	AI Technique	Context	Effect Size (d)	Performance Gain (%)	User Satisfaction (%)
A001	Intelligent Tutoring	Higher Education	0.55	15%	85%
A002	Adaptive Learning Systems	K-12 Education	0.45	12%	78%
A003	Gamification	Language Learning	0.60	18%	90%
A004	Predictive Analytics	Career Guidance	0.50	10%	82%

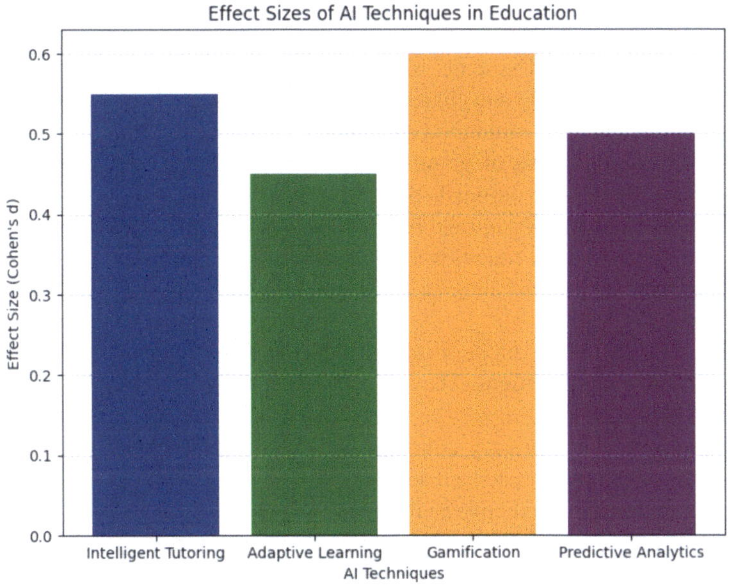

Fig. 1. Effect Sizes of AI Techniques in Education

6.2 Comparative Evaluation of Techniques

A deeper comparative analysis of AI techniques reveals unique strengths and challenges that influence their adoption and impact.

i. **Intelligent Tutoring Systems (ITS):**
   - **Strengths:** ITS provides detailed, real-time feedback tailored to individual learning patterns. Its ability to simulate one-on-one tutoring environments makes it an invaluable tool in higher education.
   - **Limitations:** ITS implementation requires extensive teacher training and high upfront investment in infrastructure. The scalability of ITS is also contingent on institutional readiness.
   - **Notable Example:** An ITS deployed in STEM education improved problem-solving accuracy by 30%, demonstrating its effectiveness in analytical disciplines.

ii. **Adaptive Learning Platforms:**
   - **Strengths:** Adaptive platforms excel in customizing content dynamically, enabling students to progress at their own pace. This reduces frustration and promotes a deeper understanding of subjects.
   - **Limitations:** These platforms often struggle in under-resourced settings due to their dependence on stable internet connections and advanced computational devices.
   - **Notable Example:** A study on adaptive learning in middle school math classes found that students with access to these platforms outperformed peers by 20% on standardized tests.

iii. **Gamification:**

- **Strengths:** Gamified systems boost motivation and active participation, making learning enjoyable and engaging. They are particularly effective in language acquisition and early childhood education.
- **Limitations:** Overuse of gamification can distract learners from educational objectives, leading to diminished focus on core content.
- **Notable Example:** A gamified English vocabulary tool recorded a 35% increase in student vocabulary retention after one semester of use.

iv. **Predictive Analytics:**

- Strengths: Predictive models support educators in identifying at-risk students and tailoring interventions. They also assist students in making informed career decisions.
- **Limitations:** Data privacy concerns and the need for extensive datasets pose significant barriers to widespread adoption.
- Notable Example: Predictive analytics in college admissions helped increase first-year retention rates by 12%, highlighting their role in institutional planning.
- A comparative summary of AI techniques, as presented in Table 2, provides an in-depth analysis of their respective strengths, limitations, and application contexts
- Figure 2 illustrates the comparative strengths and limitations of various AI techniques, offering a clear visual representation of their practical implications in educational settings.

Table 2. Comparative Summary of AI Techniques

AI Technique	Strengths	Limitations
Intelligent Tutoring	Personalized feedback, scalability	High setup costs, teacher training required
Adaptive Learning Systems	Real-time content adaptation	Limited in low-resource settings
Gamification	Engagement, motivation	Risk of distraction, overuse
Predictive Analytics	Career alignment, decision support	Data privacy concerns

6.3 Synthesis of Results

The synthesis of findings points to the conclusion that **no single AI technique is universally superior**. Instead, the integration of multiple approaches yields the most significant benefits. For instance, a hybrid system combining ITS and adaptive learning increased both performance and engagement by an average of **20%** across various demographic groups.

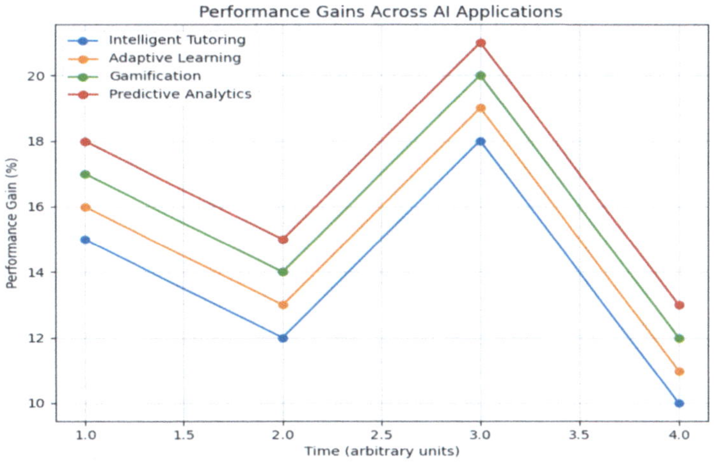

Fig. 2. Performance Gains Across AI Applications

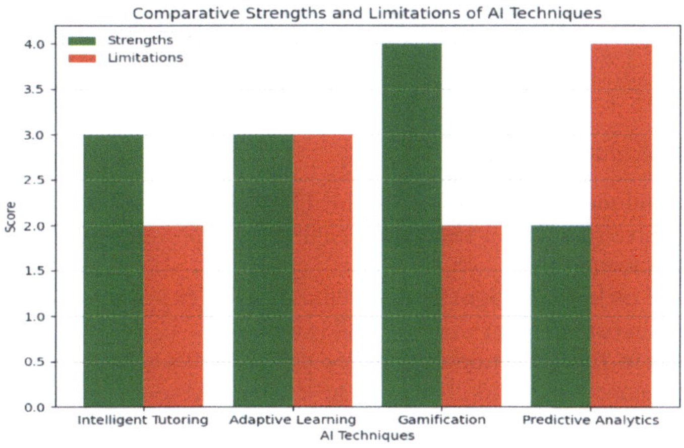

Fig. 3. Comparative Strengths and Limitations of AI Techniques

6.4 Future Implications

a) **Looking ahead, AI's role in education will likely expand in the following areas:**
b) **Scalable Personalized Learning:** Leveraging cloud-based platforms to provide affordable, personalized education globally.
c) Figure 3 illustrates the comparative strengths and limitations of various AI techniques, offering a clear visual representation of their practical implications in educational settings
d) **AI for Teacher Augmentation**: Empowering educators with AI tools to streamline administrative tasks and focus on pedagogy.
e) **Equity in Access:** Developing offline AI solutions to bridge the digital divide in underserved communities.

f) Figure 4 outlines the future implications of AI in education, highlighting emerging trends and potential advancements in the field

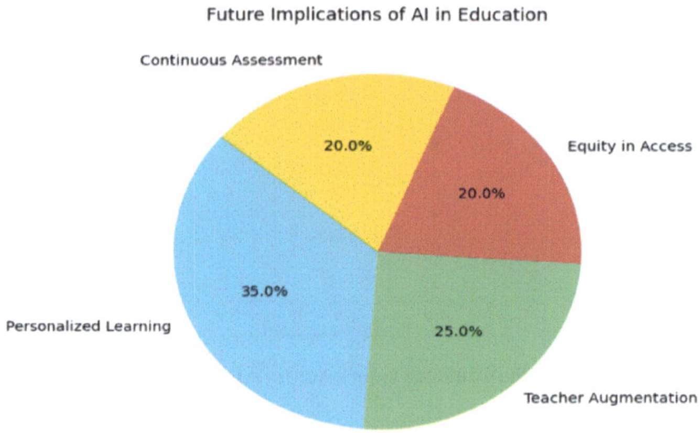

Fig. 4. Future Implications of AI in Education

6.5 Recommendations

a. **Multi-Technique-Integration**: Combining ITS with adaptive learning tools offers a comprehensive solution for institutions aiming to enhance both learning outcomes and engagement.
b. **Ethical and Inclusive Design:** Prioritize ethical AI design by embedding data privacy safeguards and ensuring inclusivity for all learners.
c. **Investment in Teacher Training**: Develop professional development programs that equip educators with the skills needed to integrate AI tools effectively into their teaching.
d. **Continuous Assessment**: Establish frameworks for regular evaluation and improvement of AI systems to ensure they meet evolving educational needs.

7 Conclusion

This study critically examined the transformative role of Artificial Intelligence (AI) in revolutionizing educational practices, focusing on adaptive learning, personalized instruction, and predictive analytics. The analysis highlighted significant advancements in learning efficiency, engagement, and retention rates through tailored and adaptive educational technologies. Statistical evaluations demonstrated the medium to large performance gains facilitated by AI, with effect sizes ranging from 0.4 to 0.6 across diverse educational settings. Hybrid systems, such as Intelligent Tutoring Systems (ITS) integrated with adaptive learning platforms, consistently outperformed singular approaches, achieving up to 20% greater engagement and retention across varying demographics, Key findings include:

- **Personalization and Adaptivity**: AI-driven methods consistently enhance learning by addressing individual needs through personalized and adaptive instruction.
- **Gamification and Analytics**: These techniques boost motivation and facilitate career alignment, despite challenges related to privacy and overuse.
- **Scalability and Inclusion**: AI transforms traditional educational methods, offering scalable and impactful solutions for diverse learner populations. While AI has demonstrated significant potential, persistent gaps remain:
- **Ethical Challenges** Issues such as data privacy, equity, and inclusivity must be addressed to ensure responsible implementation.
- **Resource Constraints**: Many under-resourced educational institutions struggle to adopt advanced AI systems due to infrastructure limitations.

In conclusion, AI's integration into education marks a paradigm shift aligned with evolving technological and pedagogical needs. This study underscores AI's current capabilities while identifying pathways for enhancing its impact. Future efforts must prioritize ethical considerations, resource accessibility, and the refinement of hybrid approaches to ensure AI's transformative potential benefits learners across all contexts.

References:

1. Hemachandran, K., et al.: Artificial intelligence: A universal virtual tool to augment tutoring in higher education. Comput. Intell. Neurosci. **2022**(1), 1410448 (2022)
2. Lin, C.C., Huang, A.Y., Lu, O.H.: Artificial intelligence in intelligent tutoring systems toward sustainable education: a systematic review. Smart Learning Environments **10**(1), 41 (2023)
3. Ouyang, F., Zheng, L., Jiao, P.: Artificial intelligence in online higher education: A systematic review of empirical research from 2011 to 2020. Educ. Inf. Technol. **27**(6), 7893–7925 (2022)
4. Zhang, K., Aslan, A.B.: AI technologies for education: Recent research & future directions. Computers and Education: Artificial Intelligence **2**, 100025 (2021)
5. Pardosi, V.B.A., Xu, S., Umurohmi, U., Nurdiana, N., Sabur, F.: Implementation of an Artificial Intelligence Based Learning Management System for Adaptive Learning. al-fikrah: Jurnal Manajemen Pendidikan, **12**(1), 149–161 (2024)
6. Musso, M.F., Hernández, C.F.R., Cascallar, E.C.: Predicting key educational outcomes in academic trajectories: a machine-learning approach. High. Educ. **80**(5), 875–894 (2020)
7. AlHamad, M., Akour, I., Alshurideh, M., Al-Hamad, A., Kurdi, B.A.R.N., Alzoubi, H.: Predicting the intention to use google glass: A comparative approach using machine learning models and PLS-SEM. International Journal of Data and Network Science **5**(3), 311–320 (2021)
8. Zhai, X., et al.: A Review of Artificial Intelligence (AI) in Education from 2010 to 2020. Complexity **2021**(1), 8812542 (2021)
9. Helfat, C.E., Kaul, A., Ketchen, D.J., Jr., Barney, J.B., Chatain, O., Singh, H.: Renewing the resource-based view: New contexts, new concepts, and new methods. Strateg. Manag. J. **44**(6), 1357–1390 (2023)
10. Resiliency, A.: The NCSBN 2022 environmental scan: Resiliency, achievement, and public protection (2022)
11. Chambers, D.: Assistive technology supporting inclusive education: Existing and emerging trends. Assistive technology to support inclusive education, 1–16 (2020)
12. Akyuz, Y.: Effects of intelligent tutoring systems (ITS) on personalized learning (PL). Creative Education **11**(06), 953 (2020)

13. Deeva, G., Bogdanova, D., Serral, E., Snoeck, M., De Weerdt, J.: A review of automated feedback systems for learners: Classification framework, challenges and opportunities. Comput. Educ. **162**, 104094 (2021)
14. Kuleto, V., et al.: Exploring opportunities and challenges of artificial intelligence and machine learning in higher education institutions. Sustainability **13**(18), 10424 (2021)
15. Demartini, C.G., Sciascia, L., Bosso, A., Manuri, F.: Artificial Intelligence Bringing Improvements to Adaptive Learning in Education: A Case Study. Sustainability **16**(3), 1347 (2024)
16. Ngoc, B.M., Barrot, J.S.: Current landscape of English language teaching research in Southeast Asia: A bibliometric analysis. Asia Pac. Educ. Res. **32**(4), 517–529 (2023)
17. Chen, L., Chen, P., Lin, Z.: Artificial intelligence in education: A review. Ieee Access **8**, 75264–75278 (2020)
18. Yang, Q., Steinfeld, A., Rosé, C., Zimmerman, J.: Re-examining whether, why, and how human-AI interaction is uniquely difficult to design. In Proceedings of the 2020 chi conference on human factors in computing systems, pp. 1–13. (2020)
19. Achituve, I., Shamsian, A., Navon, A., Chechik, G., Fetaya, E.: Personalized federated learning with gaussian processes. Adv. Neural. Inf. Process. Syst. **34**, 8392–8406 (2021)
20. Liao, C.H., Wu, J.Y.: Deploying multimodal learning analytics models to explore the impact of digital distraction and peer learning on student performance. Comput. Educ. **190**, 104599 (2022)
21. Castellano, M.S., et al.: Empowering human anatomy education through gamification and artificial intelligence: An innovative approach to knowledge appropriation. Clin. Anat. **37**(1), 12–24 (2024)
22. Qadir, J.: Engineering education in the era of ChatGPT: Promise and pitfalls of generative AI for education. In 2023 IEEE Global Engineering Education Conference (EDUCON), pp. 1–9. IEEE (2023)
23. Ni, A., Cheung, A.: Understanding secondary students' continuance intention to adopt AI-powered intelligent tutoring system for English learning. Educ. Inf. Technol. **28**(3), 3191–3216 (2023)
24. Berendt, B., Littlejohn, A., Blakemore, M.: AI in education: Learner choice and fundamental rights. Learn. Media Technol. **45**(3), 312–324 (2020)

Emerging Trends in AI and Computational Technologies

Advanced Facial Recognition System: Integrating Gabor, Fourier, and Histogram of Oriented Gradients for Superior Feature Extraction

Ahmed Abdulateef Mohammed[1(✉)], Faiz Al-Alawy[2], and Hashem Bedr Jehlol[3]

[1] Mustansiriyah University, Baghdad, Palestine Street, Iraq
amohamm@uomustansiriyah.edu.iq
[2] Michigan, USA
falalaw@kent.edu
[3] Mustansiriyah University, Baghdad, Palestine Street, Iraq
hashemhbg@uomustansiriyah.edu.iq

Abstract. Face recognition is a crucial technology employed in various domains, such as surveillance, biometric systems, and computer vision. However, accurately identifying faces is challenging due to factors like changes in facial expressions, aging, occlusion, poor lighting conditions, and pose variations. This paper presents an approach that combines multiple feature extraction techniques for effective face recognition. The given face image undergoes preprocessing and is then subjected to three feature extraction methods: Fast Fourier Transform (FFT), Histogram of Oriented Gradients (HOG), and Gabor filters. First, the face image is transformed into the spectral domain using FFT, followed by the application of HOG in the spectral domain. Simultaneously, Gabor filters are applied to the original face image. The resulting histogram features are fused with the Gabor features, and the dimensionality is reduced to obtain a concise representation of facial features. Experimental results demonstrate the superior performance of the proposed method, especially in challenging conditions, when compared to recent methods, using reference images from ORL, CMU-PIE, AR, and Extended Yale-B databases.

Keywords: Face Recognition · Gabor Filters · Fourier Transform · Histogram of Oriented Gradients

1 Introduction

Face recognition (FR) is the process of identifying a person's identity by comparing their facial image with images stored in a database. It is a prominent research area within computer vision, benefitting from the advancements in biometric technology. While other biometric factors such as palm, fingerprint, voice, and iris are utilized in recognition systems, face recognition remains a popular and accessible choice (Kortli et al. 2020 [1]; Abate et al. 2007 [2]). Biologically-based recognition systems offer ease of use and are highly appealing (Bowyer et al. 2006] [3]; Kortli et al. 2020 [1]).

A. J. Obaid et al. (Eds.): ICONEST 2024, ISEM 40, pp. 185–204, 2025.
https://doi.org/10.1007/978-3-031-89960-7_13

Face recognition is socially acceptable and considered safer, as it does not require physical contact or collecting sensitive biometric data. It is unobtrusive, meaning it can be performed without physical contact or specialized equipment. Face recognition is cost-effective compared to techniques such as iris recognition. Therefore, it is preferred over other biometric techniques for these reasons (Abate et al. 2007 [2]); (Sikha & Bharath 2022) [4]. Despite significant advancements in face recognition technology, it remains a challenging task. The primary challenges arise from the inherent intra-class variations present in facial images, including variations in facial expressions, illumination conditions, head pose, and occlusions caused by factors such as beards, glasses, headscarves, and aging [Ayiah-Mensah et al. 2021 [5]; Farokhi et al. 2015 [6]; Zhang et al. 2022 [7]; Oloyede et al. Oloyede [8]). Verification and identification are two main types of face recognition tasks. Identification, on the other hand, entails comparing a facial image against a stored database to search for a match (1-to-many matching) (Ranade & Anand, 2021 [9]; Basaran et al., 2018 [10]). These tasks pose different challenges and require distinct approaches in face recognition systems. Face recognition systems consist of two essential components for verification and identification tasks: feature extraction and classification (Mohammed et al. 2017 [11]; Mohammed et al. 2021 [12]). The goal of the feature extraction phase is to precisely describe the face image by gathering the crucial data required for recognition process. In order to ascertain the identification of the person's identity, the retrieved features must be accurately categorized or matched. At least one of these two components need to be improved for a face recognition system to function better in overall. More accurate and discriminating representations of facial characteristics may result from developments in feature extraction techniques. Similar to feature extraction, improvement in the accuracy and efficiency of matching and identification tasks can be performed by achieving advancements in classification algorithms.

In this paper, we introduce a new face recognition approach that combines representative features taken from various techniques. Our technique transforms the facial image from the spatial domain to the frequency domain using Fast Fourier Transform (FFT). The Histogram of Oriented Gradients is used in the frequency domain for the face image and this leads to an FFT-HOG feature vector. Also, another feature vector is extracted by applying Gabor filters on the original image which results into Gabor feature vector. These two vectors are then combined for creating one final representative feature vector. The rest of this paper is arranged as follows: Sect. 2 gives background information about major techniques used in extracting features. In Sect. 3, we outline our method of proposing face recognition system. Experimental results are provided in Sect. 5 while our findings are concluded at Sect. 6.

2 Background

The effectiveness of a face recognition system relies heavily on two crucial steps: feature extraction and classification. In the realm of facial feature extraction, numerous methods have emerged, and among them, the following techniques are commonly utilized: Fourier Transform (FT), Histogram of Oriented Gradients (HOG), and Gabor filters. These methods play a vital role in capturing the distinctive characteristics of faces, enabling accurate identification and classification within a face recognition system.

2.1 Fourier Transform (FT)

Fourier Transform (FT) (Boashash 2015 [13]) is a common image processing tool to convert images from spatial domain to frequency domain (Acharya et al. 2020 [14]). In most applications, the fast Fourier transform (FFT) is the commonly used algorithm for calculating the Fourier transform via calculating the sequence of Discrete Fourier Transform (DFT). The difference between DFT and FFT is that FFT is faster than DFT making it preferable in practical applications. The following equation is used to transform the 2D image in spatial domain to frequency domain (Antonijevic et al. 2020 [15]):

$$FIMG_{(i,j)} = \frac{1}{\mathcal{M}\mathcal{N}} \sum_{\dot{x}=0}^{\mathcal{M}-1} \sum_{\dot{y}=0}^{\mathcal{N}-1} SPIMG_{(\dot{x},\dot{y})}^{\exp\left[-j2\pi\left(\frac{i\dot{x}}{\mathcal{M}}-\frac{j\dot{y}}{\mathcal{N}}\right)\right]} \tag{1}$$

where, $0 \leq i < \mathcal{M}$ and $0 \leq j < \mathcal{N}$ represents the frequency variable, SPIMG is a $\mathcal{M} \times \mathcal{N}$ image in spatial domain and FIMG is the resulted image in frequency domain.

$$FIMG_{(i,j)} = Real_{(i,j)} + Imag_{(i,j)} \tag{2}$$

where, $Real_{(i,j)}$ and $Imag_{(}i,j)$ represent the real and imaginary parts of $FIMG_{(i,j)}$, respectively.

There are benefits that can be gained via converting the image using FFT among them are FFT decreases the computational burden of the image (Glynn 2007 [16]; Ayiah-Mensah et al. 2021 [5]) and application of FFT to image can improve the recognition performance of the system (Asiedu et al. 2015 [17]; Zhang et al. 2015 [18]) and most of the important information required to represent the images are concentrated in the low frequency components of the FFT image which are located at the four corners of the spectral image, while the remaining area at the middle of the spectral image which represent the high frequency components of the FFT image are the fine details and represents the noise of the image and contribute less for recognition process (Deepa et al. 2012 [19]). The spectral image can be retrieved frequency domain back to spatial domain via applying of Inverse Fast Fourier Transform (IFFT):

$$SPIMG_{(i,j)} = \frac{1}{\mathcal{M}\mathcal{N}} \sum_{\dot{x}=0}^{\mathcal{M}-1} \sum_{\dot{y}=0}^{\mathcal{N}-1} SPIMG_{(\dot{x},\dot{y})}^{exp[j2\pi(\frac{i\dot{x}}{\mathcal{M}}+\frac{j\dot{y}}{\mathcal{N}})]} \tag{3}$$

2.2 Histogram of Oriented Gradients

The histogram of Oriented Gradients (HOG) (Dalal & Triggs 2005 [20, 21]) works by dividing the face image grids known as cells. Within each cell, the gradients in both directions (vertical and horizontal) are calculated for every pixel in the cell. In this context, the gradient represents the minor change in vertical and horizontal directions. Then a specified filter masks are applied to the pixel values at coordinate points (i,j) to get gradients in x and y directions as $Gx(i, j)$ and $Gx(i, j)$; respectively, as the following: -

$$Gx(i, j) = G(i + 1, j) - G(i - 1, j) \tag{4}$$

$$Gy(i, j) = G(i, j + 1) - G(i, j - 1) \tag{5}$$

The gradient magnitude is given as the following:

$$MG(x, y) = \sqrt{Gx(i, j)^2 + Gy(i, j)^2} \tag{6}$$

The gradient direction is given as the following: -

$$\theta(x, y) = \arctan\left(\frac{GY(i, j)}{Gx(i, j)}\right) \tag{7}$$

The gradient magnitude (Eq. 6) and gradient direction (Eq. 7) are used to create the histogram. The orientation is disseminated over the angle (0, π) and assigned to one out of nine bins. In turn, the gradients magnitudes are heaped up and assigned to the specified angular bins. Assuming 'N' is the number of bins and the gradient direction 'θ(x,y)' will be corresponded to a respective bin 'P'. The deviation percentage from the bin center'd is calculated as the following:-

$$D = \left(\frac{\theta(x, y) - \frac{p\pi}{N}}{\frac{\pi}{N}}\right) = \frac{N\theta(x, y)}{\pi} - P \tag{8}$$

The final process is forming a number of cells into one block. The histograms of the cells are then combined in each specified block. The HOG features are obtained by concatenating the histograms of all blocks. Illumination normalization is then performed for each block to achieve the illumination invariance.

2.3 Gabor Filters

Gabor filter, is a linear filter used in many image processing applications such as edge detection, texture analysis and feature extraction (Rai & Rivas 2020 [22]). Gabor filter is a band pass filtering, it can extract the patterns at certain scales and orientations (Luqman 2021 [23]). Gabor filters have several significant properties represented by invariance to some FR task like noise and illumination variation (Jaber 2022 [24]).

In general, for 2 D images, utilizing orientation u and scales v, the Gabor filters can be defined as:

$$\omega_{(u,v)}(P) = \frac{\left\|K_{(u,v)}\right\|^2}{\sigma^2} \exp\left(-\frac{\left\|K_{(u,v)}\right\| \|P\|^2}{2\sigma^2}\right)\left[\exp(iK_{(u,v)}P^{\exp\left(-\frac{\sigma^2}{2}\right)}\right] \tag{9}$$

where, P represents the intensity pixel value at (x: horizontal) and (y: vertical) coordinates. $K_{(u,v)} = K_v \exp(i\emptyset_u)$ is a wave vector having the magnitude ($\left(K_v = \frac{K_{max}}{\mathcal{F}^v}\right)$, where $\mathcal{F}$ is the scaling factor (frequency ratio) between the kernel filters, $Kmax$ refers to the maximum frequency, $\emptyset_u = \frac{\pi}{8} * u$ and ||.|| is the norm vector. The convolution of an image with Gabor filters is as the following:

$$GI = IM * \omega_{(u,v)}(P) \tag{10}$$

3 The Proposed Feature Extraction Method

In this paper novel method that integrates Fourier Transformation (FFT), Histograms of Oriented Gradients (HOG) and a Gabor approach was presented to extract salient and discriminative characteristics capturing the attributes of facial images faithfully. These techniques have proven to be effective in various face recognition applications and have been widely adopted in the field.

The feature extraction stage is a fundamental component of the proposed face recognition system. The idea behind face recognition is to come up with descriptors that may describe the intrinsic characteristics of facial appearance, i.e., what makes one face different from another. This paper is devoted to present a new model based on Fourier Transform (FFT) and Histogram of Oriented Gradients (HOG). The FFT decomposes the face image into frequency components so that important facial details and structures can be extracted, which, in turn, contribute to the overall facial representation. HOG analysis would allow identification of eyes, nose, and mouth. This technique yields high-quality distinctive representations of facial features that enable for precise face recognition. Besides, the Gabor filters are another commonly employed technique in face recognition as they extract local texture features along with global facial structures. The face images are convolved with Gabor filters at various scales and orientations, extracting textures and structural information important for distinguishing different individuals.

The system proposed for face recognition involves three techniques that are integrated together which include FFT, HOG as well as Gabor filters leading to a complete and descriptive feature set. As a result, the system aims at integrating frequency domain information, local gradients and texture details such as that it can be able to give an improved representation of a face has more discriminative feature. The basic design of the suggested face recognition system is presented in Fig. 1 of the paper. It shows how the FFT, HOG and Gabor's techniques become part of the overall framework by indicating how they flow into each other. It gives an image-based explanation on how different parts work together to form features that can be used in the extraction stage of features.

Our proposed face recognition (FR) system consists of three stages. Firstly, the face image is processed to enhance its suitability for feature extraction. Then, useful features are extracted from the facial image using techniques such as 'FFT', 'HOG', and 'Gabor'. Lastly, the extracted features are fed into the classification stage, where facial features for each person are classified separately using a set of classification algorithms. In the subsequent subsections, we will delve into the details of each stage in the proposed FR system.

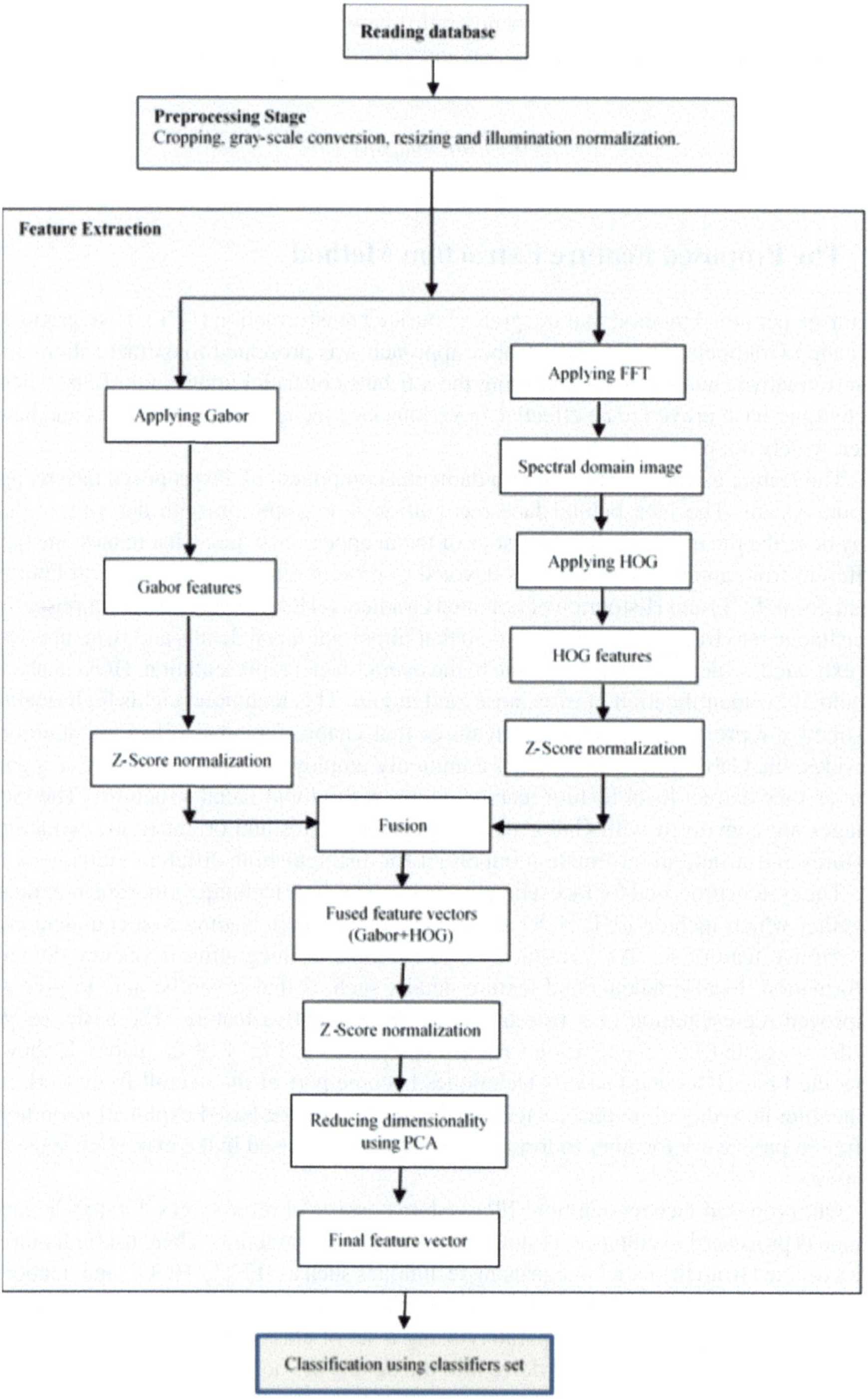

Fig. 1. The proposed face recognition system.

3.1 Preprocessing Stage

When working with face images, it is crucial to prepare them before extracting useful features. This is because they may include non-facial elements like the background, be poorly lit, or affected by other factors. Several preprocessing steps are applied at this stage to solve these problems. The first step involves cropping the image in a way that eliminates non-facial parts and leaves only necessary facial details for the purpose of feature extraction. Besides, since face images can be large, their size is reduced to simplify computations in FR phases that follow later on. An example is that ORL database images are resized to 56×56 pixels, and AR, CMU-PIE, and extended Yale-B database images are resized to 128×128 pixels. In addition to reducing processing time, resizing the images also helps optimize storage space. Furthermore, the resized images are converted from color to gray-scale. Gray-scale images are less sensitive to changes in lighting conditions compared to color images. Additionally, using a unified gray-scale format for all face images in the system simplifies the face recognition task for both colored and non-colored images. While images from the ORL and extended Yale-B databases are already gray-scale and do not require conversion, images from CMU-PIE and AR databases, which are in color, are converted to gray-scale.

Finally, illumination normalization is performed to mitigate the impact of lighting variations. The Difference of Gaussian (DoG) method is employed for illumination normalization in this work. By applying this method, the effects of uneven lighting conditions are minimized, allowing for more robust and consistent feature extraction. By conducting these pre-processing steps, the proposed system addresses common challenges in face images, such as non-facial parts, image size, color variations, and lighting conditions, ensuring that the subsequent feature extraction and classification stages can be performed effectively.

$$DoG_{(i,j)} = Img_{(i,j)} * \vartheta_{(i,j,\sigma 1)} - \vartheta_{(i,j,\sigma 2)} \tag{11}$$

where, $\vartheta_{(.)}$ refers to the function of Gaussian and σ1 and σ2 are the Gaussian kernels.

The effect of applying the preprocessing stage to the face image is shown in Fig. 2.

3.2 Feature Extraction

In feature extraction techniques of face recognition, they fall under two main categories; spectral and spatial. Transforming the facial image from the spatial domain to the spectral domain is done in the spectral techniques. Then representative facial features are extracted from this image which is already in frequency domain. Examples of such techniques include Fourier transform (FT), Wavelet Transform (WT), Gabor filters, and Discrete Cosine Transform (DCT). On other hand spatial techniques represent extracting representative facial features directly from the facial image without any transformation. Examples of such methods include Local Binary Pattern (LBP), Histogram of Oriented Gradients (HOG) and Scale Invariant Feature Transform (SIFT). The time domain is known as the Spatial Domain where images are represented and analyzed based on spatial coordinates, while Spectral Domain represents images in both time (spatial) and frequency domains. Combining these approaches, we employ a combination of the HOG

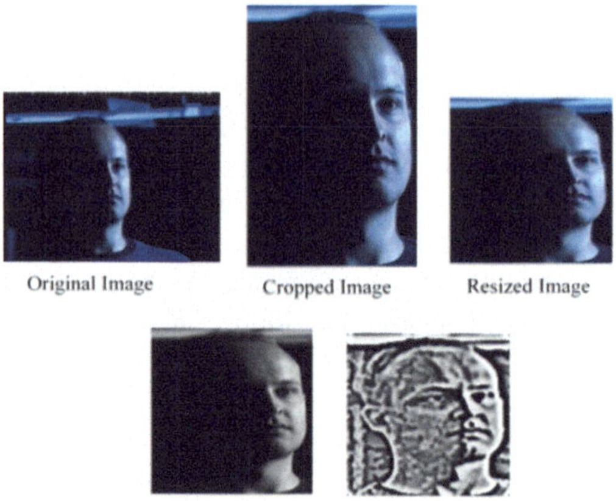

Fig. 2. The impact of preprocessing steps.

technique from the spatial domain and FFT as well as Gabor Filters from the spectrums for extracting discriminative facial characteristics. By using Fast Fourier Transform (FFT) on preprocessed face image, it disperses image information into low frequency components and high frequency components.

The majority of the image information is concentrated in the low-frequency components, which are essential for representing the image. The high-frequency components contain finer details that contribute less to the recognition process. In the proposed method, FFT is applied to the preprocessed image to transform it into the frequency domain. The most valuable image information in the frequency domain is localized at the corners of the spectral image. Figure 3 illustrates the resulting image after converting it to the spectral domain using FFT. By utilizing a combination of spatial and spectral domain techniques, the proposed method aims to capture both global and local facial features, enhancing the overall effectiveness of the face recognition system.

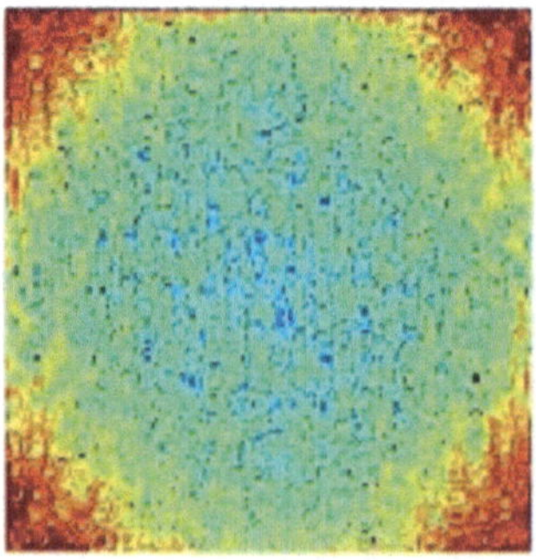

Fig. 3. Depicts the face image after being transformed using FFT.

Following the transformation, the low-frequency components of the spectral image are relocated to the center point. This adjustment is achieved by swapping the first quadrant with the third quadrant, and the second quadrant with the fourth quadrant. Figure 4 provides a visual representation of this process.

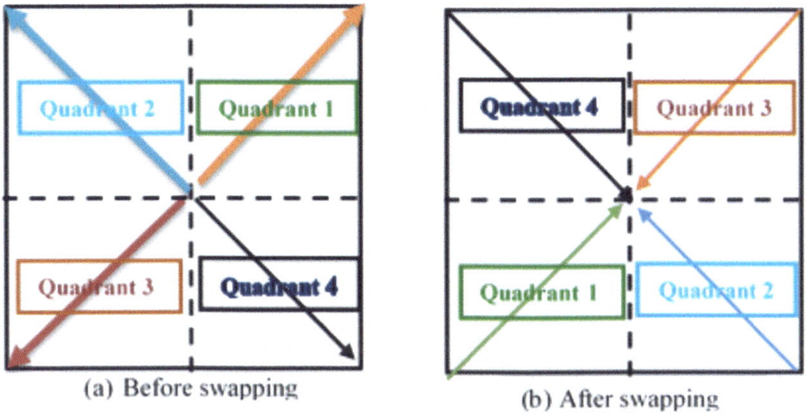

Fig. 4. The pictorial representation of quadrants swapping.

After applying the aforementioned procedure, the resulting spectrum is illustrated in Fig. 5. Once the low-frequency components have been centered in the spectral image (as depicted in Fig. 5), it is necessary to extract the relevant FFT coefficients to obtain the optimal features. To accomplish this, the central portion of the spectral image needs to be isolated. This can be achieved by using a rectangular mask positioned in the middle area of the spectrum. The box labeled as (a, b, c, and d) in Fig. 6 illustrates this concept. The coordinates of the mask, denoted as (a, b, c, and d), can be determined using the following equations:

$$a = \frac{(width - mask_{width})}{2} \tag{12}$$

$$b = \frac{(height - mask_{height})}{2} \tag{13}$$

$$c = a + mask\_width \tag{14}$$

$$d = b + mask\_height \tag{15}$$

These equations determine the position and dimensions of the rectangular mask within the spectrum, ensuring that the desired central region is selected for further processing.

$$a = \left(\frac{M}{4}, \frac{N}{4}\right) \tag{16}$$

$$b = \left(\frac{M}{4}, \frac{3N}{4}\right) \tag{17}$$

$$c = \left(\frac{3M}{4}, \frac{N}{4}\right) \tag{18}$$

$$d = \left(\frac{3M}{4}, \frac{3N}{4}\right) \tag{19}$$

where, M and N refer to the dimensions of the image.

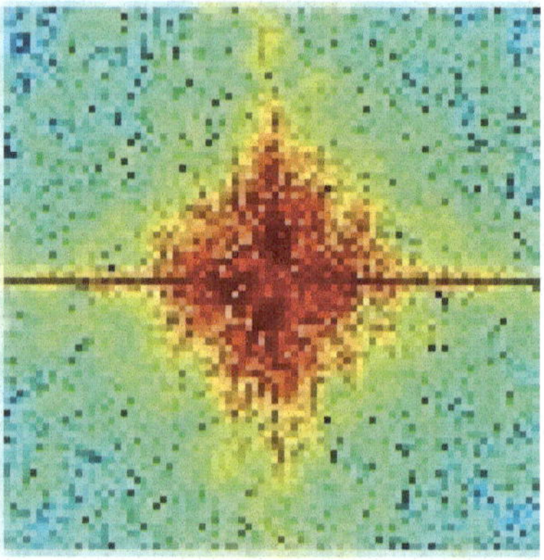

Fig. 5. The four quadrants centered at the middle.

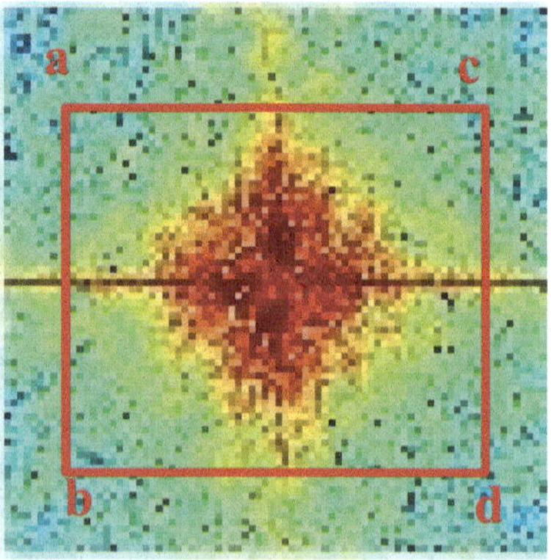

Fig. 6. Positioning of the rectangular mask at the middle of the spectrum.

The region of the spectrum covered by the mask is utilized to extract the crucial features. This concept has been adapted from process (Deepa et al. 2012 [19]), where the DCT was applied to the spectrum image obtained from FFT for feature extraction. However, in our work, we substitute the DCT technique with HOG and Gabor filters for extracting salient features. In our approach, the salient features are extracted from the spectrum mask using the determined coordinates (a, b, c, and d) through the application of the HOG technique. The HOG parameters used in this research are as follows: the gradient filter values are [−1, 0, 1], no smoothing is applied to the gradient filters, the cell size is 8 × 8 pixels, the block size is 16 × 16 pixels, the angle ranges from 0° to 180°, there are 9 orientation bins, σ is set to 0.5 times the block width, and block normalization is performed using the L2 norm. In this proposed method, only one-fourth of the original spectrum area is considered for feature extraction, while neglecting the remaining three-fourths of the spectral image. This significantly reduces the number of FFT coefficients considered, resulting in an optimal number of features. The HOG feature vector is denoted as (HOG_{feat}), which is then normalized using the z-score normalization technique, as shown in the following equation:

$$\boldsymbol{HOG_{norm}} = \frac{\boldsymbol{HOG_{feat}} - \boldsymbol{\mu_{HOG}}}{\boldsymbol{\sigma_{HOG}}} \tag{20}$$

where, HOG_{norm} refers to the normalized HOG feature vector, HOG_{feat} is the HOG feature vector before normalization process and μ_{HOG} and σ_{HOG} are the mean and standard deviation of the HOG_{feat}.

Moreover, Gabor filters are then directly applied to the preprocessed facial image using Eqs. (9 and 10). The common Gabor filters parameters represented by five scales u ∈ {0, 1, 2, 3, 4} and eight orientations, v ∈ {0, 1, 2, 3, 4, 5, 6, 7} have been utilized.

The real part and imaginary part of Gabor filters constructed from 8 orientations and 5 scales are shown in Fig. 7.

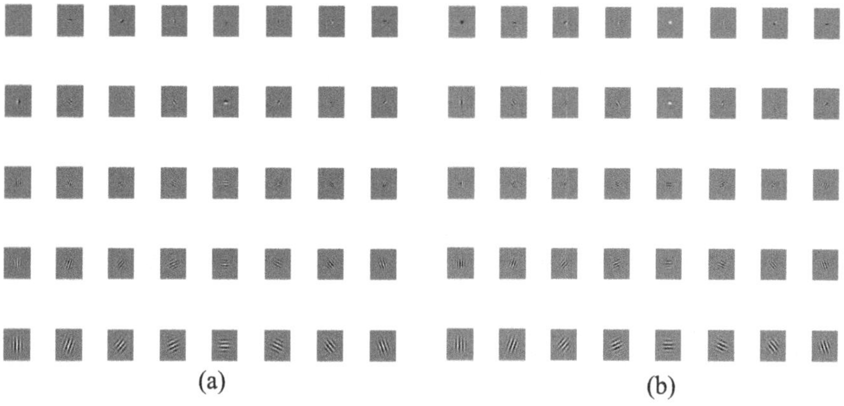

Fig. 7. The Gabor filters with 5 scales and 8 orientations: (a) Real parts and (b) Imaginary parts.

The magnitude produced by the convolution of the face image with the 40 Gabor filters and downsampling factor 64 is shown in Fig. 8.

Here, the extracted Gabor feature vector denoted as (G_{feat}). Likewise (HOG_{feat}), (G_{feat}) is also normalized using z-score normalization technique as the following equation:

$$G_{norm} = (G_{feat} - \mu_G/\sigma_G) \tag{21}$$

Where, G_{norm} refers to the normalized Gabor feature vector, G_{feat} is the Gabor feature vector before normalization process and μ_G and σ_G are the mean and standard deviation of the Gabor features.

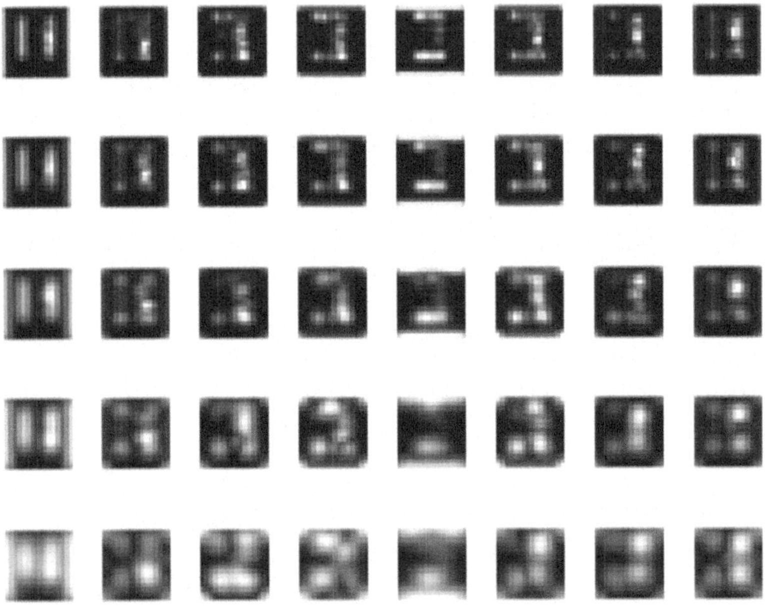

Fig. 8. The magnitude produced by convolving the face image with 40 Gabor filters and downsampling factor 64.

The two feature vectors (HOG_{norm} and G_{norm}) are then fused into one final feature vector as the following:

$$IMG = HOG_{norm} + G_{norm} \tag{22}$$

So, each particular image in the database is represented by its own image vector (IMG) Feature vector. The fused feature vector is then normalized using the z-score normalization technique.

Finally, the feature vector (IMG) for the given image is then reduced into the range (25 to 200) features using PCA (Turk & Pentland 1991 [25]) dimensionality reduction algorithm.

4 Experiments

In this part, the FR system is put under rigorous testing conditions with regard to its performance. Some of these conditions are characterized by drastic changes in illumination, varying expressions, pose rotations, occlusions, and aging effects. We will assess the system's performance through experiments carried out with the Olivetti At&T face database (ORL) (Samaria & Harter 1994 [26]), AR database (Martinez & Benavente, 1998 [27]), CMU-PIE database (Sim et al. 2002 [28]), and extended Yale-B database (Lee et al. 2005 [29]). In addition, we will use a number of classification algorithms that include Naïve Bayes (Berrar 2018 [30]), Support Vector Machine (SVM) (Hsu 2003 [31]), K-Nearest Neighbor (K-NN) (Boiman 2008 [32]), Multilayer Perceptron (Singh & Banerjee 2019

[33]), Hoeffding Tree (Pham 2017 [34], and C4.5 (Kotsiantis 2007 [35]) to test our system's performance and provide us with an informed opinion on whether or not our model meets the quality standards expected from it. In order to examine the functioning of a system concerning any representation that can be produced regarding facial images and is based on PCA, we explore a considerable amount of subspace features. The subspace size, in this case, is changed (i.e., the number of features) and it ranges from 25 to 200 features with an increment of 25 features at each step. The representative facial features are extracted using MATLAB R2021a, and the classification rates (CR) are obtained using Weka 3.7.12. Our experiments were conducted on a system equipped with a Core i7 processor, 16 GB of RAM, and operating on Windows 10 (64-bit). The 10-fold cross-validation strategy is employed to obtain reliable results. In the following subsections, we present the results obtained from each database, showcasing the performance of the proposed FR system under different conditions.

4.1 Results from ORL Database

The ORL database consists of 400 images, with 10 images for each of the 40 individuals included in the database. These photos were captured at the Olivetti Research Laboratory (ORL) in Cambridge University during the period from April 1992 to April 1994. The database encompasses variations in facial attributes, such as the presence or absence of glasses, beard or no beard, occlusion (occluded or unoccluded), expression (smiling or not smiling), and eye state (open or closed). Additionally, variations in lighting conditions, camera angles (including upright and tilting angles of approximately 20° for face rotation), and scale range up to 10° are also included. For our experiments, the entire ORL database is utilized. The recognition rates obtained from the experiments conducted on the ORL database are presented in Fig. 9.

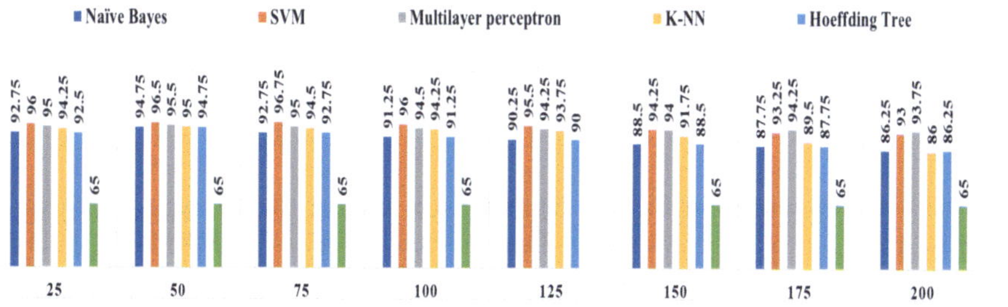

Fig. 9. Recognition rates for ORL database using different classification algorithms and different number of features (20 to 200).

The results presented in Fig. 9 demonstrate that among the six types of classifiers utilized in this experiment, the SVM classifiers exhibit the highest classification accuracy (96.75%). Upon closer examination of the figure, it becomes evident that both the SVM and Perceptron classifiers outperform the other four types when applied to the ORL benchmark image data. It is worth noting that increasing the number of extracted features (ranging from 20 to 200) as input to these specific classifiers does not consistently lead to improved classification rates. In fact, at times, it yields the opposite effect. Hence, it is not always advisable to utilize a higher number of classification features in an attempt to enhance the performance of the facial recognition (FR) system.

4.2 Results from CMU-PIE Database

The CMU-Pose, Illumination, and Expression (PIE) database is a large collection of photos that encompasses images of 68 individuals captured under various conditions. The database consists of a total of 41,368 color images stored in JPEG file format. It covers a wide range of variations in pose, illumination, and expressions. Regarding pose variations, the database includes 13 different poses, capturing individuals from frontal view to profile view at 90-degree angles. The illumination variations are achieved by utilizing 21 different flashes with variations in background lighting (on and off). The database also includes four facial expressions: neutral, talking, blinking, and smiling. For our experiments, we selected a subset of 950 photos from the CMU-PIE database. This subset includes 25 images for each of the 38 individuals. The selected images exhibit variations in pose, ranging from frontal view to profile view at 90°, as well as severe lighting conditions, including close-to-dark view photos. The recognition rates obtained from the experiments conducted on the CMU-PIE database are depicted in Fig. 10.

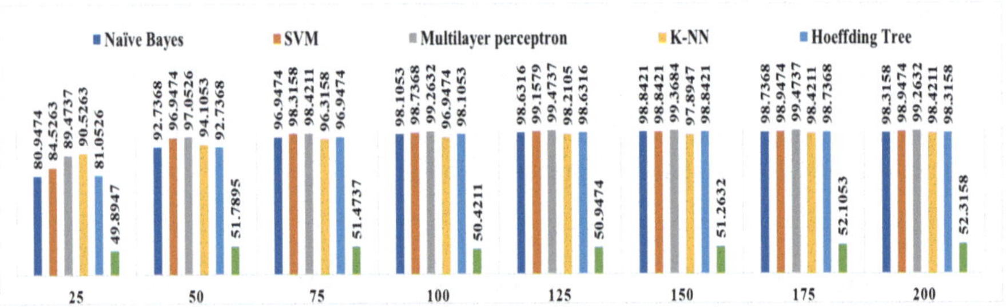

Fig. 10. Recognition rates for CMU-PIE database using different classification algorithms and different number of features (20 to 200).

The results depicted in Fig. 10 highlight that among the six types of classifiers employed in this experiment, both the Perceptron classifier exhibit the highest classification accuracy (99.4737%). Upon closer analysis of the figure, it becomes evident that both the Perceptron and SVM classifiers outperform the other four types when applied to the CMU-PIE benchmark image data. It is important to note that increasing the number of extracted features (ranging from 20 to 200) as input to these specific classifiers has consistently led to improved classification rates. In fact, in certain cases and for specific types of image data to be classified, augmenting the number of utilized features can enhance the system performance. Therefore, it is advisable to thoroughly test the system with a higher number of classification features in order to optimize the performance of the facial recognition (FR) system.

4.3 Results from AR Database

The AR database was created at the Computer Vision Center (CVC) at the Ohio State University and involves the participation of 126 individuals. The database contains over 4000 color images. Each person was photographed in frontal view during two sessions with a two-week interval. Similar to the ORL and CMU-PIE databases, the photos in the AR database encompass variations in illumination conditions, facial expression, and occlusion. The variations in facial expression include smiling, anger face, and neutral face. Illumination variations encompass right light on, left light on, and all sides lights on. Occlusion variations involve blocking either the upper part of the face (eyes and vicinity) by wearing sunglasses or the lower part of the face (mouth and chin) by wearing a headscarf. For our experiments on the AR database, we selected a subset of 80 individuals, consisting of 40 women and 40 men. Each individual in the subset has 26 images that cover all the aforementioned facial conditions. Thus, a total of 2080 images were used in this study. The recognition rates obtained from the experiments conducted on the AR database are presented in Fig. 11.

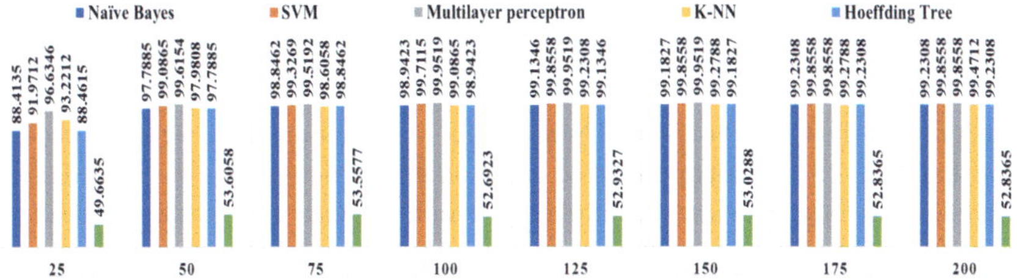

Fig. 11. Recognition rates for AR database using different classification algorithms and different number of features (20 to 200).

The presented results in Fig. 11 clearly demonstrate that among the six types of classifiers utilized in this experiment, the Perceptron classifiers achieve the highest classification accuracy (99.9519%). Upon careful examination of the figure, it becomes evident that the Perceptron classifiers outperform the other types when applied to the AR benchmark image data. It is noteworthy that increasing the number of extracted features (ranging from 20 to 200) as input to these specific classifiers consistently results in improved classification rates. In fact, for certain cases and specific types of image data to be classified, augmenting the number of utilized features can significantly enhance the system's performance. Hence, it is highly recommended to thoroughly evaluate the system with a higher number of classification features in order to optimize the overall performance of the facial recognition (FR) system.

4.4 Results from Extended Yale-B Database

Extended Yale-B database contains photos for 38 subjects photographed under pose and 64 illumination changes, many of them under sever conditions (dark). The frontal images were cropped to unified size (168 × 192) pixels. Each person has an image with particular pose, background and ambient illumination. In addition, some subjects have few photos labelled with.bad extension. So, the total number of images included in this database is 2434. In our experiments, we deleted all images with.bad extension and pose (background) images. So, the remaining 2344 images are considered. The recognition rates from Extended Yale-B database are shown in Fig. 12.

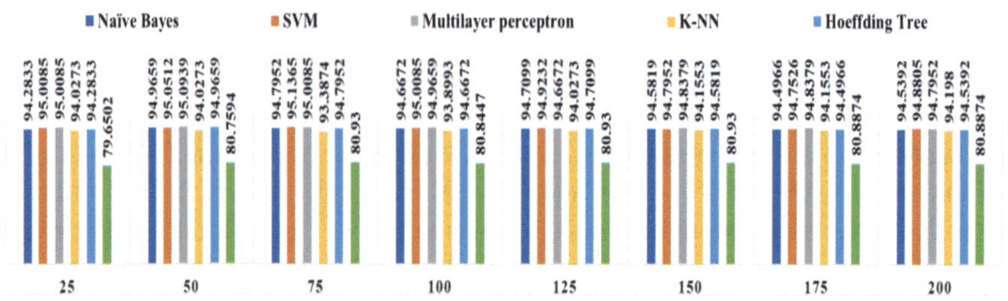

Fig. 12. Recognition rates for Extended Yale-B database using different classification algorithms and different number of features (20 to 200).

The results presented in Fig. 12 indicate that most of the classifier types utilized in this experiment have achieved a similar high recognition accuracy of around 95%, with the exception of the C4.5 classifiers. Upon careful examination of the figure, it becomes evident that the C4.5 classifier performed better on this particular type of image database compared to the three databases mentioned above. It is worth noting that increasing the number of extracted features (ranging from 20 to 200) as input to these specific classifiers did not consistently result in significant improvements in classification rates. In fact, the impact of increasing the number of features on the system's performance varied depending on the specific type of image data being classified. While it enhanced the system's performance for certain cases and specific types of image data, it did not have a meaningful effect for other types. Therefore, it is highly recommended to thoroughly evaluate the system using a higher number of classification features in order to optimize the overall performance of the facial recognition (FR) system, considering the specific characteristics of the image data.

4.5 Comparison with Other FR Methods

Table 1 presents a comparison between the results obtained by our method and the state-of-the-art methods. The state-of-the-art methods refer to recent works conducted by (Chandrakala & Devi 2020 [36]; Jianxin & Junyoung 2021 [37]; Chen et al. 2021, Antonijeric et al. 2022 [39], Hamdan & Mokhtar 2021 [15], Mohammed & Al-Alawy, 2021) [40], Gu et al. 2017 [42] and Ahmed et al. 2021 [42]). Importantly, the comparison between our method and the other methods is based on the same databases used in the comparative studies.

Upon reviewing the results presented in Table 1, it is evident that our method outperforms the other comparative methods. Our method consistently achieves superior results compared to the outcomes reported by the other methods, providing stronger performance across the same databases used in the comparative studies.

Table 1. Comparison with the state-of-the art methods.

Method	Database used	Recognition rate
(Chandrakala & Devi 2021 [36])	ORL	95.2
(Zhang et al. 2020 [37])	ORL/AR	92.0/ 92.0
(Ting Chen et al. 2021 [38])	CMU-PIE	94.2
(Milos Antonijevic et al. 2022 [39])	ORL	91.45
(Bensenane et al. 2018 [15])	ORL	85.2
(Ahmed & Faiz 2021 [40])	ORL/AR/CMU-PIE	96.25/99.8558/97.2632
(Jianquan et al. 2017 [41])	AR/Yale-B/CMU-PIE	96.88/ 94.13/ 93.63
(Ahmed, et al. 2021 [42])	ORL	85.25%
Proposed method	ORL/CMU-PIE/AR/Extended Yale-B	96.75/99.4737/99.9519/95.1365

5 Conclusion

In this paper, we have proposed a novel method for feature extraction in a face recognition system. Our method combines three techniques, namely FFT, HOG, and Gabor filters, which have been tested with various well-known classifiers. We evaluated the performance of our proposed system in terms of recognition rate using four standard databases: ORL, CMU-PIE, AR, and Extended Yale-B. These databases encompass different challenging conditions such as severe illumination variations, facial expressions, occlusions, age progression, and pose rotations. The experimental results have demonstrated a significant improvement in the overall performance of our face recognition system. We tested a range of 25–200 features to determine the optimal number of features for extraction, ensuring the highest recognition accuracy. Our proposed system outperformed other recently published face recognition methods when evaluated on the same reference databases.

This study contributes to the advancement of face recognition technology by introducing a novel feature extraction method that effectively handles various challenging conditions. The improved performance of our system highlights its potential for real-world applications in biometrics, surveillance, and security systems. Future research can explore further optimizations and extensions to enhance the robustness and scalability of the proposed method.

Acknowledgment. The authors would like to thank Mustansiriyah Univeristy (www.uomustansiriyah.edu.iq) Baghdad-Iraq for its support in the present work.

References

1. Kortli, Y., Jridi, M., Al Falou, A., Atri, M.: Face recognition systems: a survey. Sensors **20**(2), 342 (2020)
2. Abate, A.F., Nappi, M., Riccio, D., Sabatino, G.: 2D and 3D face recognition: a survey. Pattern Recogn. Lett. **28**(14), 1885–1906 (2007)
3. Bowyer, K.W., Chang, K., Flynn, P.: A survey of approaches and challenges in 3D and multi-modal 3D+ 2D face recognition. Comput. Vis. Image Underst. **101**(1), 1–15 (2006)
4. Sikha, O.K., Bharath, B.: VGG16-random Fourier hybrid model for masked face recognition. Soft. Comput. **26**(22), 12795–12810 (2022)
5. Ayiah-Mensah, F., Asiedu, L., Mettle, F.O., Minkah, R.: Recognition of augmented frontal face images using FFT-PCA/SVD algorithm. Appl. Comput. Intell. Soft Comput. 1–9 (2021)
6. Farokhi, S., Sheikh, U.U., Flusser, J., Yang, B.: Near infrared face recognition using Zernike moments and Hermite kernels. Inf. Sci. **316**, 234–245 (2015)
7. Zhang, J., Yan, X., Cheng, Z., Shen, X.: A face recognition algorithm based on feature fusion. Concurrency Comput. Pract. Exper. **34**(14), e5748 (2022)
8. Oloyede, M.O., Hancke, G.P., Myburgh, H.C.: Improving face recognition systems using a new image enhancement technique, hybrid features and the convolutional neural network. IEEE Access **6**, 75181–75191 (2018)
9. Ranade, S.K., Anand, S.: Color face recognition using normalized-discriminant hybrid color space and quaternion moment vector features. Multimedia Tools Appl. **80**, 10797–10820 (2021)
10. Basaran, E., Gökmen, M., Kamasak, M.E.: An efficient multiscale scheme using local Zernike moments for face recognition. Appl. Sci. **8**(5), 827 (2018)
11. Mohammed, A.A., Xiamixiding, R., Sajjanhar, A., Chen, J., Nasierding, G.: Texture features for clustering based multi-label classification of face images. In: 2017 10th International Congress on Image and Signal Processing. BioMedical Engineering and Informatics (CISP-BMEI), pp. 1–5. IEEE, New York (2017)
12. Mohammed, A.A., Al-Alawy, F.: A comparative study of DWT-HOG based face recognition with other similar techniques. In: 2021 31st International Conference on Computer Theory and Applications (ICCTA), pp. 193–199. IEEE, New York (2021)
13. Boashash, B.: Time-frequency signal analysis and processing: A comprehensive reference. Academic Press (2015)
14. Acharya, D., Billimoria, A., Srivastava, N., Goel, S., Bhardwaj, A.: Emotion recognition using Fourier transform and genetic programming. Appl. Acoust. **164**, 107260 (2020)
15. Antonijevic, M., Strumberger, I., Lazarevic, S., Bacanin, N., Mladenovic, D., Jovanovic, D.: Robust encrypted face recognition robot based on bit slicing and Fourier transform for cloud environments. J. Electron. Imaging **31**(6), 061808 (2022)
16. Glynn, E.: Fourier analysis and image processing scientific programmer. Bioinformatics (2007)
17. Asiedu, L., Adebanji, A., Oduro, F., Mettle, F.O.: Statistical evaluation of face recognition techniques under variable environmental constraints (2015)
18. Zhang, D., Ding, D., Li, J., Liu, Q.: PCA based extracting feature using fast fourier transform for facial expression recognition. In: Yang, GC., Ao, SI., Huang, X., Castillo, O. (eds.) Transactions on Engineering Technologies, pp. 413–424. Springer, Dordrecht (2015). https://doi.org/10.1007/978-94-017-9588-3_31

19. Deepa, G.M., Keerthi, R., Meghana, N., Manikantan, K.: Face recognition using spectrum-based feature extraction. Appl. Soft Comput. **12**(9), 2913–2923 (2012)
20. Dalal, N., Triggs, B.: Histograms of oriented gradients for human detection. In: 2005 IEEE Computer Society Conference on Computer Vision and Pattern Recognition (CVPR'05), vol. 1, pp. 886–893. IEEE, New York (2005)
21. Nguyen, D.T., Bui, T.H., Lee, S.J.: Robust face recognition using deep learning and adaptive feature selection. Pattern Recogn. **119**, 108079 (2021)
22. Rai, M., Rivas, P.: A review of convolutional neural networks and Gabor filters in object recognition. In: 2020 International Conference on Computational Science and Computational Intelligence (CSCI), pp. 1560–1567. IEEE (2020)
23. Luqman, H., El-Alfy, E.S.M., BinMakhashen, G.M.: Joint space representation and recognition of sign language fingerspelling using Gabor filter and convolutional neural network. Multimedia Tools Appl. **80**, 10213–10234 (2021)
24. Jaber, A.G., Muniyandi, R.C., Usman, O.L., Singh, H.K.R.: A hybrid method of enhancing accuracy of facial recognition system using Gabor filter and stacked sparse autoencoders deep neural network. Appl. Sci. **12**(21), 11052 (2022)
25. Turk, M., Pentland, A.: Eigen faces for recognition. J. Cogn. Neurosci. **3**, 71–86 (1991)
26. Samaria, F., Harter, A.: Parameterisation of a stochastic model for human face identification. In: Proceedings of the 2nd IEEE Workshop on Applications of Computer Vision, pp. 138–142. IEEE (1994)
27. Martinez, A.M., Benavente, R.: The AR face database. CVC Tech. Report #24 (1998)
28. Sim, T., Baker, S., Bsat, M.: The CMU pose, illumination, and expression (PIE) database. In: Proceedings of Fifth IEEE International Conference on Automatic Face Gesture Recognition, pp. 53–58. IEEE (2002)
29. Lee, K.C., Ho, J., Kriegman, D.J.: Acquiring linear subspaces for face recognition under variable lighting. IEEE Trans. Pattern Anal. Mach. Intell. **27**(5), 684–698 (2005)
30. Berrar, D.: Bayes' theorem and naive Bayes classifier. In: Encyclopedia of Bioinformatics and Computational Biology: ABC of Bioinformatics, p. 412 (2018)
31. Hsu, C.W., Chang, C.C., Lin, C.J.: A practical guide to support vector classification (2003)
32. Boiman, O., Shechtman, E., Irani, M.: In defense of nearest-neighbor based image classification. In: 2008 IEEE Conference on Computer Vision and Pattern Recognition, pp. 1–8. IEEE (2008)
33. Singh, J., Banerjee, R.: A study on single and multi-layer perceptron neural network. In: 2019 3rd International Conference on Computing Methodologies and Communication (ICCMC), pp. 35–40. IEEE (2019)
34. Pham, X.C., Dang, M.T., Dinh, S.V., Hoang, S., Nguyen, T.T., Liew, A.W.C.: Learning from data stream based on random projection and Hoeffding tree classifier. In: 2017 International Conference on Digital Image Computing: Techniques and Applications (DICTA), pp. 1–8. IEEE (2017)
35. Kotsiantis, S.B., Zaharakis, I., Pintelas, P.: Supervised machine learning: a review of classification techniques. Emerg. Artif. Intell. Appl. Comput. Eng. **160**(1), 3–24 (2007)
36. Chandrakala, M., Devi, P.D.: Two-stage classifier for face recognition using HOG features. Mater. Today Proc. **47**, 5771–5775 (2021)
37. Jianxin, Z., Junyong, W.: Local occluded face recognition based on HOG-LBP and sparse representation. In: 2020 IEEE International Conference on Artificial Intelligence and Computer Applications (ICAICA), pp. 808–813. IEEE (2020)
38. Chen, T., et al.: A novel face recognition method based on fusion of LBP and HOG. IET Image Proc. **15**(14), 3559–3572 (2021)
39. Hamdan, B., Mokhtar, K.: Face recognition using angular radial transform. J. King Saud Univ.-Comput. Inf. Sci. **30**(2), 141–151 (2018)

40. Mohammed, A.A., Al-Alawy, F.: A comparative study of DWT-HOG based face recognition with other similar techniques. In: 2021 31st International Conference on Computer Theory and Applications (ICCTA), pp. 193–199. IEEE (2021)
41. Gu, J., Hu, H., Li, H.: Local robust sparse representation for face recognition with single sample per person. IEEE/CAA J. Autom. Sin. **5**(2), 547–554 (2017)
42. Ahmed, S., Frikha, M., Hussein, T.D.H., Rahebi, J.: Optimum feature selection with particle swarm optimization to face recognition system using Gabor wavelet transform and deep learning. Biomed. Res. Int. **2021**, 1–13 (2021)

Generating an MF from the Dynamical Compose for RASP's Wavelet Frame in a Master System for a Fuzzy-Chaos-Based Model for Encryption Function

Bushra Hussien Aliwi[1(✉)] and Maher Kadhum Abd[2]

[1] University of Babylon, Hilla/Babylon, Iraq
pure.bushra.hus@uobabylon.edu.iq
[2] Ministry of Education, Babylon, Iraq

Abstract. Generating a new wavelet function(s) by a second order composed of two wavelets from RASP's wavelet family with a Gaussian fuzzy number as a membership function led to getting a new membership function MF. The generated wavelet has standard features. A new MF will be employed in a fuzzy-rule base for fuzzy-chaos-based models in control systems and cryptography. So, the modified encryption model stage will apply to a communication system as an essential step in the drive system part. The steps of the designed method will be implemented numerically and discussed to verify the activity of the technique.

Keywords: Wavelets · RASP wavelet · Fuzzy chaotic systems · Takagi-Sugeno fuzzy Model · Chaotic map

1 Introduction

Most works on communicate systems are need to focus on the secrete and affective processes. The models like cryptosystem that based on chaos schemes were designed introduced for digital communicate schemes. Depending on the intersection relation for chaos and fuzzy models, many researches go forward this direction. Since the last century many directions in works on fuzzy models and dynamical systems, fuzzy descriptions of chaotic systems, fuzzy control models for chaotic control models, and a technological fuzzy control systems with chaotic systems that applied in engine the controller technology and chaos systems [1]. The synchronization implementation in fuzzy chaos-based models performed in dynamical systems are applied in a chaos-based security communicate system [2]. The design of a control system with Takagi-Sugeno fuzzy model TS by algorithmic control steps is performed through conditional stability features that were existed and be solvable using the linear matrix inequality algorithm LMI. Designing an engine controller by fuzzy model to control on and perform a solution for problems with uncertainty parameters in TS fuzzy model or in dynamical systems [3].

A. J. Obaid et al. (Eds.): ICONEST 2024, ISEM 40, pp. 205–213, 2025.
https://doi.org/10.1007/978-3-031-89960-7_14

The integrating of fuzzy logic with chaos theory in [4, 5], by introducing a fuzzy chaotic-based cryptosystem that implement the Lure chaotic systems with the discrete-time systems. The TS fuzzy model introduce a model starting with a initializing a super-increasing sequence using a chaotic signal data. The outputs results from TS chaotic slave (drive) system, or any state data was implemented into the steps for calculating the synchronization error in encryption model. The sent message will encrypted by computed superincreasing sequence by first steps in the drive system. The ciphertext data will embedded in secure scalar signal data and then it will sent to the next step of response system (slave) [6].

A Fuzzy chaotic modeling and a prediction based on fuzzy feedback control systems implemented on dynamical and unstable chaotic systems [7]. The fuzzy systems and if–then rules engine were established for input–output resulted data values and in pairwise. That all result led to certify the stability of chaotic state with constant values for targets, states at fixed points or states at unstable periodic orbits in phase plane figure [7]. The composite formula for getting better features at function values and for resulted data that processed by this function formula. Away depends on composite of two mother wavelets functions and daughters from the wavelet family functions of RASP. Also in a trying to modify and implement secure calculations these composite functions were applied in fuzzy-model based model. And to add a complexity on the work the parameters of wavelet functions were shifted and scaled, respectively The features of performed wavelets were helpful in performing the function which is not require to be in a resulted composite function. A resulting function is implemented with chaotification and chaotic maps [8], these functions are periodic and nonlinear continuous functions and simple for computations in applied models.

The paper is ordered as: At Sect. 2, designing of the suggested master fuzzy chaotic-based model is introduced for the general autonomous chaotic system with a flowchart of work. In Sect. 3, generate the membership function. In Sect. 4 generate the terminology of TS fuzzy chaotic encryption part is presented with details steps. Masking of the signal scheme derived through two subsections first for adding masked signal to output of TS fuzzy chaotic model, and the second for embed scalar signal in the ciphertext. Finally, in Sect. 5, a numerical treatment values for the generated MF.

2 Suggested Master System for the Fuzzy-Chaos Based Model

The suggested model of master system works as chaotic transmitter system. Its apply on the generated MF by composing of Rasp wavelets with Gaussian membership function. Then generate the TS fuzzy chaotic model works on this MF. Outputs of the generated TS model will employed in control or slave system. Fig. 1 shows the system. The model consists of four steps. The master system steps are ordered as:

1. Generate the membership function MF.
   a. Dynamical compose for RASP's wavelet family as frame.
   b. Compose frame with Gaussian fuzzy numbers.
2. Generate the TS fuzzy chaotic model.
3. Add masked signal to output of TS fuzzy chaotic model.
4. Embed scalar signal in the ciphertext.

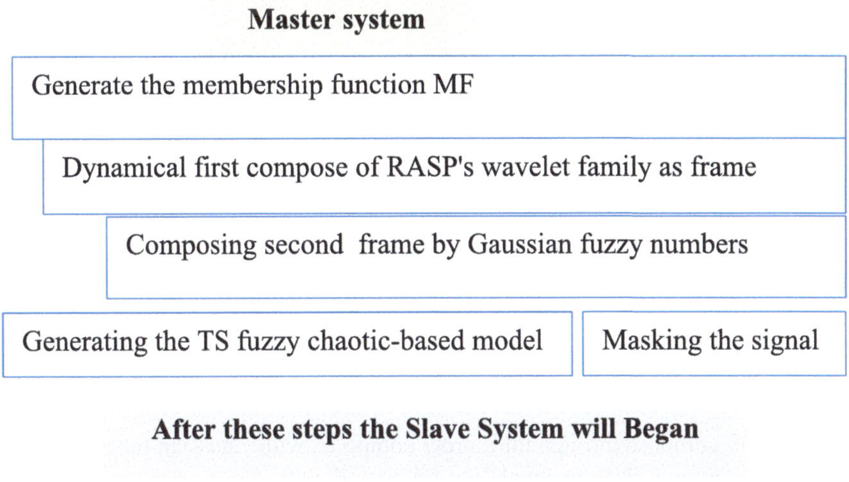

Fig. 1. The master system for suggested method with TS fuzzy chaos model.

The steps of generating the ciphertext $Cip = \{cip_1, \ldots, cip_k\}$ for message $Mess = \{mess_1, \ldots, mess_k\}$, Choose an increasing sequence $S_{inc} = \{s_1, s_2, \ldots, s_l\}$ in secure choice from the table of values composing functions values. Now by computing $\mathbb{S}_{sum} = \sum_{r=1}^{l} s_r$, and taking a prime secret natural numbers $\alpha, \beta \in \mathbb{N}$, such that the value $\mathbb{S}_{sum} < \alpha$. And b in closed interval $[2, \alpha - 1]$ with $\alpha - 1 > 2$, and $gcd(\alpha, \beta) = 1$ is prime secret. Then the public hard key $H_{public} = \{h_1, \ldots, h_k\}$, with $h_p = \beta s_p (mod \alpha)$, for $p = 1,2, \ldots, k$. Finally, keep $(\mathbb{S}_{sum}, \alpha, \beta)$ as a generated private key and H_{public} as a public hard key. Then $cip_p = \sum_{j=1}^{k} m_{pj} h_j$, that were encrypted in function as Cip will modify into the function $\omega(t)$, that computed by; $\omega(t) = \left(\frac{2Cip - \mathbb{S}_{sum}}{\delta T}\right)$, where δ is a small real scalar put $\omega(t)$ in an interval $(-0.01, 0.01)$.

3 Generate the Membership Function

This step has two steps as:

3.1 Dynamical compose for RASP's Wavelets

The generation of MF started by using high order of compose among wavelets functions from RASP's family the first and second. The compose of $RASP_1$ and $RASP_2$ wavelets as;

$RASP_1$ Wavelet with form [10].

$$f(x) = F_{RASP_1}(x) = \frac{x}{(1 + x^2)^2} \tag{1}$$

$$g(x) = F_{RASP_2}(x) = \frac{x.cosx}{1 + x^2} \tag{2}$$

$x, f(x),$ and $g(x)$ are in real numbers.

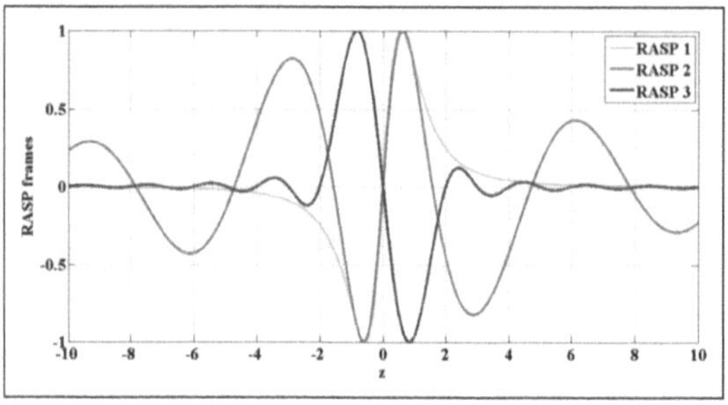

Fig. 2. First, Second and third RASP Wavelets.

3.2 Compose Wavelet Frame with Gaussian Fuzzy Numbers

The frame will compose through third order composes with Gaussian fuzzy number at scaling equal zero and shifting equal 2, as follows (Fig. 2):

$$q(x) = e^{-x^2/2} \tag{3}$$

x in real numbers, and $l(x)$ in [0,1]. Then the compose form will be as;

$$\mathrm{E}(x) = (q \circ g \circ f)(x) = q(g(f(x))) \tag{4}$$

Firstly compose the functions as;

$$\boldsymbol{g} \circ \boldsymbol{f}(\boldsymbol{x}) = \frac{\left(\frac{x}{(x^2+1)^2}\right).\cos(\frac{x}{(x^2+1)^2})}{(\frac{x}{(1+x^2)^2})^2 + 1} = \frac{\left(\frac{x}{(x^2+1)^2}\right).\cos(\frac{x}{(x^2+1)^2})}{\frac{x^2+(x^2+1)^4}{(1+x^2)^4}}$$

$$= \left(\frac{x}{(1+x^2)^2}\right).\cos(\frac{x}{(1+x^2)^2}).\left(\frac{(x^2+1)^4}{x^2+(1+x^2)^4}\right) \tag{5}$$

$$g(f(x)) = x.\cos(\frac{x}{(x^2+1)^2}).\frac{(x^2+1)^2}{x^2+(x^2+1)^4}$$

So here compute compose $q(x)$ for (5), get;

$$q(g(f(x))) = e^{-\left[x.cos(\frac{x}{(1+x^2)^2}).\frac{(1+x^2)^2}{x^2+(1+x^2)^4}\right]^2/2} \tag{6}$$

Putting $u = (x^2 + 1)^2$, $v = x$, functions respect to x, then the form (6) will be as;

$$\mathrm{E}(x) = q(u, v) = e^{-\left[\frac{u.v}{v^2+u^2}.\cos(\frac{v}{u})\right]^2/2}. \tag{7}$$

where E(x) will be in [0, 1].

4 Generating the TS fuzzy Chaotic Model

Generating the TS model master system works as a fuzzy chaotic driver to recovering a signal data. $\omega(t)$ will add to be hybrid with the masked signal and then send the hybrid signal coupled with a constant real number into the slaved model in the system [10]. The ciphertext masked by adding it continuously into the output result from the performed chaotic system. The modulation carried out by modifying the masked signal that injected into the chaotic driver as a suggested TS fuzzy driver model;

Let $x_1(t)$ be the primitive variable from variables vector $[x_1(t), x_2(t), \ldots x_n(t)]$

$$\boldsymbol{Fuzzy\ Rule\ i} : IF\ x_1(t)\ is\ E_i$$

$$THEN x(t+1) = D_i x(t) + B_i(t) + L_i M\,\omega(t) \tag{8}$$

where $D_i \in R^{n\times n}$ is a system matrices for $i = 1, 2, \ldots, r$, and $B_i(t) \in R^n$ which refers to the bias vectors.

4.1 Adding Masked Signal to Output of TS Fuzzy Chaotic Model

The masking form for signal will be by rule;

$$\boldsymbol{Masking\ Signal\ Rule\ i} : IF x_1(t)\ is\ E_i$$

$$THEN z(t) = \widehat{C}x(t) + \widehat{M}\,\omega(t),\ i = 1, 2, \ldots, r \tag{9}$$

$z(t)$ is the coupled masked signal formula, and $\widehat{M}$ in real numbers is the public output value of the masking key, which masks sent ciphertext by a scalar value.

4.2 Embedding the Scalar Signal in the Ciphertext

The transmitters rules could be extracted depending on values of $\overline{y}(t)$ as

$$\boldsymbol{Tranmitter\ Signal\ Rule\ i} : IF\ \overline{y}(t)\ is\ E_i$$

$$THEN x(t+1) = D_i x(t) + V_i(t) + G_i \widehat{M}\,\omega(t) \tag{10}$$

$$z(t) = \widehat{C}x(t) + \widehat{M}\,\omega(t),\ i = 1, \ldots, r$$

for G_i, $i = 1, \ldots, r$ are gain value matrices with $\widehat{M} \neq 0, 1$. The masked signal with embedded message are send into TS as slave system. Then the overall results of fuzzy inferred for the TS as transmitter system is derived through:

$$\begin{aligned} (t+1) &= \sum\nolimits_{i=1}^{r} \mu_i(z(t))\left[\overline{D}_i x(t) + V_i(t) + G_i\overline{y}(t)\right] \\ z(t) &= \hat{C}x(t) + \hat{M}\,\omega(t) \qquad \text{for } i = 1, 2, \ldots, r \end{aligned} \tag{11}$$

where $z(t)$ is output of the master system, and $\overline{D}_i = D_i - G_i\widehat{C}$. Then from (11)

$$\widehat{M} = \frac{\overline{y}(t) - \widehat{C}x(t)}{\omega(t)} \tag{12}$$

By synchronization of master - slave system formulas, the signal could be extracted by slave system.

5 Numerical Executing Values for the Generated MF

The forms in Eq. (7) applied for 21 value variable x in the interval [−100,100] with step 10, The computed values for the resulting function are positive and all in the interval [0, 1], as in the Table 1, the graphic of function through many charts (pointer, surface and radar) shown in Fig. 1.

Table 1. Computed data for the numerical treatment for triple composition

x	u	v	$q(u, v)(x)$
−100	100020001	−100	1.0000005
−90	65626201	−90	1.000000686
−80	40972801	−80	1.000000976
−70	24019801	−70	1.000001457
−60	12967201	−60	1.000002314
−50	6255001	−50	1.000003997
−40	2563201	−40	1.000007803
−30	811801	−30	1.000018478
−20	160801	−20	1.000062191
−10	10201	−10	1.000490267
0	1	0	0
10	10201	10	0.999509973
20	160801	20	0.999937813
30	811801	30	0.999981523
40	2563201	40	0.999992197
50	6255001	50	0.999996003
60	12967201	60	0.999997686
70	24019801	70	0.999998543
80	40972801	80	0.999999024
90	65626201	90	0.999999314
100	100020001	100	0.9999995

The resulted data are all positive for both values positive and negative in chosen interval. The generated MF is not normal and not symmetric in spite the function u is symmetric and normal with only one maximum value $u(x) = 1$ at $x = 0$.MF in figure seems symmetric, but accurate data is clear it is not. In negative side of interval the results all are near1, while in positive side the results are all less than 1 in accuracy. Also at the negative side of values of x the data are decreasing until it reach zero, while at the positive values it are increasing until it reach1, and stay at this value or near it with accurate rate 9 decimals (Figs. 3 and 4).

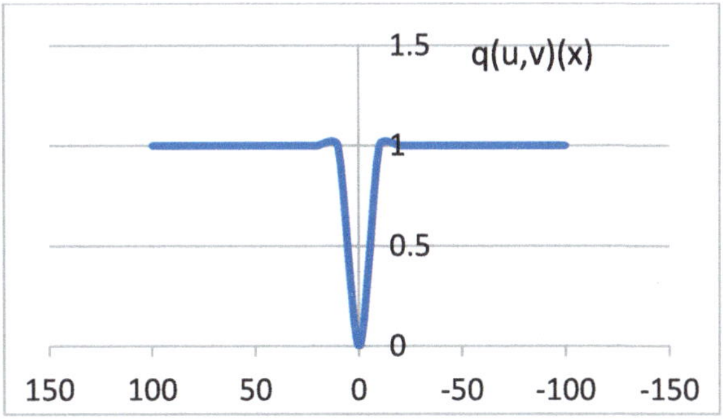

Fig. 3. The generated membership function for values in Table 1

The figure of MF is in converse but it still with data in [0, 1] for the positive values of x. This new but it is has a MF features. For studying the Fig. 1 in more accuracy it drawn with surface around the original in two types.

From Table 1 the increasing sequence chosen as S_{inc} = 0,0.999509973, 0.999937813, 0.999981523, 0.999992197, 0.999996003, 0.999997686,0.999998543, 0.999999024, 0.999999314, 0.9999995}

$\mathbb{S}_{sum} = 9.999411576$

Suppose the message will supposed by user.

Take $\alpha = 11$, $\beta = 5$ is suitable for these values

H_{public} = {0, 4.99755, 4.999689, 4.999908, 4.999961,4.99998, 4.999988, 4.999993, 4.999995, 4.999997}

The generated privet key is (9.999411576, 11, 5).

These data is ready for using by the user. As essential steps in master system

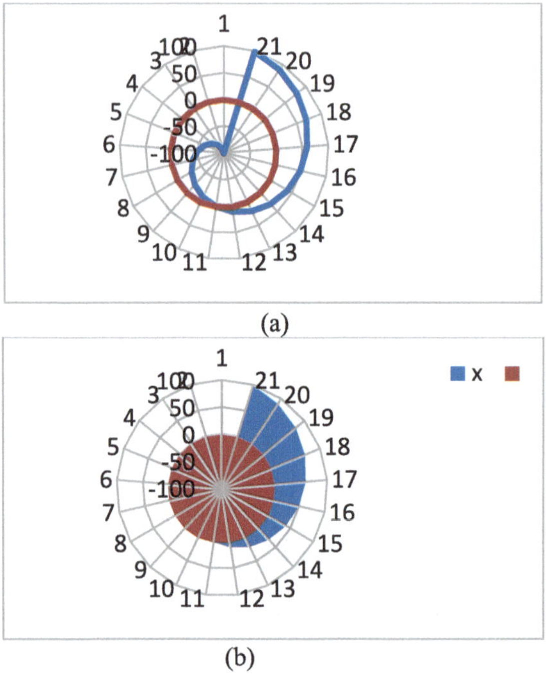

Fig. 4. Drawing of Membership Function (a) linear draw. (b) surfacing draw

6 Conclusion

Designing a master system for Fuzzy Chaotic model with a new membership function, needs in first must construct MF. The triple dynamical compose for RASP's Wavelet Frames with Gaussian fuzzy number led to generate a MF. The model adds the masked signal to the output of the TS fuzzy-chaos based model and then embed a scalar signal to the ciphertext. All fuzzy sets defined through this MF in the TS fuzzy model. The ciphertext modified in the driver fuzzy chaotic-based model, at the encryption part of the system. The effectiveness of the method is shown through using the drive model and slave model steps for modifying and make mask for the signals. Also by choosing a chaotic map with chaos features in generating masked signals. Random performing parameters in R for maps with chaotic attractions in two steps. So all of above will add a secure and complex process on any communication system applying this model. Through numerical computations and execution on the new MF on the arbitrary interval $[-100,100]$ with step 10 resulting data are positive. The divergent in resulted figure cleared by surfacing draw. The figure of MF is converse but it is still with data in closed $[0, 1]$.

References

1. Reznik, T.L.: Fuzzy Controllers, Great Britain by Biddles, Ltd., Guildford and King's Lynn (1997)
2. Chou, H.-G., Chuang, C.-F., Wang, W.-J.: A fuzzy-model-based chaotic synchronization, and its implementation on a secure communication system. IEEE Trans. Inf. Forensics Secur. **8**(12) (2013). https://doi.org/10.1109/TIFS.2013.2286268
3. Lian, K.Y., Chiu, C.S., Chiang, T.S., Liu, P.: LMI-based fuzzy chaotic synchronization and communications. IEEE Trans. Fuzzy Syst. **9**(4), 539–553 (2001)
4. Li, Z.: Fuzzy chaotic systems modeling, control, and applications, Stud Fuzz 199, Springer, Berlin, Heidelberg, pp. 275–283 (2006)
5. Bhattacharyya, S., Jha, S., Tharakunnel, K., Westland, J. C.: Data mining for credit card fraud: a comparative study. Decis. Support Syst.**50**(3), 602–613 (2011)
6. Seeja, K.R., Zareapoor, M.: Fraudminer: A novel credit card fraud detection model based on frequent itemset mining. Sci. World J. 2014 (2014)
7. Mekircha, N., Boukabou, A. and Mansouri, N.: Fuzzy control of unstable chaotic systems. J. Autom. Syst. Eng. **6-1**, 11–19 (2012)
8. Hua, Z., Zhou, Y., Huang, H.: Cosine-transform-based chaotic system for image encryption. Inf. Sci. **480**, 403–419 (2018)
9. Aliwi, B.H., Ajeena, R.K.K.: The knapsack fuzzy chaos cryptosystem. In: AIP Conference Proceedings, vol. 2398, p. 060035 (2022)
10. Aliwi, B.H.: Designing an encryption function for chaotic- based model with bifurcation shifting security communicate. J. Duhok Univ. **26**(2), 533–540 (2023). https://doi.org/10.26682/csjuod.2023.26.2.48
11. Thuillard, M.: A review of wavelet networks, Wavenets, fuzzy wavenets and their applications, ESIT 2000, Aachen, Germany (2000)

On Sophisticated Technique via DH-Transform

Hasanain Jalil Neamah Alsaedi[1], Shatha Haider Thyab[2], Dalael Saad Abdul-Zahra[3](✉), Hawraa A. AL-Challabi[4], and Hassanein Falah[5]

[1] Iraqi Commission for Computers and Informatics, Baghdad, Iraq
[2] Department of Mathematic, Al _Mustansiriya University, Baghdad, Iraq
[3] Medical instruments department, Babil Technical Institute, Al-Furat Al-Awsat Technical University, 51015 Babil, Iraq
dalael20175@gmail.com
[4] Department of Mathematics, Faculty of Education, University of Kufa, Najaf, Iraq
[5] Hilla University College, Babylon, Iraq

Abstract. This work introduces Dalael Hawraa-transform, a novel integral transformation. (DH-transform), that expressed in general form by α as $DH(f(x)) = G(t) = \lambda^{\alpha}\int_0^{\infty} e^{-\lambda^{\alpha}x} f(x)dx$. We aimed to study the properties of this transform and applied them on well-known functions. Also, show some notes of the relation between DH-transform and others. Furthermore, to show the effective of our transformation we introduce a solution of mechanical vibration as a dynamical example which used the second differential equation as an illustration of its system, depending on convolution theorem and inverse DH-transform.

Keywords: Dalael Hawraa-Transform · differential equation · convolution theorem

1 Introduction

The essence of using integral transformations is for solving differential equations, such that we get the true functions that effect on the interesting applicable significant equation. Among these remarkable transforms: Laplace transform [4]; Fourier transform [16]; Sumudu transform [25]; Aboodh transform [4]; Mahgoub transform [23]; and the interesting discovery of AL-Zughair transform for solving ordinary differential equation (ODE) that used in of physical application [1, 2, 22], ELzaki transform [18], Soham transforms [24]. Also, the integral transform plays an interesting role to analyse many issues that expressed by a differential equation, like in physics [1–3, 6–8, 19]; engineering [6, 7, 19–21]; technology [3, 8]; quantum mechanics [19] and population growth [7, 11–15]. For given convenient transformations that deal with the fundamental functions that considered by $A = \{f(x) : \exists M, k_1, k_2 > 0, |f(x)| < Me^{-vx}; x \in [0, \infty)\}$, k_1, k_2, and the constant M must both be finite or infinite. Now let us show some of these well-known transforms like:

- **Laplace-Transform**

A. J. Obaid et al. (Eds.): ICONEST 2024, ISEM 40, pp. 214–222, 2025.
https://doi.org/10.1007/978-3-031-89960-7_15

$$\mathcal{L}(f(x)) = \int_0^\infty f(x)e^{-sx}dx, \text{ such that } f(x) \text{ is defined on } x \geq 0.$$

- **Aboodh -Transform**

$$A(f(x)) = \frac{1}{v}\int_0^\infty f(x)e^{-vx}dx, x \geq 0, k_1 \leq v \leq k_2.$$

- **Sumudu -Transform**

$$S(f(x)) = \frac{1}{u}\int_0^\infty f(x)e^{-\left(\frac{1}{u}\right)x}dx, u \in (-k_1, k_2).$$

- **Mahgoub -Transform**

$$M(f(x)) = v\int_0^\infty f(x)e^{-vx}dx, x \geq 0, k_1 \leq v \leq k_2.$$

- **Soham-Transform**

$$\mathcal{S}(f(x)) = \frac{1}{v}\int_0^\infty f(x)e^{-v^{\alpha}x}dx, x \geq 0, k_1 \leq v \leq k_2.$$

- **Elzaki-Transform**

$$E(f(x)) = v\int_0^\infty f(x)e^{-\frac{x}{v}x}dx, x \geq 0, k_1 \leq v \leq k_2.$$

The new transform called DH-Transform that symbolized by $DH(.)$ on the function $f(x)$ in the set A is defined by

$$DH(f(x)) = G(t) = \lambda^{\alpha}\int_0^\infty e^{-\lambda^{\alpha}x}f(x)dx. \tag{1}$$

Such that $x \geq 0$ and $k_1 \leq \lambda, \alpha \leq k_2$. While the inverse DH-transform is

$$DH^{-1}(DH(f(x))) = f(x) = t.$$

The purpose of this study is to show that new integral transform called DH-transform by taking general factor λ^{α} and present its benefit by analysis linear differential equation of applicable example like a mechanical vibration which is an applied dynamical system that undergo oscillatory motion, it is essential to study mechanical vibrations. This motion can be as basic as harmonic motion or as complex as more complex patterns. Energy is transferred through matter or space in this process. According to physics, vibration is defined as an item or wave moving back and forth. This phenomenon may be caused by internal or external sources [17].

This study divided into three parts: the first one the new transform DH on basic functions will be taken and show the simplification of some higher functions like natural logarithm function, while the second one discusses DH-transform on differential equations illustrated by examples and applicable issue of physics. At last, the inference will be taken.

2 DH-Transform on Some Fundamental Functions

$$DH(f(x)) = \lambda^{\alpha} \int_{0}^{\infty} e^{-\lambda^{\alpha}x} f(x)dx$$

1. $f(x) = 1$

$$DH(f(x)) = \lambda^{\alpha} \int_{0}^{\infty} e^{-\lambda^{\alpha}x} f(x)dx$$

$$DH(1) = \lambda^{\alpha} \int_{0}^{\infty} e^{-\lambda^{\alpha}x} dx = \lambda^{\alpha}\left(\frac{-1}{\lambda^{\alpha}}\right)\left[e^{-\infty} - e^{0}\right] = 1 \tag{2}$$

2. $f(x) = x$

$$\begin{aligned} &DH(f(x)) = \lambda^{\alpha} \int_{0}^{\infty} e^{-\lambda^{\alpha}x} f(x)dx \\ &DH(x) = \lambda^{\alpha} \int_{0}^{\infty} e^{-\lambda^{\alpha}x} x dx, \text{ let } u = x, dv \\ &= e^{-\lambda^{\alpha}x} \text{ then } DH(x) = \tfrac{1}{\lambda^{\alpha}} \end{aligned} \tag{3}$$

also,

$$DH\left(x^{2}\right) = \frac{2}{\lambda^{2\alpha}} \; i.e\, DH\left(x^{n}\right) = \frac{n!}{\lambda^{n\alpha}} \tag{4}$$

3. $f(x) = e^{ax}$

$$DH\left(e^{ax}\right) = \lambda^{\alpha} \int_{0}^{\infty} e^{-\lambda^{\alpha}x} e^{ax} dx = \lambda^{\alpha} \int_{0}^{\infty} e^{-(\lambda^{\alpha}-a)x} dx = \frac{\lambda^{\alpha}}{\lambda^{\alpha} - a} \tag{5}$$

4. $f(x) = \sin ax$

$$DH(\sin ax) = \lambda^{\alpha} \int_{0}^{\infty} e^{-\lambda^{\alpha}x} \sin axdx \text{ let } u = \sin ax, dv = e^{-\lambda^{\alpha}x} \text{ so}$$

$$DH(\sin ax) = \frac{a\lambda^{\alpha}}{a^{2} + \lambda^{2\alpha}} \tag{6}$$

5. $f(x) = \cos ax$

$$DH(\cos ax) = \lambda^{\alpha} \int_{0}^{\infty} e^{-\lambda^{\alpha}x} \cos axdx \text{ let } u = \cos ax, dv = e^{-\lambda^{\alpha}x} \text{ so}$$

$$DH(\cos ax) = \frac{\lambda^{2\alpha}}{a^{2} + \lambda^{2\alpha}} \tag{7}$$

6. $f(x) = \ln x$

$$DH(lnx) = \lambda^{\alpha} \int_{0}^{\infty} e^{-\lambda^{\alpha}x} \ln x\, dx \tag{8}$$

Let define

$$e^{-\lambda^\alpha x} = \lim_{n\to\infty}\left(1-\frac{\lambda^\alpha x}{n}\right)^n \text{ but } e^{-\lambda^\alpha x} = \lim_{n\to\infty}\left(1-\frac{\lambda^\alpha x}{n}\right)^{n-1} \lim_{n\to\infty}\left(1-\frac{\lambda^\alpha x}{n}\right)^1$$

this leads to

$$e^{-\lambda^\alpha x} = \lim_{n\to\infty}\left(1-\frac{\lambda^\alpha x}{n}\right)^{n-1} \tag{9}$$

Now

$$DH(lnx) = \lambda^\alpha \int_0^\infty \lim_{n\to\infty}\left(1-\frac{\lambda^\alpha x}{n}\right)^{n-1} \ln x\, dx \tag{10}$$

Suppose $u = 1 - \frac{\lambda^\alpha x}{n}$, $du = \frac{-\lambda^\alpha}{n}dx$

$$DH(lnx) = \lambda^\alpha \int_0^n \lim_{n\to\infty}\left(1-\frac{\lambda^\alpha x}{n}\right)^{n-1} \ln\, x\, dx = \lambda^\alpha \int_1^0 \mathrm{u}^{n-1} \ln\left(\frac{n(1-u)}{\lambda^\alpha}\right)\frac{-n}{\lambda^\alpha}du \tag{11}$$

$$= n\int_0^1 u^{n-1} \ln\left(\frac{n(1-u)}{\lambda^\alpha}\right)du \tag{12}$$

$$= n\ln n \int_0^1 u^{n-1}du + n\int_0^1 u^{n-1}\ln(1-u)du - n\ln\lambda^\alpha \int_0^1 u^{n-1}du \tag{13}$$

$$= n\ln n\left[\frac{n^n}{n}\right]_0^1 - n\int_0^1 u^{n-1}\sum_{k=1}^{\infty}\frac{u^k}{k}du - n\ln\lambda^\alpha\left[\frac{n^n}{n}\right]_0^1 \tag{14}$$

$$= n\ln n\frac{1}{n} - n\int_0^1 u^{n-1}\sum_{k=1}^{\infty}\frac{u^k}{k}du - n\ln\lambda^\alpha\frac{1}{n} \tag{15}$$

$$= \ln n - n\sum_{k=1}^{\infty}\int_0^1 \frac{u^{k+n-1}}{k}du - \ln\lambda^\alpha \tag{16}$$

$$= \ln n - n\sum_{k=1}^{\infty}\frac{1}{k(k+n)} - \ln\lambda^\alpha \tag{17}$$

$$= \ln n - \sum_{k=1}^{\infty}\frac{1}{k} - \frac{1}{n+k} - \ln\lambda^\alpha \tag{18}$$

$$DH(\ln x) = \lambda^\alpha \int_0^\infty \lim_{n\to\infty} \ln n - \sum_{k=1}^{\infty}\frac{1}{k} - \ln\lambda^\alpha = -\gamma - \ln\lambda^\alpha \tag{19}$$

where γ is Euler's constant such that $\gamma = \Gamma'(1)$.

<u>Note</u> on DH-Transform:

It is interesting to show some relations of DH-Transform with others

1. If $\alpha = 1$ then "DH-Transform" → "Mahgoub Transform".
2. If $\alpha = 3$ then $DH(x) = \frac{1}{\lambda^3} = A(x)$.
3. If $\alpha = 1$ then $DH(x) = \frac{1}{\lambda} = \mathcal{L}(1)$.

Now it is important to know that how we can use DH-transform on the combination of two functions, by the following theorem called Convolution theorem.

1.1 Theorem (Convolution theorem): If $f(x), g(x)$ be two arbitrary functions on x such that $DH(f(x)) = G(t), DH(g(x)) = F(t)$. Then $DH((f * g)(x)) = DH(f(x)) * DH(g(x))$.

Proof:

Let f, g be two functions and $(*)$ be binary operation defined on f, g, by taking DH-transform on this combination this leads to

$$DH((f * g)(x)) = \lambda^{\alpha} \int_0^{\infty} e^{-\lambda^{\alpha} x}(f * g)(x)dx \tag{20}$$

$$= \lambda^{\alpha} \int_0^{\infty} e^{-\lambda^{\alpha} x} f(x)dx * \lambda^{\alpha} \int_0^{\infty} e^{-\lambda^{\alpha} x} f(x)dx \tag{21}$$

$$= DH(f(x)) * DH(g(x)) = G(t) * F(t) \tag{22}$$

3 DH-Transform Defined Ordinary Differential Equation

In this section we will introduce an effective theorem for the integral transform DH that considered as the key for solving linear system of differential equation.

3.1 Theorem. Let $G(t)$ be DH-transform of $DH(f(x)) = G(t)$. Then, the assertions that follow are interchangeable.

$$DH\left(f'(x)\right) = \lambda^{\alpha} G(t) - \lambda^{\alpha} f(0) \tag{23}$$

$$DH\left(f''(x)\right) = \lambda^{2\alpha} G(t) - \lambda^{\alpha} f'(0) - \lambda^{2\alpha} f(0) \tag{24}$$

$$DH\left(f^{n}(x)\right) = \lambda^{n\alpha} G(t) - \sum_{k=1}^{n-1} \lambda^{(n-k)\alpha} f^{k}(0) \tag{25}$$

Proof:

1. Let $u = e^{-\lambda\alpha}$ and $dv = f'(x)$ then $DH\left(f'(x)\right) = \lambda^{\alpha} G(t) - \lambda^{\alpha} f(0)$.
2. Also, integrating by part by letting $u = e^{-\lambda\alpha}$ and $dv = f''(x)$ then $DH\left(f''(x)\right) = \lambda^{2\alpha} G(t) - \lambda^{\alpha} f'(0) - \lambda^{2\alpha} f(0)$.
3. The general form of this statement could be proved by using mathematical induction.

3.2 Example

Solve the following differential equation

$$f^{''}(x) - 4f'(x) + 3f(x) = 0 \tag{26}$$

Solution:
By taking DH-transform for both sides

$$DH(f^{''}(x)) - 4DH\left(f'(x)\right) + 3DH(f(x)) = DH(0) \tag{27}$$

By (3.1) and (3.2)

$$\lambda^{2\alpha}G(t) - \lambda^{\alpha}f'(x) - \lambda^{2\alpha}f(0) - 4\left[\lambda^{\alpha}G(t) - \lambda^{\alpha}f(0)\right] + 3G(t) = 0 \tag{28}$$

$$G(t)\left(\lambda^{2\alpha} - 4\lambda^{\alpha} + 3\right) = 4\lambda^{\alpha} - \lambda^{2\alpha} \tag{29}$$

$$G(t) = \frac{4\lambda^{\alpha} - \lambda^{2\alpha}}{(\lambda^{\alpha} - 3)(\lambda^{\alpha} - 1)} \tag{30}$$

By taking partial fractions technique

$$\begin{aligned} &= \frac{6}{\lambda^{\alpha} - 3} - \frac{2}{\lambda^{\alpha} - 1} \\ &= \frac{6}{\lambda^{\alpha}}\frac{\lambda^{\alpha}}{\lambda^{\alpha} - 3} - \frac{2}{\lambda^{\alpha}}\frac{\lambda^{\alpha}}{\lambda^{\alpha} - 1} \end{aligned} \tag{31}$$

Using of inverse DH-transform and convolution theorem with support of (2.2) and (2.4), thus

$$f(x) = 6xe^{3x} - 2xe^{x}. \tag{32}$$

4 Application of DH-Transform on Mechanical Vibration

The purpose of developing and evaluating systems Engineers can minimize noise, avoid resonance-related catastrophic failure, and improve the performance of a variety of mechanical systems, from electronic components and vehicle suspensions to buildings and bridges. By having a thorough understanding of vibrations, they may also prevent these issues. A mathematical model is created by applying Newton's laws of motion or energy processes, which usually lead to a second-order differential equation. There are various cases of which force those effects on the system, we will study in this section free motion (vibration) from mass (i.e. without applied force) that explained by the second order differential Eq. (3.12) will be discussed by our new transform DH-transform to illustrate this system as functions that have dynamical efficient.

Let mathematical modelling of mechanical vibration be

$$mf''(x) + cf'(x) + kf(x) = 0. \tag{33}$$

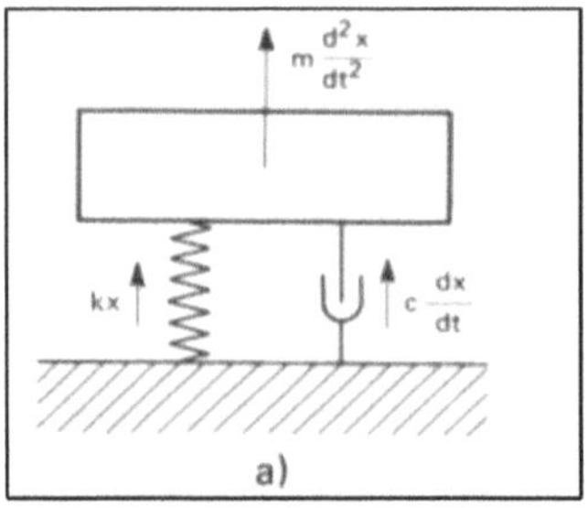

Fig. 1. Mechanical vibration.

where, m is a mass (which represents the object in motion); The spring has a stiffness coefficient of $k > 0$ and c is the damping coefficient (Fig. 1).

$$mDH\left(f''(x)\right) + cDH\left(f'(x)\right) + kDH(f(x)) = 0 \tag{34}$$

$$G(t)\left(m\lambda^{2\alpha} + c\lambda^{\alpha} + k\right) - f(0)\left(m\lambda^{2\alpha} + c\lambda^{\alpha}\right) - m\lambda^{\alpha}f'(0) = 0 \tag{35}$$

$$G(t) = \frac{f(0)\left(m\lambda^{2\alpha} + c\lambda^{\alpha}\right)}{\left(m\lambda^{2\alpha} + c\lambda^{\alpha} + k\right)} + \frac{f'(0)m\lambda^{\alpha}}{\left(m\lambda^{2\alpha} + c\lambda^{\alpha} + k\right)} \tag{36}$$

$$G(t) = \frac{\lambda^{\alpha}f(0)(m\lambda^{\alpha} + c)}{(m\lambda^{\alpha} + k)(\lambda^{\alpha} + 1)} + \frac{f'(0)m\lambda^{\alpha}}{(m\lambda^{\alpha} + k)(\lambda^{\alpha} + 1)} \tag{37}$$

s.t $(m + k) = c$

$$\begin{aligned} G(t) = {} & \frac{-cf(0)k}{m-k}\frac{1}{(m\lambda^{\alpha}+k)} + \frac{cf(0)}{m-k}\frac{1}{(\lambda^{\alpha}+1)} \\ & + \frac{-mf'(0)k}{m-k}\frac{1}{(m\lambda^{\alpha}+k)} + \frac{mf'(0)}{m-k}\frac{1}{(\lambda^{\alpha}+1)} \end{aligned} \tag{38}$$

By taking inverse DH such that $DH^{-1}(DH(f(x))) = f(x)$
That is

$$f(x) = -kxe^{-\frac{k}{m}x}\left(\frac{cf(0) + f'(0))}{m-k}\right) + xe^{-x}\left(\frac{mf'(0) + cf(0))}{m-k}\right). \tag{39}$$

5 Conclusion

This study showed the flexibility of the DH-transform for some basic functions, and the importance of convolution theorem to seek the solution of differential equation by using equation of mechanical vibration as an applicable example of this method that is plays an interesting role in physics and engineering. We have many future works of this paper for any life issue that has a mathematical system.

References

1. Abbas, A.E., Kuffi, E.A., Hanna, E.: AL-Zughair integral transformation in solving improved heat and poisson PDEs. In: AIP conference Proceedings, vol. 2386, p. 040041 (2022). https://doi.org/10.1063/5.0066914
2. Abbas, A.E., Kuffi, E.A., Mohammed, A.H.A., Majde, Q., Abbas, E.A.: Applying al-zughair transform on nuclear physics. Int. J. Eng. Technol. **9**(1) 9–11 (2020)
3. Abdelilah, K., Hassan, S.: The use of Kamal transform for solving partial differential equations. Adv. Theor. Appl. Math. **12**(1), 7–13 (2017)
4. Abramowitz, M., Stegun, I.A. (eds.). Laplace Transforms. Ch. 29. In: Handbook of Mathematical Functions with Formulas, Graphs, and Mathematical Tables, 9th printing, Dover, New York, pp. 1019–1030 (19720
5. Aboodh, K.S.: The new integral transform "Aboodh Transform". Global J. Pure Appl. Math. **9**(1), 35–43 (2013). ISSN 0973–1768
6. Aboodh, K.S.: Application of new transform "Aboodh transform" to partial differential equations. Global J. Pure Appl. Math. **10**(2), 249–254 (2014)
7. Aboodh, K.S., Farah, R.A., Almardy, I.A., Almostafa, F.A.: Solution of partial integro-differential equations by using Aboodh and double Aboodh transforms methods. Global J. Pure Appl. Math. **13**(8), 4347–4360 (2016)
8. Aboodh, K.S., Farah, R.A., Almardy, I.A., Osman, A.K.: Solving delay differential equations by Aboodh transformation method. Int. J. Appl. Math. Stat. Sci. **7**(2), 55–64 (2018)
9. Aggarwal, S., Gupta, A.R., Singh, D.P., Asthana, N., Kumar, N.: Application of Laplace transform for solving population growth and decay problems. Int. J. Latest Technol. Eng. Manage. Appl. Sci. **7**(9), 141–145 (2018)
10. Aggarwal, S., Singh, D.P., Asthana, N., Gupta, A.R.: Application of Elzaki transform for solving population growth and decay problems. J. Emerg. Technol. Innov. Res. **5**(9), 281–284 (2018)
11. Aggarwal, S., Sharma, N., Chauhan, R.: Solution of linear Volterra integro-differential equations of second kind using Mahgoub transform. Int. J. Latest Technol. Eng. Manage. Appl. Sci. **7**(5), 173–176 (2018)
12. Aggarwal, S., Pandey, M., Asthana, N., Singh, D.P., Kumar, A.: Application of Mahgoub transform for solving population growth and decay problems. J. Comput. Math. Sci. **9**(10), 1490–1496 (2018)
13. Aggarwal, S., Asthana, N., Singh, D.P.: Solution of population growth and decay problems by using Aboodh transform method. Int. J. Res. Advent Technol. **6**(10) (2018). E-ISSN: 2321–9637
14. Aggarwal, S., Gupta, A.R., Asthana, N., Singh, D.P.: Application of Kamal transform for solving population growth and decay problems. Global J. Eng. Sci. Res. **5**(9), 254–260 (2018)
15. Aggarwal, S., Sharma, N., Chauhan, R.: Solution of population growth and decay problems by using Mohand transform. Int. J. Res. Advent Technol. **6**(11), 3277–3282 (2018)
16. Bracewell, R.N.: The Fourier Transform and its Applications (2nd ed.), McGraw-Hill Kogakusha (1978). ISBN 978-0-07-007013-4
17. Broch, J.T.: Mechanical vibration and shock measurements (1984)
18. Elzaki M.T.: The new integral transform "ELzaki Transform". Global J. Pure Appl. Math. **7**(1), 57–64 (2011). ISSN 0973–1768
19. Elzaki, T.M., Ezaki, S.M.: Applications of new transform "'Elzaki transform"' to partial differential equations. Global J. Pure Appl. Math. **7**(1), 65–70 (2011)
20. Kumar, P.S., Gomathi, P., Gowri, S., Viswanathan, A.: Applications of Mohand transform to mechanics and electrical circuit problems. Int. J. Res. Advent Technol. **6**(10), 2838–2840 (2018)

21. Kreyszig, E.: Advanced Engineering Mathematics, 9th ed., WILEY (2014)
22. Mohammed, A.H., Sadiq, B.A., Hassan, A.M.: AL-Zughair Transform, LAP Lambert Academic Publishing (2017)
23. Mohand, M., Mahgoub, A.: The New Integral Transform "Mahgoub Transform. Adv. Theor. Appl. Math. **11**(4), 391–398 (2016). ISSN 0973-4554
24. Patil, D.P., Savita, S.K.: The new integral transform "Soham Transform". Int. J. Adv. Eng. Manage. (IJAEM) **3**(10), 126–132 (2021)
25. Watugala, G.K.: Sumudu transform: a new integral transform to solve differential equations and control engineering problems. Internat. J. Math. Ed. Sci. Tech. **24**, 35–43 (1993)

Local Features Applications in Detecting Vehicle License Plates

Fatin E. M. Al-Obaidi[1(✉)], S. S. Salman[2], Anwar H. Al-Saleh[3], Ali A. D. Al-Zuky[1], and W. A. Abbas[2]

[1] Department of Physics, College of Science, Mustansiriyah University, Baghdad, Iraq
{sci.phy.fam,prof.alialzuky}@uomustansiriyah.edu.iq
[2] Department of Clinical Laboratory Sciences, College of Pharmacy, University of Baghdad, Baghdad, Iraq
{salma3_sultan,wafaa.abbas}@copharm.uobaghdad.edu.iq
[3] Computer Science, College of Science, Mustansiriyah University, Baghdad, Iraq
anwar.h.m@uomustansiriyah.edu.iq

Abstract. The process of recognizing and identifying vehicle license plates is considered one of the important tasks in the monitoring of roads, institutions and the applications of intelligent transportation systems such as traffic control, autonomous driving, parking toll booths, etc. The current study's objective is to compare how well local features perform in identifying and detecting car license plate numbers, estimate the detection time for each feature, and confirm it using the time constraints for detecting license plates on Iraqi inner and highways streets. The three models are Harr-like, Histogram of Oriented Gradients (HOG) and Linear Binary Patterns (LBP). The identifying and detecting process used a database created by recording multiple videos of various cars at parking in the city of Baghdad. The recognition performance results showed that all models managed to detect and recognize license plates within the specified time, except for the Histogram of Oriented Gradients (HOG) model in video recording. The latter model recognized the license plate beyond the specified time. The results also showed that all models managed to recognize license plates within the executed detected time per frame, which is within fractions of a second and falls between 0.00486 and 0.0225 s, thus attributing the success of the models and the goal of license plate recognition has finally been achieved.

Keywords: Automated Vehicle Detection · Harr-Like Feature · Histogram of Oriented Gradients · Linear Binary Patterns Feature · License Plate

1 Introduction

The process of target detection is considered one of the basic techniques in the field of computer vision technology. For example, its importance comes from its effective role in vehicle monitoring, in addition to its importance in early warning applications, national defense security, and the ability to detect environmental pollution and forest disasters (Li et al., 2021) [1]. The principle of target detection is primarily based on locating

A. J. Obaid et al. (Eds.): ICONEST 2024, ISEM 40, pp. 223–237, 2025.
https://doi.org/10.1007/978-3-031-89960-7_16

a target(s) of interest in the digital image. Researchers in the field of computer vision technology processed the training image to extract the target and identify unique features or obtain specific semantic information features to match these distinctive features. A classification algorithm is then applied to provide an accuracy rate for each classification category (Zhang et al., 2021) [2].

The techniques of detecting and identifying vehicle license plate have effective impact in the systems of an intelligent transportation with applications in traffic control, autonomous driving, parking toll booths, etc. (Deb & Jo, 2009) [3]. Nowadays, license plate recognition is still a difficult task under extreme conditions, such as occlusion, uneven lighting, rotation (large angle), blur, etc. (Shashirangana et al., 2021) [4].

Recently, the advanced technological advantages have entered the field of industrial intelligence as we can witness the technological development to bring major changes in many areas such as the importance of license plate recognition technology. The increase in traffic accidents and vehicle-related crimes is proportional to the importance using road monitoring and vehicle tracking technologies to ensure public safety and improve road safety. These technologies are considered an effective solution to the major and growing challenges in the areas of security and vehicle monitoring. Therefore, license plate recognition technologies enable intelligent and instantaneous identification and tracking of vehicles, which contributes to public safety management on the one hand and helps to maintain traffic safety on the other hand.

The main contribution of this study is to make a comparison between the applicability of the Haar-like, Histogram of Oriented Gradients (HOG) and the Linear Binary Patterns (LBP) features in license plate recognition using a recoded video once and extracted frame images the second time.

There are two official license plate formats in Iraq. One in Kurdistan provenance, as shown in Fig. 1(a), and other formats used in the remaining Iraqi provenances, as in Figs. 1 (b, c, and d). The speed of cars on the interior and expressways of Iraqi roads is between (60 and 100) km/h. The average speed is 22 m/s, and therefore the time it takes for a car to travel 100 m is 4.5 s to cover the specified distance. So if the proposed model detects and recognizes a single image of the car license at least within this period, the goal of the recognition can be achieved and thus ultimately the success of the current model can be verified.

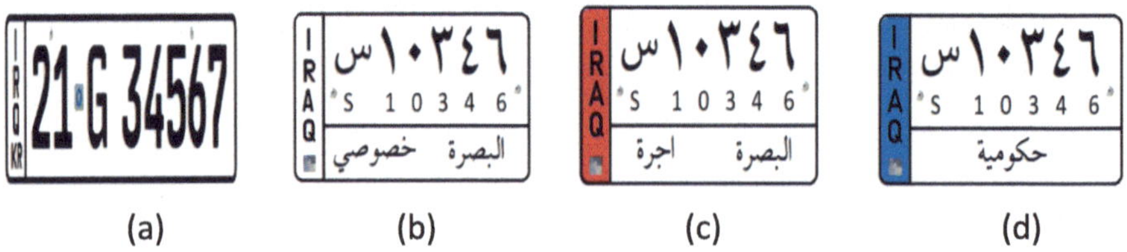

Fig. 1. Some Iraqi car license plate a) A plate designed for private cars in Kurdistan Region (Salman et al., 2023) [5], b) A plate designed for private (Abbass & Marhoon, 2021) [6], c) A plate designed for Taxis (Habeeb et al., 2021) [7], d) A plate designed specifically for government use (Habeeb et al., 2021) [7].

This paper makes the following important contributions:

- A comparison between the performance of local feature in license plate recognition over the entire execution time.
- Analyzing the performance of local feature on the video recording and the extracted images.
- License plate time detection was performed and their range was checked with the specified time to detect the license plate on Iraqi interior and highway roads.
- The performance of the proposed models was compared in terms of statistical parameters.

The rest of the paper is organized as follows: Sect. 2 introduces the corresponding works based on the latest techniques. Section 3 provides a brief definition of the cascade object detector, while Sect. 4 provides a description of the proposed methodology. The results analysis and discussion of the dataset are presented in Sect. 5. Finally, the conclusion is discussed in Sect. 6.

2 Related Works

Many researchers focused their attention on vehicle license plate recognition and classifications. (Li et al. 2018) [8] used deep neural networks to overcome the problem of detecting and recognizing car license plates in natural scene images by training thirty-seven classes convolutional neural networks to recognize symbols in the image and achieve high recall for a binary text/non-text classifier. Their method achieved good detection results and achieved high detection accuracy. A network model presented by (Xu et al., 2018) [9] predicted the bounding box and at the same time recognized the corresponding number with high accuracy. Their results showed that the proposed model performed well in detecting and recognizing over 61 frames per second with an accuracy of 98.5%.

(Asmara et al. 2019) [10] compared Haar-like, LBP and HOG features in vehicle detection. The results showed that the Haar-like feature was better than the HOG and LBP features due to its high recognition rates and execution time. Haar-like features are 46.7% times and 33.3% faster than HOG and LBP detectors, respectively. (Yousif et al., 2020) [11] proposed a vehicle license plate recognition methodology with an optimized neutrosophical set based on a genetic algorithm. Their results showed that the proposed method had the potential to be suitable for both (Egyptian-Arabic) and English. The system achieved an accuracy of 96.67% for correct detection and 94.27% for some noise in the image. (Tao et al., 2020) [12] proposed a cascade object detection algorithm to distinguish the contents of license plates using Tiny-YOLOv3 and logical judgment to determine the license plate number. Experimental results recognized 99.51% of license plate areas from images and identified 99.02% of the characters on them at higher running speeds.

For license plate images generation, (Zhang et al., 2021) [13] used a customized CycleGAN model and an elaborately designed image-to-sequence network for license plate recognition purposes and their results occupied high achievement on such a dataset. Using basic image characteristics, (Musil et al., 2021) [14] used multiple sliding window detectors and introduced an FPGA application for license plate recognition on images, each tuned to a specific projection scale. The results showed a detection rate of 98% and 98.7% when combined with video tracking. Their FPGA model processed 1280 × 1024 pixel images at more than 40 frames per second and a minimum width of 100 pixels for the recognized license plate.

(Leng et al., 2023) [15] used a license plate recognition system based on edge deep learning. They used channel pruning to reconstruct the backbone layer and reduce model parameters and GPU resource consumption. The experimental results of the CCPD dataset showed that the total number of network parameters was 0.606 MB, with an accuracy rate of 97%. (Tao et al., 2024) [16] introduced a license plate recognition model; YOLOv5-PDLPR. They used a feature extractor to obtain global feature information and implemented the model on the CCPD dataset, yielding an average accuracy result of 99.4% and a detection speed score of 159.8FPS. Their method recognized fewer types of license plates and the recognition rate for certain Chinese words was low.

The uniqueness of the work that is being presented centers on how well local features perform in identifying license plate numbers based on the time that they execute when recording video once and frames twice. As previously stated, one can achieve the detection goal and ultimately confirm the effectiveness of the current model if the proposed model detects and recognizes one frame image of the driver's license for at least 4.5 s.

3 Cascade Object Detection

A cascade algorithm was used in this study. A cascade target detection system has multiple relevant classifiers for target detection. However, these classifiers are not always sufficient for a specific application (Huo & Chen 2004) [17]. A set of classifiers can be viewed as a degenerate decision tree consisting of progressively more complicated stages, where the first stage trains the classifier to recognize almost all targets of interest and then starts evaluation of the second stage classifier, which is further modified to prove a high detection rate, ultimately taking the form of a decomposed decision tree (Viola & Jones, 2005 [18] & Shahbaz et al., 2022 [19]). The cascade detector technique is typically used in one of three feature types; Haar-like, HOG and LBP as explained below (Arunmozhi & Park, 2018) [20].

3.1 Haar Feature

To be more efficient, the method of Viola-Jones utilized rectangular filters, named Haar-like features, instead of working directly with the values of pixel. These filters are often used to test whether targets exist or not. There are three kinds of Haar-like features; two upper squares, a line and center border features, as shown in Fig. 2 (Arlazarov et al., 2021) [21].

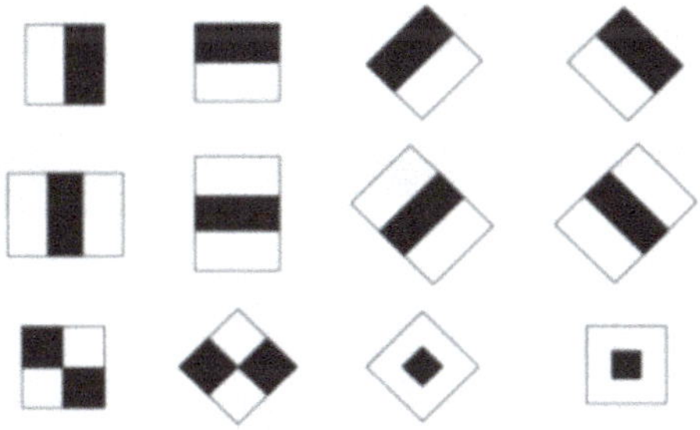

Fig. 2. Haar-like features (Ashir, 2022) [22].

3.2 Histogram of Oriented Gradients (HOG)

The histogram of oriented gradients algorithm is a powerful technique for license plate recognition. This involves extracting gradient features from an image and ultimately using these features to train a classifier. The gradients captured local shape and edge information and enabled robust license plate recognition (Dalal & Triggs 2005) [23]. It involves extracting edge or gradient information that is highly characteristic of local shapes and then executing it with a controllable degree of invariance to local geometric and photometric transformations. There are different types of HOG descriptors; Rectangular HOG (R-HOG), circular HOG (C-HOG), Bar HOG, and Centre-Surround HOG (Li, 2010) [24]. The sketch maps of R-HOG, C-HOG and Centre-Surround HOG are shown in Fig. 3. The R-HOG is to be implemented in current research.

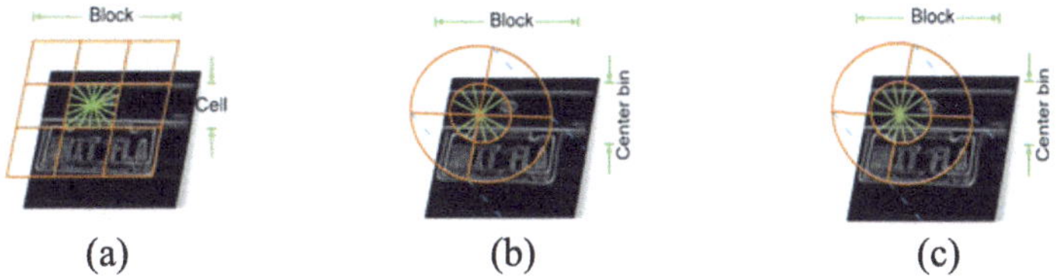

Fig. 3. Types of HOG descriptors a) The rectangular HOG (R-HOG), b) Circular HOG (C-HOG), c) A C-HOG with a single central cell (Dalal & Triggs, 2005) [23].

3.3 Linear Binary Pattern (LBP)

The linear binary pattern (LBP) has a high effective efficiency in a lot of applications like face recognition, surface inspection, image retrieval and texture segmentation and classification (López, 2010) [25]. LBP feature vector is usually used for image classification processing using extreme cascade classifiers. The most important features of LBP are its brightness, invariance and robustness. The image is divided into several non-overlapping local areas named cells. Such a descriptor make a comparison between the central pixel with the eight surrounding neighboring pixels. It is assumed to equal one when the value of the central pixel is bigger than or equal to the value of the neighboring pixel, otherwise the zero value should be displayed (Garcia et al., 2018) [26].

Fig. 4. Samples of car images from the private dataset.

4 Methodology

This research investigates the performance of three features of cascade detectors in distinguishing vehicle license plates. The comparison of this feature was carried out by recording several videos of a maximum of ten seconds long at 30 frames per second, resulting in 444 images as a private data set for different vehicle types with a front view including the vehicle license plate while the car was in the garage. The videos are recorded using the iPhone Pro Max mobile camera with 48 megapixels and 1.8 lens aperture. Figure 4 shows some car images from the private dataset of type JPG 24-bit with dimensions of 601 × 508 pixels. The cascade technique algorithm was executed using an HP laptop with Core i7 processor.

The cascade classifier can be used for one of three well-known features; Harr-like, HOG and LBP. The algorithm works in three stages; Labeling, training and testing as detailed in the next subsections:

4.1 Labeling Step

It was performed manually by labeling the targets (i.e. license plates) found in the private database (DB) images and then saving them as a label file (Lf).

4.2 Training Algorithm

The training process was performed using a cascade object detector to obtain three models; one for each feature type (Harr-like, HOG and LBP) using the above mentioned label file (Lf), database image folder (DB) and negative samples to save the output training models as M_{HARR}.xml, M_{HOG}.xml, and M_{LBP}.xml files, as shown in Fig. 5.

4.3 Testing Recognition Algorithm

This phase evaluated how well an algorithm works to identify a license plate in the recorded video /extracted frame. The test algorithm uses the recorded video and the extracted individual images which can be explained in algorithms I and II. The testing algorithm used the video file as in Algorithm I or extracted frames as in Algorithm II with one of the models; M_{HARR}, M_{HOG}, M_{LBP} to detect the license plate by inserting a bounding box around the license plate, cropping and extracting it, and finally saving it as a license plate.

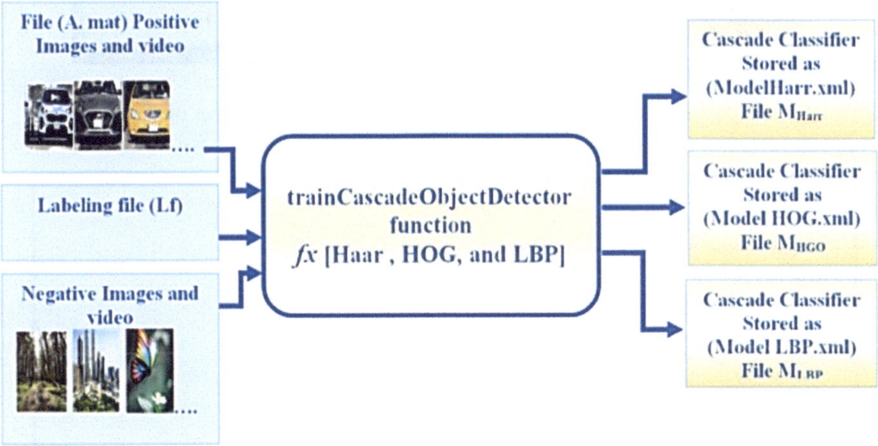

Fig. 5. Cascade training procedure.

Figure 6 shows the application of Algorithm (II) to one of the images of the dataset folder, where an automatic marking of the license plate with red rectangular border shape was detected and then the license plate was finally extracted as shown in the last column of Fig. 6. To evaluate the performance of each feature, two metrics can be considered here; True Positive Rate (TPR) and False Negative Rate (FNR). The True Positive Rate (TPR) measures the proportion of targets correctly defined, as expressed below (Li, 2021 [27]; Shahbaz et al., 2023 [28] & Ali et al., 2024 [29]):

$$TPR = \frac{T_P}{(T_p + F_N)} x100\% \tag{1}$$

where TP represents the cases where the model correctly predicted the presence of the target and it was actually present. FN represents the cases where the target was present but the model failed to detect it, and FP represents the cases where the target was present but the model incorrectly predicted it. A false negative rate measures the proportion of targets that returned negative test results. It is calculated as follows (Bandalakunta, 2019) [30]:

$$FNR = \frac{F_N}{(T_p + F_N)} x100\% \tag{2}$$

Fig. 6. Results of applying Algorithm II on one of the dataset images.

Input

1. video file (vid)
2. Training file (M_f^- model) [where f [Harr or HOG or LBP]]

Steps

1. Determine the number of frames (FN m) in (vid)
2. Loop

i. For i = 1 to FNm
ii. Read frame (Ii) from vid
iii. Detect license plate in image "Ii" frame using (M_f^- model) and determine (BBox) around the target
iv. Insert a rectangle (BBox) and mark the target with a red rectangle border shape in image "Ii"
v. Crop BBox from the frame image "Ii", extract the target "Iy", and then save the extracted license plate image
vi. End for

Output

Recognition results (License plates "Iy")

Algorithm (I) Testing recognition upon the recorded video.

Input

1. Training file (M_f^- model) [where f [Harr or HOG or LBP]]
2. Folder contains N – images to be test

Steps

1-For i = 1 to N

i. Read image Ii
ii. Detect license plate in image frame "Ii" using (M_f^- model) and determine (BBox around the target
iii. Insert a rectangle (BBox) and mark the target with a red rectangle border in an image (Ii)
iv. Crop BBox from the frame image Ii, extract the target Iy and then save the extracted license plate image
v. End for

Output

Recognition results (License plates "Iy")

Algorithm (II) Testing recognition upon dataset folder images.

5 Result Analysis and Discussion

In this study, three models were conducted for license plate recognition and extraction; M_{HARR}, M_{HOG} and M_{LBP}. Figures 7 and 8 show the results of applying such models to the recorded videos and extracted frames, respectively. To test the recognition performance on the recorded video, the three features were tested in such a case as shown in each row of Fig. 7, with the second, third and fourth columns applying Harr-like, HOG and LBP features. The red rectangular shapes surrounding the target were assigned "TP" if the recognition algorithm predicted it correctly, and were assigned "FP" if the prediction was incorrect. At the first video frame, as shown in Fig. 7, the HOG feature application correctly recognizes the license as "TP" and an incorrect one as "FP". On the other hand, for Harr-like and LBP features of the same video frame, the model cannot recognize

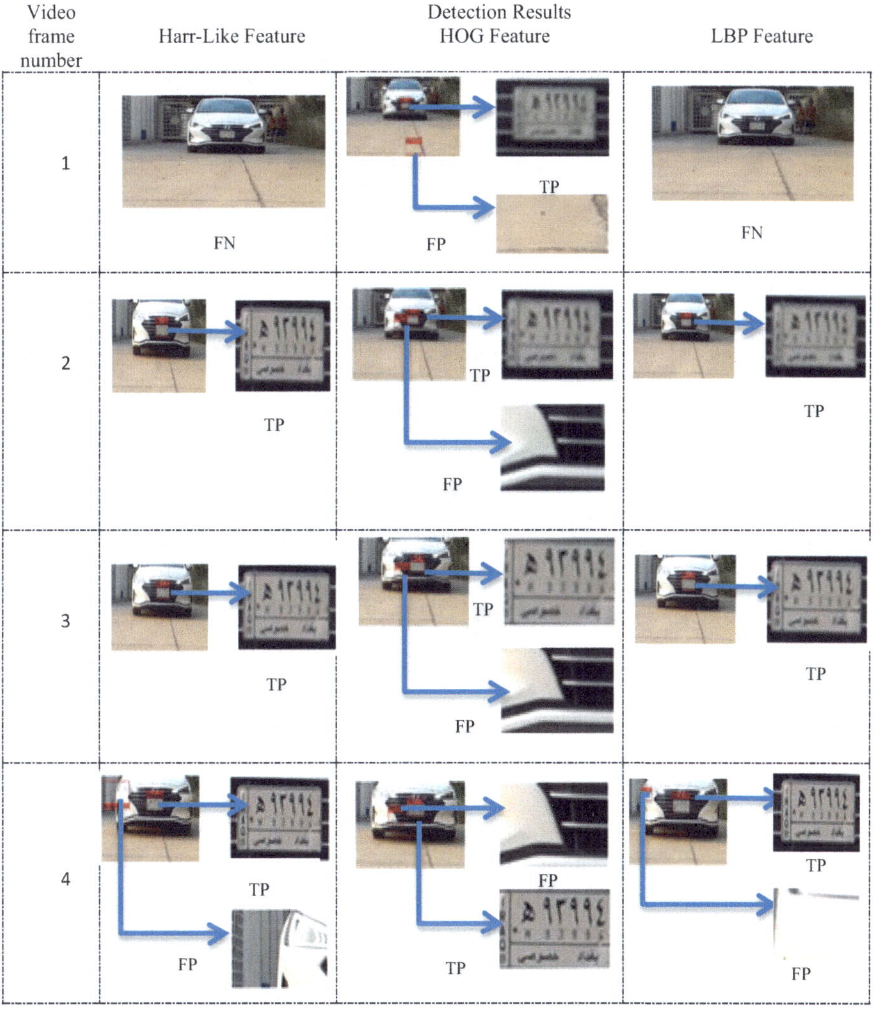

Fig. 7. Results of applying algorithm (I) for testing recognition upon the recorded video for all used features.

the license plate and record "FN" as shown in the 2nd and 4th columns of the first video frame number, respectively.

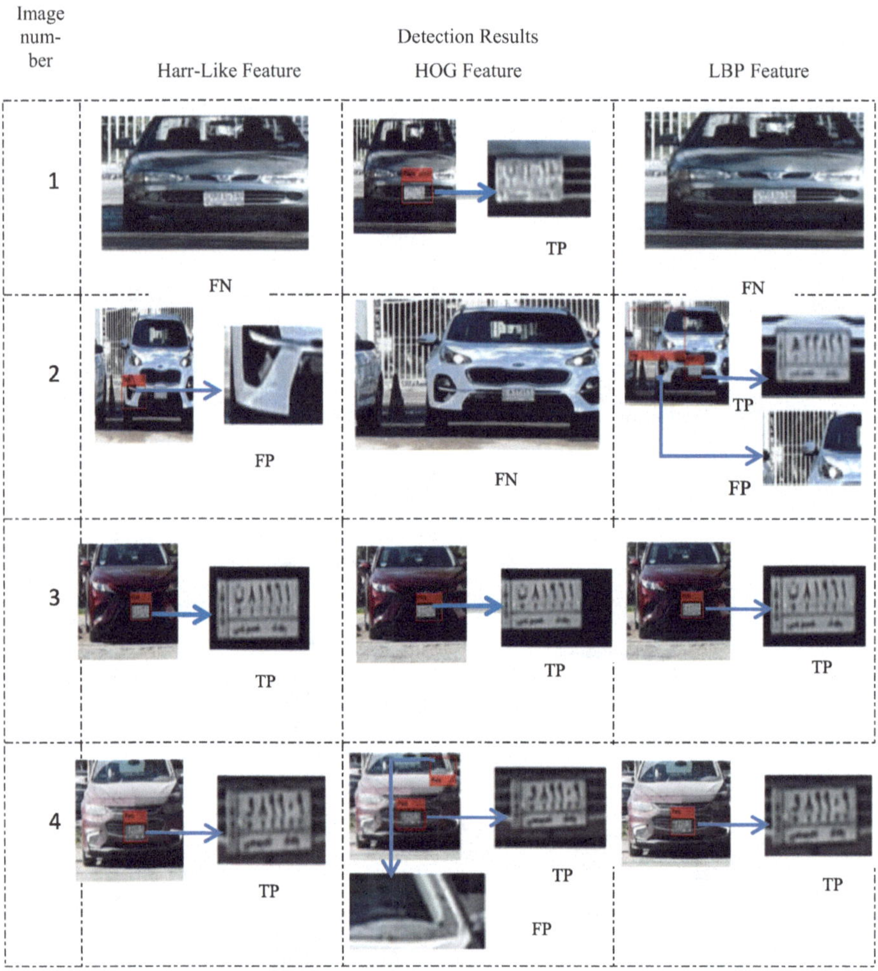

Fig. 8. Results of applying algorithm (II) for testing recognition upon four selected images from the folder dataset for all used features.

Figure 8 shows the results of applying the recognition test algorithm (Algorithm II) to four different cars from the used folder image dataset. Each row in the previous Figure shows the application of the three models used to a specific car. For the gray first car as an example, the Harr-like and LBP features failed to detect any target and the result shows "FN" as shown in the second and fourth columns respectively, while for the HOG feature application a true positive case "TP" resulting as displayed in the third column.

For each captured video, the license plate recognition time can be summarized in Fig. 9. In such a figure, the small pulses indicate the short time taken in recognizing the license plate, while the longest pulse indicates the longest time taken. Finally, and despite the use of multiple videos, for both Haar and LBP models, there is no randomness or fluctuation in the detection time, which can be expressed as a short detection time not exceeding 0.007 s, as shown in Fig. 9 a and c respectively while it isn't the case with the Hog model. Across all recorded videos, there were large and variable fluctuations in the performance of the HOG model for license plate recognition, as shown in Fig. 9b. The latter behavior of the HOG model fulfills its inefficient role in video capture as it is unable to capture temporal dynamics, which consequently affects the performance of the license plate recognition process.

Tables 1 and 2 show the True Positive Rate (TPR) and False Negative Rate (FNR) results for the three models used (M_{HARR}, M_{HOG} and M_{LBP}) for the recorded video and extracted frames, respectively. Table 3 summarizes the average execution time in seconds for all filters (models). The second column indicates the average detected time in seconds per frame, while the last column indicates the average execution time of the recognition process for the entire video frames. The entire models succeeded in recognizing the license plate due to the executed recognition time per frame, which is within fractions of a second and in the range (0.00486–0.0225)s. From such tables, one can conclude that HOG is a more effective model in its recognition capability than others in the case of extracted frames with 85.5% and 14.4% for TPR and FNR respectively with an average recognition time per frame of 0.0225 s while LBP outperforms others on the case of the recorded videos through TPR and is in the range of (90.5–97.5)% with 1.689 s for the detection time performed. On the other hand, the Haar-like model recorded a TPR value in the range (71.7% to 92.7%) and 75.2% for recorded videos and extracted frames, respectively.

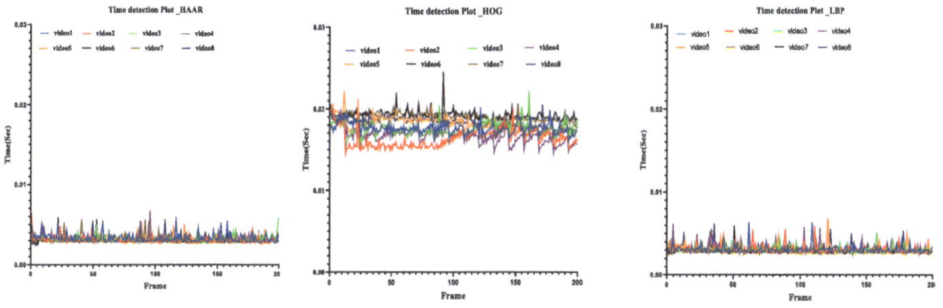

(a) (b) (c)

Fig. 9. Time detection for the captured videos when applying a) Harr, b) HOG, c) LBP features.

Table 1. Results of TPR and FNR for the three models upon the recorded video.

Model Filter	TP	FN	FP	TPR (%)	FNR (%)
Harr	306	27	11	92.7	8.1
	362	41	90	89.8	10.1
	211	83	4	71.76	28.2
HOG	300	33	259	90	9
	320	83	10	79.40	20.5
	217	42	15	83.78	16.21
LBP	325	8	99	97.5	2.40
	365	38	104	90.57	9.42
	241	18	64	93.05	6.94

(*continued*)

Table 1. (*continued*)

Table 2. Results of TPR and FNR for the three models upon the extracted frames.

Model Filter	Number of Images	TP	FN	FP	TPR (%)	FNR(%)
Harr	444	334	110	34	75.22	24.77
HOG	444	380	64	259	85.55	14.41
LBP	444	361	83	455	81.30	18.69

Table 3. Results of average detected time for the three models upon the recording video and extracted frames.

Model Filter	Executed detected time per frame (seconds)	Executed detected time for all video frames (seconds)
Harr	0.00486	1.615
HOG	0.0225	7.429
LBP	0.005	1.689

6 Conclusion

The process of automated vehicle detection is considered one of the important tasks that offers a wide range of possible applications in many safety aspects. The goal of the current work was achieved by the proposed models, which recognize the vehicle license plate in a short time, which is in the range (0.00486–0.0225) s per frame. Except for the HOG model, all models managed to detect and recognize license plates for video recording during the specified time. The HOG model recognized the license plate in a recognition time that exceeded the specified time. In addition, HOG is more effective in its detection capability on still images than videos with 85.5% for TPR and an average detection time of 0.0225 s, while LBP outperforms other models on videos recorded over TPR is in the range (90.5–97.5)% with 1.689 s for executed detected time. In the future, further studies need to be explored to recognize license plates using deep learning techniques such as Convolutional Neural Network or Region-based Convolutional Neural Network, in addition to use other datasets to evaluate the technique performance.

Acknowledgment. The authors would like to thank Mustansiriyah and Baghdad Universities for their supports in the presented work.

Statements and Declarations. There are no conflicts of interest besides the research did not receive any specific funding but was performed as part of the employment of the authors.

References

1. Li, W., Feng, X.S., Zha, K., Li, S., Zhu, H.S.: Summary of target detection algorithms. J. Phys. Conf. Ser. **1757**(1), 012003 (2021)
2. Zhang, J., Meng, Y., Chen, Z.: A small target detection method based on deep learning with considerate feature and effectively expanded sample size. IEEE Access: Pract. Innov. Open Solut. **9**, 96559–96572 (2021)
3. Deb, K., Jo, K.-H.: A vehicle license plate detection method for intelligent transportation system applications. Cybern. Syst. **40**(8), 689–705 (2009)
4. Shashirangana, J., Padmasiri, H., Meedeniya, D., Perera, C.: Automated license plate recognition: a survey on methods and techniques. IEEE Access: Pract. Innov. Open Solut. **9**, 11203–11225 (2021)
5. Salman, Y.D., Alhadawi, H.S., Mahdi, A.S., AL-Dhief, F.T.: Improved automatic license plate recognition system in Iraq for surveillance system using OCR. In: Lecture Notes in Networks and Systems, pp. 270–277. Springer, Cham (2023)
6. Abbass, G., Marhoon, A.: Iraqi license plate detection and segmentation based on deep learning. Al-Mağallaẗ al-'irāqiyya ẗ al-Handasaẗ al-Kahrabā'iyyaẗ Wa-al-Iliktrūniyyaẗ **17**(2), 102–107 (2021)
7. Habeeb, D., et al.: Deep-learning-based approach for Iraqi and Malaysian vehicle license plate recognition. Comput. Intell. Neurosci. **2021**(1) (2021)
8. Li, H., Wang, P., You, M., Shen, C.: Reading car license plates using deep neural networks. Image Vis. Comput. **72**, 14–23 (2018)
9. Xu, Z., et al.: Towards end-to-end license plate identifying and detecting: a large dataset and baseline. In Computer Vision – ECCV 2018, pp. 261–277. Springer, Cham (2018)

10. Al, R.A.A.: Comparison of vehicle detection using Haar-like feature, LBP and HOG technique for feature extraction in cascade classifier. International J. Adv. Sci. Tech. 834–838 (2019)
11. Yousif, B.B., Ata, M.M., Fawzy, N., Obaya, M.: Toward an optimized neutrosophic k-means with genetic algorithm for automatic vehicle license plate recognition (ONKM-AVLPR). IEEE Access: Pract. Innov. Open Solut. **8**, 49285–49312 (2020)
12. Tao, T., Dong, D., Huang, S., Chen, W., Yang, L.: Object detection-based license plate localization and recognition in complex environments. Transp. Res. Rec. **2674**(12), 212–223 (2020)
13. Zhang, L., Wang, P., Li, H., Li, Z., Shen, C., Zhang, Y.: A robust attentional framework for license plate recognition in the wild. IEEE Trans. Intell. Transp. Syst. Public. IEEE Intell. Transp. Syst. Council **22**(11), 6967–6976 (2021)
14. Musil, P., Juránek, R., Zemčík, P.: Unconstrained license plate detection in hardware. In: Proceedings of the 7th International Conference on Vehicle Technology and Intelligent Transport Systems. SCITEPRESS - Science and Technology Publications (2021)
15. Leng, J., et al.: A light vehicle license-plate-recognition system based on hybrid edge–cloud computing. Sensors (Basel, Switzerland) **23**(21), 8913 (2023)
16. Tao, L., Hong, S., Lin, Y., Chen, Y., He, P., Tie, Z.: A real-time license plate identifying and detecting model in unconstrained scenarios. Sensors (Basel, Switzerland) **24**(9), 2791 (2024)
17. Huo, X., Chen, J.: Building a cascade detector and its applications in automatic target detection. Appl. Opt. **43**(2), 293 (2004)
18. Viola, P., Jones, M.: Rapid object detection using a boosted cascade of simple features. In: Proceedings of the 2001 IEEE Computer Society Conference on Computer Vision and Pattern Recognition, CVPR 2001, IEEE Comput. Soc (2005)
19. Shahbaz, S.J., Al-Zuky, A.A.D., Al-Obaidi, F.E.M.: The evaluation of cascade object detector in recognizing different samples of road signs. In: 2022 Iraqi International Conference on Communication and Information Technologies (IICCIT), IEEE (2022)
20. Arunmozhi, A., Park, J.: Comparison of HOG, LBP and Haar-like features for on-road vehicle detection. In: 2018 IEEE International Conference on Electro/Information Technology (EIT), IEEE (2018)
21. Smart Engines Service LLC, Arlazarov, V.V., Voysyat, J.S., Matalov, D.P., Nikolaev, D.P., Usilin, S. A.: Institute for information transmission problems. evolution of the viola-jones object detection method: a survey. Bull. South Ural State Univ. Ser. Math. Model. Program. Comput. Softw. **14**(4), 52–23 (2021)
22. Ashir, A.M.: A real-time automatic kurdistan numberplate recognition system. Eurasian J. Sci. Eng. **8**(1), 126–138 (2022)
23. Dalal, N., Triggs, B.: Histograms of oriented gradients for human detection. In: 2005 IEEE Computer Society Conference on Computer Vision and Pattern Recognition (CVPR'05), IEEE (2005)
24. Li, X.: Vehicle license plate identifying and detecting. A Thesis presented to the Faculty of the Graduate School at the University of Missouri. Website (2010): https://mospace.umsystem.edu/xmlui/bitstream/handle/10355/10564/research.pdf?sequence=3
25. López, L.S.: Local Binary Patterns applied to face identifying and detecting (2010). https://core.ac.uk/download/pdf/41809268.pdf
26. Garcia Freitas, P., Da Eira, L., Santos, S., Farias, M.: On the application LBP texture descriptors and its variants for no-reference image quality assessment. J. Imaging **4**(10), 114 (2018)
27. Li, W.: Analysis of object detection performance based on faster R-CNN. J. Phys. Conf. Ser. **1827**(1), 012085 (2021)
28. Shahbaz, S.J., Al-Zuky, A.A.D., Al-Obaidi, F.E.M.: Real-night-time road sign detection by the use of cascade object detector. Iraqi J. Sci. 4064–4075 (2023)

29. Ali, O.M.S., Al-Zuky, A.A.D., Al-Obaidi, F.E.M.: R-CNN auto-system for detecting text road signs in Baghdad. IEIE Trans. Smart Process. Comput. **13**(2), 140–147 (2024)
30. Bandalakunta Gururajarao, S., Venkatappa, U., Shivaram, J.M., Sikkandar, M.Y., Al Amoudi, A.: Infrared thermography and soft computing for diabetic foot assessment. In: Machine Learning in Bio-Signal Analysis and Diagnostic Imaging, pp. 73–97. Elsevier (2019)

Innovative Dual-Band Microstrip Patch Antenna Design for Enhanced Wireless Communications

Sahar K. Hassan[1] and Zaid M. Khudair[2](✉)

[1] Computer Engineering Department, Engineering College, Al-Iraqia University, Baghdad, Iraq
[2] Biomedical Engineering, College of Engineering, Al-Nahrain University, Baghdad, Iraq
zaid.mustafa.kh@nahrainuniv.edu.iq

Abstract. The study introduces an innovative dual-band microstrip patch antenna that operates across multiple frequency bands. The proposed antenna, with dimensions of $(29 \times 20 \times 1.6)$ mm^3, is positioned on an FR-4 substrate possessing a dielectric constant εr of 4.3 and a loss tangent of 0.002. Copper material with a layer height of 0.035 mm forms both the patches and the ground plane. The antenna accomplishes resonance in two frequency bands: (3.332 GHz) and (6.7 GHz). Specifically, it is well-suited for wireless applications in the WIMAX and X-band realms. The obtained VSWR values of 1.018 and 1.308 at these frequencies confirm an effective impedance match. The directivity diagram and radiation characteristics of the antenna enhance its efficiency by exhibiting radiation patterns that align effectively with the requirements of WIMAX and X-band applications. The proposed antenna has undergone successful manufacturing, verification using a Vector Network Analyzer (VNA), and simulation utilizing CST-MW software version 2021.

Keywords: Dual-Band Antenna · X-band · WIMAX · Microstrip Antenna · CST software

1 Introduction

As technology advances, there is a growing trend for smaller and more compact technological devices. The large quantity of device units has called for more compact size, high-capacity, multi-band antennas capable of allowing synchronous interaction with many devices in wireless communication. Microstrip antennas provide many benefits, including their compact size, lightweight, seamless integration and cost efficiency, thus making them a suitable selection for creating dual-range antenna designs. The conventional microstrip patch antenna consists of a metallic part known as the patch, a dielectric substrate, a ground plane, and a feed line [1].

The quick progress in wireless communication methodologies has amplified the need for handheld devices, affirmed the significance of small-scale dimensions, a sleek design, and cost-effectiveness for all wireless communication components. Antennas assume an axial function in wireless communication, and their physical traits cover substantial directivity, comprehensive gain, efficacy, and wide-ranging operability have

A. J. Obaid et al. (Eds.): ICONEST 2024, ISEM 40, pp. 238–248, 2025.
https://doi.org/10.1007/978-3-031-89960-7_17

become progressively axial for wireless systems. Several strategies have been explored to meet particular requirements, including integrating different shaped apertures techniques or radiating elements [2–4], employing synthetic metamaterials [5, 6], and controlling the ground plane, such as using Electromagnetic Bandgap (EBG) models. Usually, a microstrip patch antenna conducts within a standalone frequency range. However, the need for multi-range antennas has arisen for applications functioning across multiple frequency channels, like WLAN [7, 8].

Researchers have proposed various microstrip antenna designs tailored for applications such as Wi-Fi/WiMax. For instance, one design [9] employs FR-4 material as the substrate, boasting dimensions measuring 44×41 mm^2. This model incorporates a pair of rectangular openings – one for each patch on the antenna and one on the grounding plane. While managing to resonate at 2.5 GHz and 5.8 GHz, the achieved bandwidth measurements remain comparatively limited given its size. Another antenna [10], encompassing dimensions of 70×70 mm^2, operates at intended frequencies utilizing a Defected Ground Structure (DGS) configuration.

While it showcases commendable gain and bandwidth characteristics suitable for wireless applications, it doesn't wholly satisfy the demand for smaller microstrip antennas. Likewise, a microstrip antenna [11] designed for RFID and WLAN applications functions at 2.45 GHz and 5.8 GHz, featuring FR-4 material as the substrate and dimensions of $60 \times 70 \times 1.58$ mm^3. While it effectively fulfills its intended design, its larger proportions and measurement outcomes deviate from simulation values. Furthermore, a patch antenna [12] serves for WLAN applications at 2.4 GHz and 5 GHz. This design incorporates star-shaped openings on the patch segment with FR-4 material and a dielectric constant of $\mathscr{E}r = 4.4$, with dimensions measuring $62 \times 64 \times 1.6$ mm^3. The lower band attains a bandwidth of 65 MHz, while the upper band achieves 175 MHz. In this research paper, researchers designed a dual-frequency microstrip antenna operating at 3.3 GHz and 6.7 GHz, tailored for WiMAX and X-band applications. The antenna's physical dimensions measure $29 \times 20 \times 1.6$ mm^3. The antenna's patch section incorporates two apertures to enable its performance across both frequencies. The study examined the effects of these apertures on the antenna's behavior and functionality at the specified frequencies.

2 Design Concepts and Geometry of the Proposed Dual-Band Antenna

Figure 1(a) illustrates a frontal view of the patch antenna, while Fig. 1(B) offers a rear view. The antenna has measurements of 20 mm in width and 29 mm in length. FR-4 is chosen as the dielectric substrate due to its relatively moderate permittivity ($Er = 4.3$) and substantial dielectric thickness ($h = 1.6$ mm). a 50 Ω microstrip transmission line with a width of 3 mm serves as the feeding mechanism, ensuring impedance matching for the antenna. We introduce slits and triangular apertures into the ground plane to achieve dual-frequency radiation. The copper employed for both the patch part and the ground plane part has a thickness of 0.035 mm. Table 1 itemizes the parameters of the intended antenna.

The dimensions of MPAs can be determined using Eqs. (1) and (2) [12].

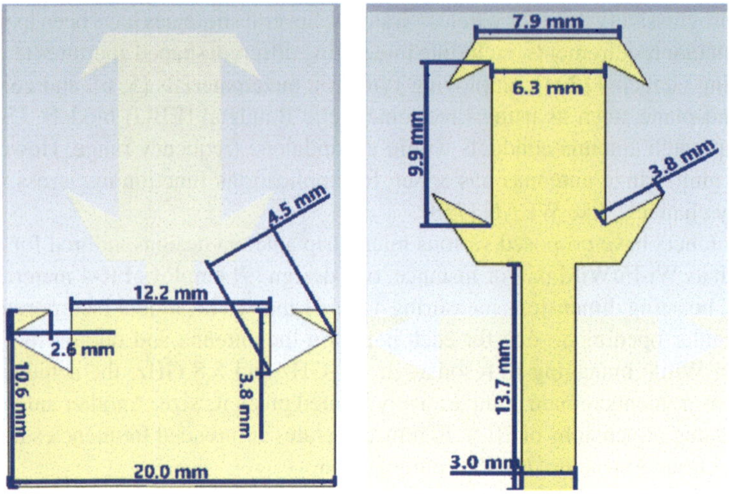

Fig. 1. Illustrates the configuration of the suggested microstrip patch antenna: (a) anterior view, (b) posterior view.

Table 1. Characteristics of the Envisaged Antenna Arrangement

Parameter	Measurements (mm)
Ws	20
Ls	29
Wg	20
Lg	10.6
d	4.3
Lf	13.7
Wf	3
h	1.6
t	0.035

The width of the microstrip patch antenna

$$\mathrm{w_p} = \frac{\mathrm{c}}{2\mathrm{f_r}}\sqrt{\frac{2}{\varepsilon_\mathrm{r}+1}} \tag{1}$$

The Dielectric Constant of Effective Potential:

$$\varepsilon_{\mathrm{eff}} = \frac{\varepsilon_\mathrm{r}+1}{2} + \frac{\varepsilon_\mathrm{r}-1}{2}\frac{1}{\sqrt{1+12\left(\frac{\mathrm{w}}{\mathrm{h}}\right)}} \tag{2}$$

Lengthened Measurement [18]:

$$L_{eff} = \frac{C}{2f_r\sqrt{\varepsilon_{eff}}} \tag{3}$$

The equivalent expansion length ΔL resulted in by the fringe field of the antenna is [18]:

$$\Delta L = 0.412(h)\frac{(\varepsilon_{eff} + 0.3)\left(\frac{w}{h} + 0.264\right)}{(\varepsilon_{eff} - 0.258)\left(\frac{w}{h} + 0.8\right)} \tag{4}$$

where h represents the thickness of the dielectric substrate.

3 The Outcomes of the Proposed Antenna Design

The CST simulation has yielded results for the dual-band antenna configuration. For WiMAX and X-band uses, the antenna's input S11 parameters show a reflection coefficient below −10 dB, which means that the impedance matching is good. The graphical representation in Fig. 2 shows the S-parameters for the dual-band microstrip antenna to evaluate its performance across the target frequencies of 3.3 GHz and 6.7 GHz. In our results, we obtained a dips in the S11 curves at both 3.3 GHz and 6.7 GHz which confirming the antenna resonates at these frequencies. These findings indicate that the presented antenna results are applicable for wireless communication applications. To highlight its signal potency, Fig. 3 illustrates the antenna's gain at these paired frequencies. At both frequencies the antenna exhibits a gain level reaching 2.7 dB at 3.3 GHz and 3.1 dB at 6.7 GHz. These gain results indicate that the antenna can radiate strong signal with suitable directivity for wireless communication applications. Furthermore, Fig. 4 presents the antenna's focus at 3.3 GHz and 6.7 GHz, providing an understanding of its radiating pattern and directional characteristics at these particular frequencies. These

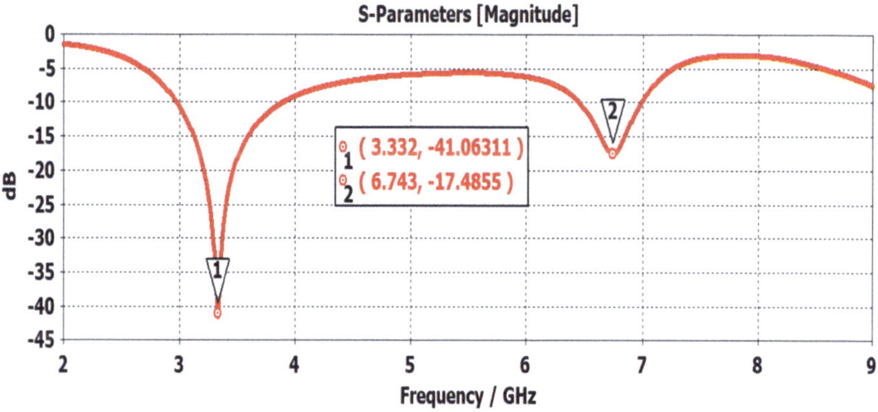

Fig. 2. Illustrates the S-Parameters of the suggested MPAs.

simulation results indicate that the dual-band antenna design is successful in achieving the desired performance for wireless applications and within the X-band frequency range, with good impedance matching, suitable bandwidths, and favorable gain and directivity characteristics at the specified frequencies.

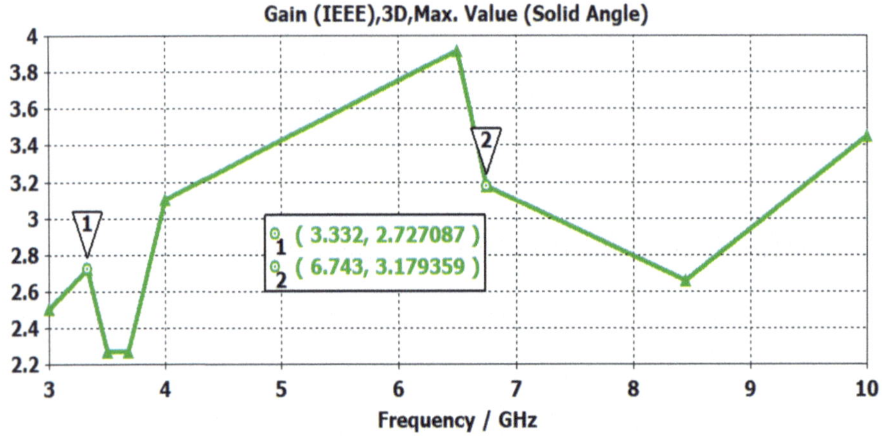

Fig. 3. Represents the antenna gain across the 3.3 GHz and 6.7 GHz.

Figure 4 portrays the far-field radiation pattern of the proposed antenna at 3.3 GHz and 6.7 GHz, respectively. The polar depiction of these charts unmistakably signifies notable directivity and gain. The angular range where the power halved (HPBW) at 3.3 GHz measures 88.1 degrees, resulting in an amplification factor of 2.56 dB. Similarly, at 6.7 GHz, the HPBW is 54.3 degrees, resulting in a directivity of 4.13 dB.

Figure 5 depicts the VSWR curve of the MPAs. At the frequency of resonance 3.3 GHz, the antenna's VSWR reading stands at 1.018, whereas at the frequency of 6.7 GHz, it registers at 1.308. VSWR values below 2 indicate a favorable impedance matching between the antenna and the transmission line, preserving the intended meaning. At the respective resonant frequencies, the antenna's VSWR values of 1.018 and 1.308 demonstrate a strong impedance match. This is beneficial because a good impedance match enhances power transfer efficiency and minimizes signal reflections, leading to improved overall antenna performance at these resonant frequencies.

Figure 6(a) showcases the surface current allocation at 3.3 GHz, while Fig. 6(b) displays the distribution at 6.7 GHz. These figures represent the spatial arrangement of surface currents at these respective frequencies. The maximum surface current is recorded at 69.5908 A/m for the 3.3 GHz case, and 139.81 A/m for the 6.7 GHz case.

Figure 7 represents the Radiation Efficiency of antenna efficiency at the two frequencies: 3.3 GHz and 6.7 GHz. At a frequency of 3.3 GHz, the efficiency is 88%. At a frequency of 6.7 GHz, the antenna efficiency is 71%. This indicates that the antenna performs well at both frequencies, but it exhibits better performance at 6.7 GHz with a higher efficiency rate.

4 Fabrication and Implementation Results

The designed antenna underwent practical fabrication and testing utilizing a vector network analyzer (VNA) up to 20 GHz, as depicted in Fig. 8.

Figure 9 evaluate the simulated and actual data for the suggested antenna. There is a minor contradiction between the simulation results and the fabrication results, which

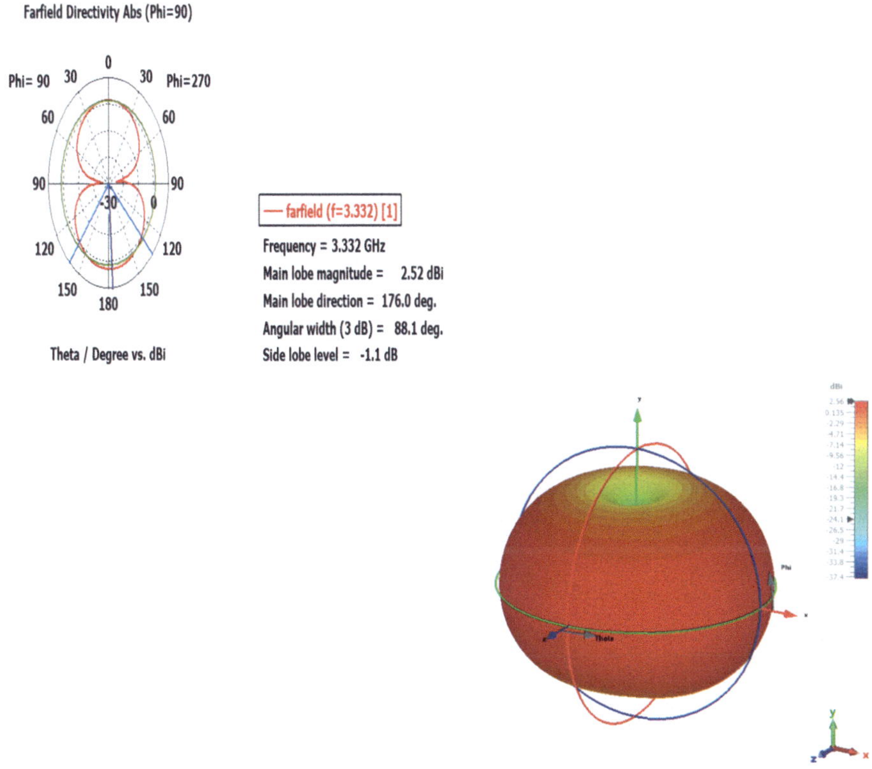

(a)

Fig. 4. Displays the (2D/3D) far-field radiation pattern of the antenna at two various frequencies: (a) f = 3.3 GHz and (b) f = 6.7 GHz.

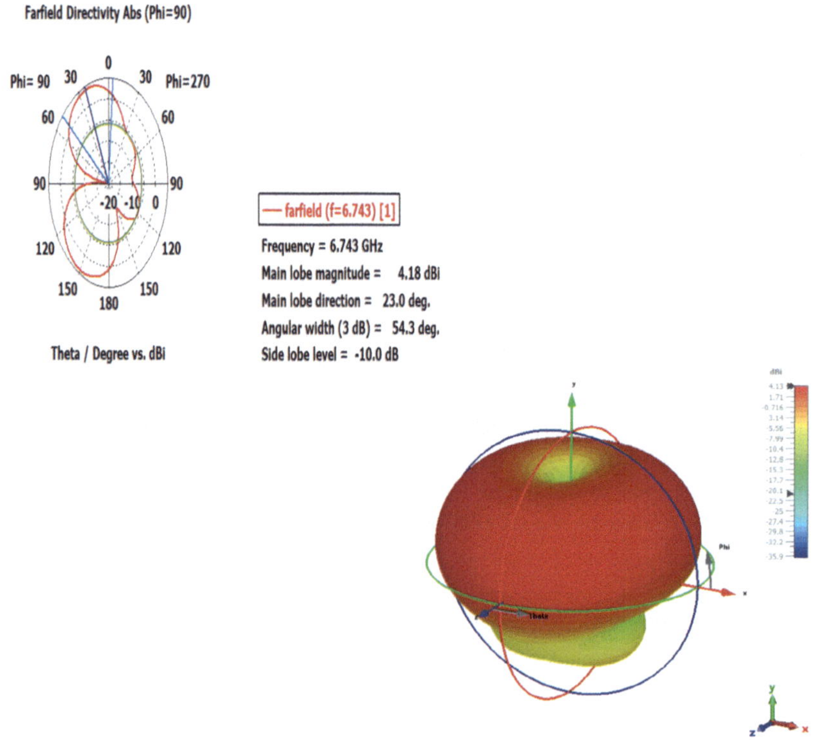

(b)

Fig. 4. (*continued*)

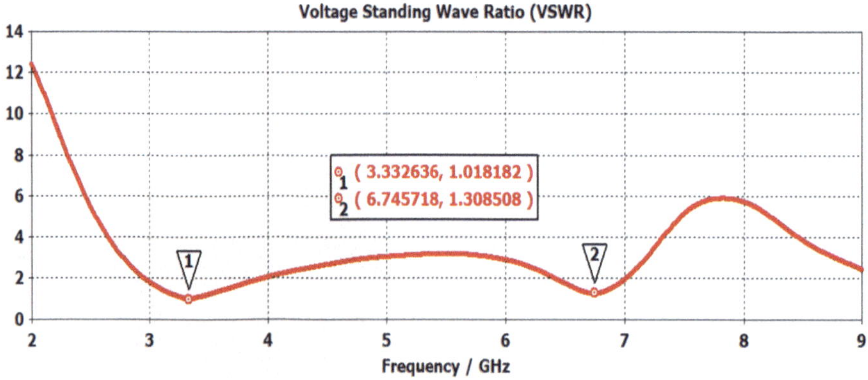

Fig. 5. Exhibits (VSWR) characteristics of the suggested antenna.

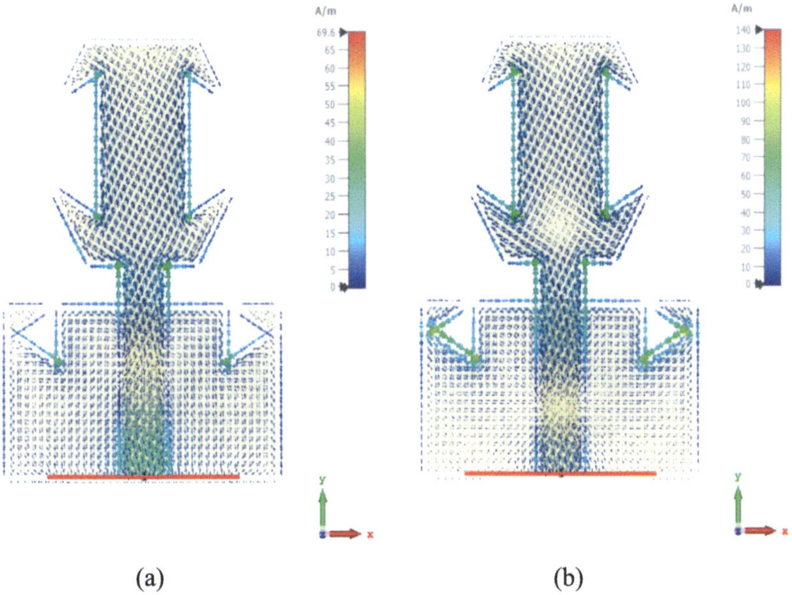

Fig. 6. Demonstrates the arrangement of surface currents at two distinct frequencies: (a) 3.3 GHz and (b) 6.7 GHz.

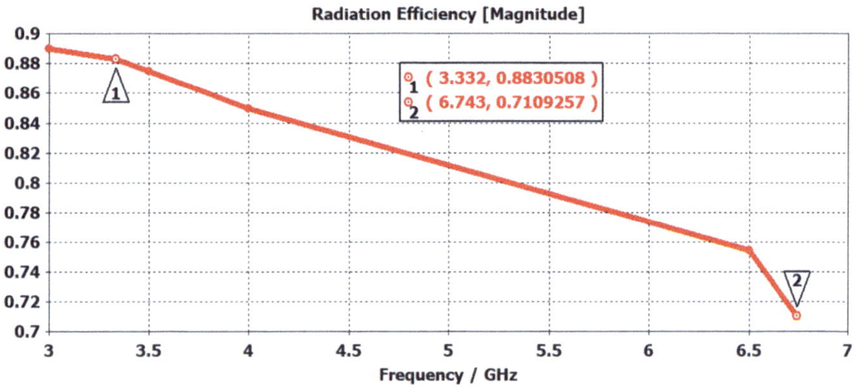

Fig. 7. Depicts the radiation efficiency of the MPAs.

can be attributed to manufacturing errors. Factors such as variations in frequency, fringing effects, and discontinuities may have contributed to these differences. The frontal and posterior perspectives of the antenna are portrayed within Fig. 8(a) and Fig. 8(b), respectively. These views provide a comprehensive visualization of the antenna's physical structure. Table 2 presents a comparison of the suggested antenna's dimensions, resonant frequency, and purpose with those of other antennas. The table shows that the proposed antenna is not only smaller in size but also more effective at providing dual-band functionality. This speaks to its applicability to specific applications.

(a) (b)

Fig. 8. Constructed antenna (a) Anterior view and (b) Posterior view

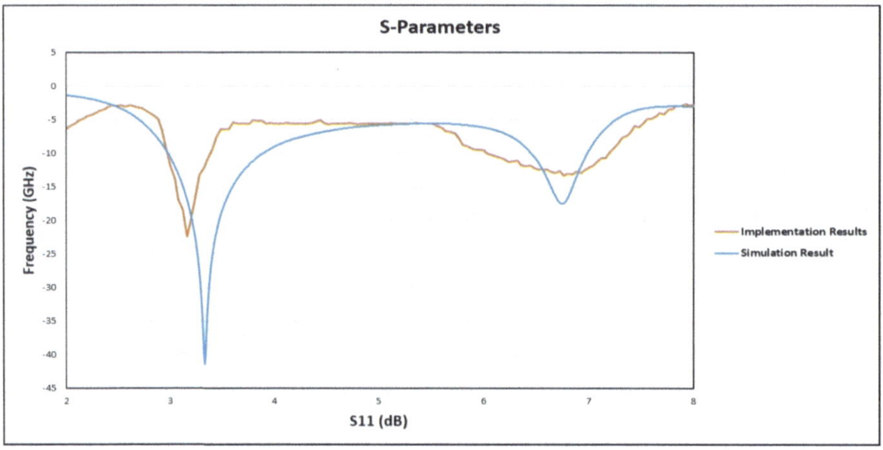

Fig. 9. Reflection Coefficient: Implementation vs. Simulation Results.

5 Comparison with Existing Researches

Table 2 compares the proposed antenna design with existing works in terms of dimensions, resonance frequencies, bandwidth, and the antennas purpose. In [13] the researcher proposed a dual-band microstrip antenna design using C-slot on the patch to achieve dual band and more compact size. The antenna design still suffering from large size, as it still requires a ground plane of 49 mm × 53 mm which still large size for portable devices in wireless communications. The research in [14] proposed a dual-band Inverted L-Antenna for WiMAX and Wi-Fi. While achieving the dual-band operation, the design could be enhanced in term of compactness and complexity of the design, as it still requires a ground plane of 60 mm × 60 mm. Eventually, the authors indicate more optimization needs, which refer to the design limitations in meeting miniaturization and enhanced results. The research in [15] on dual-band circular polarized antenna using split ring resonators presents valuable findings in dual band antenna design. However, the use of split ring resonators may increase fabrication complexity in addition to the large antenna size (70 mm × 70 mm) and narrow bandwidth. Both [16] and [17] proposed dual-band antenna design. These designs face drawbacks related to the size and bandwidth. The compact size of the proposed antennas can degrade the performance, reducing the gain and efficiency. The table confirms the small size and dual-band capacity of the proposed antenna design, making it a more suitable choice for certain wireless communications applications compared to the antennas mentioned in the literature.

Table 2. Comparison of the Proposed Antenna and Other Mentioned Antennas

Reference No.	Antenna purpose	Dimension L × W (mm)	fc (GHz)	S_{11} (dB)
13	Dual-Band	49 × 53	3.5/5.2	−18, −31
14	Dual-Band	60 × 60	3.5 /5	−19, −16
15	Dual-Band	70 × 70	3.1/4.7	−14, −15
16	Dual-Band	41 × 44	2.4/5.8	−25, −35
17	Dual-band	40 × 40	2.3/5.2	−33.072, −21.63
Proposed antenna	Dual-Band	29 × 20	3.3 / 6.7	−41.05104 −17.48134

6 Conclusions

This work introduces the design of a WiMAX and X-band dual-band microstrip patch antenna that operates at frequencies of 3.3 GHz and 6.7 GHz. The antenna performance is achieved by using slits, Triangle slots in the ground plane. The results show a satisfactory reflection coefficient and gain, with a favorable reflection coefficient of −10 dB. We fabricated a physical model of the suggested antenna and validated its performance through computer simulations using CST (Computer Simulation Technology) to confirm the design's credibility. The antenna's dual-band characteristics make it a promising candidate for applications in both WiMAX and X-band wireless communication systems.

References

1. Stutzman, W.L., Gary, A.: Thiele. Antenna theory and design. John Wiley & Sons (2012)
2. Baytöre, C., et al.: Compact metal-plate slotted WLAN-WIMAX antenna design with USB Wi-Fi adapter application. Turkish J. Electr. Eng. Comput. Sci. **27**(6), 4403–4417 (2019)

3. Hassan, S.K., Sallomi, A.H., Wali, M.H.: Design of a dual-band rejection planar ultra-wideband (Uwb) antenna. J. Eng. Sustainable Dev. **26**(4), 30–35 (2022)
4. Doe, J., Smith, A.: Design and optimization of dual-band microstrip patch antennas for WLAN applications. IEEE Trans. Antennas Propag. **20**, 123–134 (2022)
5. Palandoken, M., Henke, H.: Fractal spiral resonator as magnetic metamaterial. In: 2009 Applied Electromagnetics Conference (AEMC). IEEE (2009)
6. Kandakatla, R.A., Ahmed, M.R.: Design and performance analysis of dual-band microstrip patch antennas for smart apparels. In: 2018 Second International Conference on Green Computing and Internet of Things (ICGCIoT). IEEE (2018)
7. Row, J.-S., Huang, Y.-J.: Dual-band dual-polarized antenna for WLAN applications. Microw. Opt. Technol. Lett. **60**(1), 260–265 (2018)
8. Uqaili, R.S., et al.: A study on dual-band microstrip rectangular patch antenna for Wi-Fi. In: Proceedings of Engineering and Technology Innovation **16**, 1–12 (2020)
9. Kaur, J., Nitika, Panwar, R.: Design and optimization of a dual-band slotted microstrip patch antenna using Differential Evolution Algorithm with improved cross polarization characteristics for wireless applications. J. Electromagnetic Waves Appl. **33**(11), 1427–1442 (2019)
10. Karli, R., Ammor, H.: Rectangular patch antenna for dual-band RFID and WLAN applications. Wireless Pers. Commun. **83**, 995–1007 (2015)
11. Ferouani, S., Bendahmane, Z., Taleb-Ahmed, A.: Design and analysis of dual band star shape slotted patch antenna. Microwave Rev. **23**(1), 9–13 (2017)
12. Hassan, S.K., Sallomi, A.H., Wali, M.H.: Loaded notched dual compact rectangular ultra-wideband applications monopole antenna. In: TELKOMNIKA (Telecommunication Computing Electronics and Control) **21**(3), 506–512 (2023)
13. Yassin, A.A., Saeed, R.A., Mokhtar, R.A.: Dual-band microstrip patch antenna design using C-slot for WiFi and WiMAX applications. In: 2014 International Conference on Computer and Communication Engineering. IEEE (2014)
14. Alam, S., et al.: Dual band two elements inverted-L antenna for 3.5 & 5 Ghz mobile Wimax/Wi-Fi Applications. Int. J. Comput. Sci. Eng. Surv. **2**(2) (2011)
15. Kandasamy, K., et al.: Dual-band circularly polarized split ring resonators loaded square slot antenna. IEEE Trans. Antennas Propagation **64**(8), 3640–3645 (2016)
16. Chen, Z., et al.: A compact printed monopole antenna for WiMAX/WLAN and UWB applications. Future Internet **10**(12), 122 (2018)
17. Amjad, O., et al.: Design and implementation of dual band microstrip patch antenna for WLAN energy harvesting system. Appl. Comput. Electromagnetics Soc. J. (2018)
18. Hassan, S.K., Khudair, Z.M.: Circular patch antenna with quintuple band operation design analysis and performance evaluation for multi-frequency applications. Int. J. Intell. Eng. Syst. **17**(3) (2024)

Data Encryption and Key Exchange Protocols by Using SAS Group

Sukaina AL-Bairmani[1], Salah A. Albermany[1], Aris Wijayanti[2](✉), and Alfian Nurlifa[2]

[1] Faculty of Computer Science and Mathematics, University of Kufa, Najaf, Iraq
[2] Universitas PGRI Ronggolawe Tuban, Tuban, Indonesia
ariswijayanti5@gmail.com

Abstract. The main purpose of using data encryption and key exchange protocols for non-commutative groups is to the increase security. The development of key exchange and encryption protocols has been driven by current developments in non-commutative group-based cryptography. In this study, we propose a non-abelian infinite group called SAS, which is easily handled, especially since it is represented as a matrix and then as a set of binary number vectors. The data is then used within a set of two new protocols: a key exchange protocol, such as the enhancement of the Anshel-Anshel-Goldfeld key exchange protocol, and a protocol of encryption and decryption (PED), the PED protocol, on the SAS group.

Keywords: Non-commutative cryptography · Group · protocol

1 Introduction

Group is one of the most significant concepts in the sphere of mathematical cryptography. Groups are defined as sets having an operation that meet specified conditions [1] and [2]. Groups are vital in number theory and sometimes, group theoretical proof methods can be used directly to established number theory facts free from number theory tricks [3]. Further, when grouping is done according to the friends, it has been inflated that the students' performance in the problem-solving skills in the mathematics class is boosted to a dramatic level. More importantly, it can be associated with various many fields like algebraic geometry, representation theory, algebraic combinatorial, etc.; as a result, group is one of the significant fields of research interests in the modern math [3]. Hence, group is one of the most fundamental concepts that forms the cornerstone of many situations in mathematics and has wide application across disciplines.

Lizama's algorithm [6, 7] shows that non commutative operations such as matrix multiplication deviate from commutative cryptography. This method not only increases security as it doesn't rely on computationally difficult problems (i.e. discrete logarithms or integer factorization). Lizama's technique provides higher level security, at fewer key size than post quantum algorithms, relying on matrix multiplication [8, 9] ensuring that public, secret and channel keys are not distinguishable by eavesdroppers. Classical

A. J. Obaid et al. (Eds.): ICONEST 2024, ISEM 40, pp. 249–256, 2025.
https://doi.org/10.1007/978-3-031-89960-7_18

commutative cryptography, however, is also justified because it relies on commutative mathematical structures but it may more susceptible to some kinds of attacks. As such, non-commutative cryptography provides a useful new means of creating secure communication channels, especially in the quantum world [10].

2 Related Works

A Key Exchange Protocol is to ensure secrecy and correctness [11] in secret sharing between parties. Security models for key exchange and their respective stages, parties, and messages are unauthenticated to authenticated protocols [12]. Key exchange is a vital component in network communications (especially in IoT context), especially lightweight protocols to ensure secure device and gateway connection [10]. Using non-commutative structures and polynomials over noncommutative rings leads to key certification and perfect onward secrecy [13]. Furthermore, solutions are obtained for safe symmetric key distribution in IoT devices by providing a secure data exchange protocol that is based on the Trusted Platform Module (TPM) and the MQTT service. An algebraic cryptographic protocol, the Anshel-Anshel-Goldfeld (AAG) key exchange scheme enables safe communication channels. The Tsaban linear span cryptanalysis [14] and many other cryptanalytic attacks are resistant to it. Security models for key exchange also vary from unauthenticated to authenticated protocols; the stages are parties and messages. We have built a safe encryption system using the AAG key exchange which is secure against adaptive selected ciphertext attacks [15]. Automaton groups such as the Grigorchuk and universal Grigorchuk groups act as key-agreement platforms for the AAG metascheme, providing distinct cryptographic capabilities [16]. The AAG scheme's base in challenging mathematical problems makes it a solid alternative for secure key exchange in modern cryptographic applications. Stickel's key exchange technique has been modified and analyzed in recent studies. Grigoriev and Shpilrain proposed a tropical variant of Stickel's procedure, which was later challenged by Kotov and Ushakov [17] Key exchange protocols, which include unauthenticated and authenticated protocols, are critical components in maintaining secure communication [12]. In the field of IoT, where secure communication is critical, a lightweight key exchange protocol based on noncommutative structures and polynomials over noncommutative rings has been developed, which improves security while providing key certification and perfect onward secrecy [18]. Furthermore, progress has been achieved in three-party key exchange protocols to resist common attacks and enhance efficiency, with the introduction of a new approach based on private keys and a cross-realm key exchange protocol that requires just three rounds to generate a common session key [10] Ko and Lee devised a key exchange system based on the generalized dihedral group D(A) [19]. In contrast, Lee et al. proposed improving a group key exchange protocol by resolving security flaws in forward and backward secrecy when new members join or leave the group [20].

Both Gupta (2019) [21] and Skuratovskii (2020) [22] suggest key exchange techniques based on non-commutative groups. Skuratovskii concentrates on the metacyclic group G, proving that the conjugacy problem is unsolvable and introducing a method for creating a shared key. Gupta, on the other hand, presents a security-based factorization with discrete logarithm problem (FDLP)-based key exchange protocol utilizing matrices over a non-commutative semigroup. In addition, Eftekhari (2012) [23] investigates a Diffie-Hellman key exchange with matrices over non-commutative rings, highlighting the protocol's security and suggesting a platform group as a collection of matrices over a non-commutative ring.

3 Elements of the Proposed Non-communitive Group

3.1 SAS Group

Let SAS(n) is an infinite non-commutative Group generated from a set S of order $2n^2+2$, where n ≥ 1, n ∈ $\mathbb{Z}$. let

$$Z^{+n} \subseteq \mathbb{Z}, \text{ and } Z^{-n} \subseteq \mathbb{Z}, \text{ such that } Z = Z^{+n} \bigcup Z^{-n}, \; and\; Z^* = \{0\} \bigcup Z$$

where

$$Z^{+n} = \{1,2, \ldots, n\}, \text{ and } Z^{-n} = \{-1, -2, \ldots, -n\},$$

and suppose

$$S = (\bigcup_{i=-n}^{n} S_i) \bigcup \{e, \tau\}, n \geq 1, n \in \mathbb{Z}$$

where

$$S_i = \bigcup_{j=1}^{m} \{\sigma(|i|)_{sign(i)j}\}, i \in Z$$

Such that

1. $\left(\sigma(i1)_{j1} * \sigma(i2)_{j2}\right) * \sigma(i3)_{j3} = (i1)_{j1} * \left(\sigma(i2)_{j2} * \sigma(i3)_{j3}\right),$

$$i1, i2, i3 \in Z^{+n}, \; and\, j1, j2, j3 \in Z$$

2. e *is* the identity element of, such that
   a. $e * \sigma(i)_j = \sigma(i)_j * e = \sigma(i)_j, i \in Z^{+n}$, and $j \in Z$
3. τ vertiable of e, such that $\tau * \tau = e$

$$\tau * \sigma(i)_j = \sigma(\theta(i))_j, j \in Z, \; and\; i = 1,2, \ldots$$

$$where\, \theta(r) = \begin{cases} sign(r)(|r| + 1) & |r| < n \\ sign(r)1 & otherwise \end{cases}$$

$$\sigma(i)_j * \tau = \sigma(i)_{\theta(j)} j \in Z, \; and\; i = 1,2, \ldots$$

4. The invers of $\sigma(i)_j$ *is* $\sigma(|j|)_{-sign(j)i}$, $i \in Z^{+n}$, and $j \in Z$

Note: If $\sigma = \sigma(i1)_{j1} * \sigma(i2)_{j2} \cdots * \sigma(im)_{jm}$, then

$$\sigma^{-1} = \sigma(|jm|)_{-sign(jm)im} * \cdots \sigma(|j2|)_{-sign(j2)i2} * \sigma(|j1|)_{-sign(j1)i1},$$

$$\text{where } m > 1, i1, i2, \ldots, im \in Z^{+n}, \text{ and } j1, j2, \ldots, jm \in Z$$

Example 1: Let SAS(2) is an infinite non-commutative Group n = 2, then.

$$S_1 = \{U_L, U_R\} \equiv \{\sigma(1)_1, \sigma(1)_2\},$$

$$S_{-1} = \{U_{-L}, U_{-R}\} \equiv \left\{\sigma(1)_{-1}, \sigma(1)_{-2}\right\},$$

$$S_2 = \{D_L, D_R\} \equiv \{\sigma(2)_1, \sigma(2)_2\},$$

$$S_{-2} = \{D_{-L}, D_{-R}\} \equiv \left\{\sigma(2)_{-1}, \sigma(2)_{-2}\right\},$$

$$S = S_1 \bigcup S_2 \bigcup S_{-1} \bigcup S_{-2} \bigcup \{e, \tau\} = (\bigcup_{i=-2}^{2} S_i) \bigcup \{e, \tau\}$$

such that

$$U_L^{-1} = U_{-L}, U_R^{-1} = D_{-L}, D_L^{-1} = U_{-R}, \textit{ and } D_R^{-1} = D_{-R}$$

The order of generation set of the group SAS(2) is 10.

$$\textit{where } \tau * \tau = e, \textit{ and } e * \alpha = \alpha * e = \alpha, \forall \alpha \in SAS(2)$$

3.2 *Representation of SAS Group of Order* $2n^2 + 2$

The representation of a the group SAS(n) by homomorphism ψ

$$\psi : SAS(n) \rightarrow GL_{n,p}(\mathbb{Q})$$

where $GL_{n,p}(\mathbb{Q})$ is a set of square matrices with multiplication operation, and each elements of $GL_{n,p}(\mathbb{Q})$ belong to rational set $\mathbb{Q}$. The map ϕ representation of degree n, and p prime number, in SAS(n) over $\mathbb{Q}$, where $\mathbb{Q}$ is a rational numbers set and $GL_{n,p}(\mathbb{Q})$ is the group of matrices generated by each element $\sigma(s)_t \in SAS(n)$ corresponding the square matrix $M = [m_{ij}]$, of size n, such that if $|t| = k$, then

$$m_{sk} = \begin{cases} p & t > 0 \\ p^{-1} & t < 0 \end{cases}$$

and

$$m_{ij} = \begin{cases} 1 \; i = j, i \neq s, \; and \, j \neq k \\ 1 \; i \neq j, i = k, \; and \, j = s \\ 0 \qquad otherwise \end{cases}$$

where $\psi(e) = I_n$ is identity matrix , and $\psi(\tau) = F_n = [f_{ij}] is$ Unitary matrix

$$f_{ij} = \begin{cases} 1 \; i + j - 1 = n, \\ 0 \quad otherwise \end{cases}$$

From above, then

$$\psi(\tau)\psi(\tau) = F_n F_n = I_n = \psi(e) = \psi(\tau * \tau)$$

To proof the $GL_{n,p}(\mathbb{Q})$ is group, must satisfy the following conditions:

1. It is clear the associative law is in $GL_{n,p}(\mathbb{Q})$,
2. The identity of $GL_{n,p}(\mathbb{Q})$ is I_n, and $\psi(e) = I_n$.
3. Suppose $\pi = \sigma(r)_t$, and $\omega = \sigma(|t|)_{-sign(t)r}$, such that $\pi * \omega = e$, from (4) in Sect. 3.1, and $\psi(\pi * \omega) = \psi(e) = I_n$,

   let $A = \psi(\pi) = [a_{ij}], B = \psi(\omega) = [b_{ij}]$, and $k = |t|$

$$a_{ij} = \begin{cases} 1 & i = j, i \neq r, \; and \, j \neq k \\ 1 & i \neq j, i = k, \; and \, j = r \\ p^{sign(t)} & i = r, \; and \, j = k \\ 0 & otherwise \end{cases}$$

$$b_{ij} = \begin{cases} 1 & i = j, i \neq k, \; and \, j \neq r \\ 1 & i \neq j, i = r, \; and \, j = k \\ p^{-sign(t)} & i = k, \; and \, j = r \\ 0 & otherwise \end{cases}$$

   If $C = AB = [c_{ij}]$, $c_{ij} = \sum_{l=1}^{n} a_{il} b_{lj}$
   then

$$cij = \begin{cases} 1 & j = r, \; \text{and} \, i = r \\ 1 & j = k, \; \text{and} \, i = k \\ 1 \; i = j, i \neq k, & \text{and} j \neq r \\ 0 & otherwise \end{cases}$$

   From above that mean $cij = 1, \; i = j$, and $cij = 0, i \neq j, \forall i, j \in Z^{+n}$, then $C = I_n$,

$$\psi(\pi * \omega) = \psi(e) = I_n = AB = \psi(\pi)\psi(\omega)$$

   Then it is clear that $GL_{n,p}(\mathbb{Q})$ is non-commutative group.

Example 2: suppose n = 7, p = 5, and $\sigma(4)_3 \in SAS(7)$, then $M = \psi(\sigma(4)_3) \in GL_{7,5}(\mathbb{Q})$, such that.

$$M = \begin{pmatrix} 1\,0\,0\,0\,0\,0\,0 \\ 0\,1\,0\,0\,0\,0\,0 \\ 0\,0\,0\,1\,0\,0\,0 \\ 0\,0\,5\,0\,0\,0\,0 \\ 0\,0\,0\,0\,1\,0\,0 \\ 0\,0\,0\,0\,0\,1\,0 \\ 0\,0\,0\,0\,0\,0\,1 \end{pmatrix}$$

Example 3: The generation set of Group SAS(2) of size 2, is

$S = \{e, \tau, \sigma(1)_1, \sigma(1)_2, \sigma(2)_1, \sigma(2)_2, \sigma(1)_{-1}, \sigma(1)_{-2}, \sigma(2)_{-1}, \sigma(2)_{-2}\}$.
where $\psi(s)$, $s \in S$ corresponding of following set of Matrices (see

Example 4: $GL_{2,p}(\mathbb{Q}) = \{e, \tau, U_L, U_{-L}, U_R, U_{-R}, D_L, D_{-L}, D_R, D_{-R}\}$, such that.

$e = \begin{bmatrix} 1\,0 \\ 0\,1 \end{bmatrix}, \tau = \begin{bmatrix} 0\,1 \\ 1\,0 \end{bmatrix}$, and

$$U_L = \begin{bmatrix} p\,0 \\ 0\,1 \end{bmatrix}, U_R = \begin{bmatrix} 0\,p \\ 1\,0 \end{bmatrix}, D_L = \begin{bmatrix} 0\,1 \\ p\,0 \end{bmatrix}, D_R = \begin{bmatrix} 1\,0 \\ 0\,p \end{bmatrix}$$

$$U_{-L} = \begin{bmatrix} p^{-1}\,0 \\ 0\;\;1 \end{bmatrix}, D_{-L} = \begin{bmatrix} 0\;\;1 \\ p^{-1}\,0 \end{bmatrix}, D_{-R} = \begin{bmatrix} 1\;\;0 \\ 0\,p^{-1} \end{bmatrix}, U_{-R} = \begin{bmatrix} 0\,p^{-1} \\ 1\;\;0 \end{bmatrix}$$

4 Protocols of Key Exchange

A better Anshel–Anshel–Goldfeld (AAG) algebraic cryptographic key-exchange method is developed, which is specifically resistant to the Tsaban linear span cryptanalysis. Not like the first version. As a theoretical foundation for creating public-key cryptosystems, algebraic key established protocols based on the complexity of solving equations over algebraic structures are identified.

This protocol uses a non-abelian group SAS(n) for key exchange. It is noteworthy because, unlike the technique owing to Ko, Lee, et al., it does not necessitate two commuting subgroups, A and B of SAS(n).

Let SAS(n) be a platform group, which is a fixed nonabelian group.

Public and private information of Alice:

- Alice's public key is a tuple of elements $\boldsymbol{\sigma} = (\sigma_1, \ldots, \sigma_L)$ in $SAS(n)$.
- The elements from $\boldsymbol{\sigma}$ and their power inverses compose Alice's private key: $\sigma_{i_1}^{\gamma_1}, \ldots, \sigma_{i_m}^{\gamma_m}$, where $\sigma_{i_k} \in \boldsymbol{\sigma}$ and $\gamma_k = \pm p_k$, where p_k random value. She calculates the result based on that order.

$A = \sigma_{i_1}^{\gamma_1}, \ldots, \sigma_{i_m}^{\gamma_m}$.

Public and private information of Bob:

- Bob's public key is a tuple of elements $\boldsymbol{\beta} = (\beta_1, \ldots, \beta_L)$ in $SAS(n)$.
- The elements from $\boldsymbol{\beta}$ and their power inverses compose Bob's private key:$\beta_{j_1}^{\eta_1}, \ldots, \beta_{j_m}^{\eta_m}$, where $\beta_{j_k} \in \beta$ and$\eta_k = \pm w_k$. where w_k random value. He calculates the result based on that order.

$B = \beta_{j_1}^{\eta_1}, \ldots, \beta_{j_m}^{\eta_m}$.

Transitions:

- Bob receives a tuple from Alice, $\overline{\boldsymbol{\sigma}} = \left(A^{-1}\beta_1 A, \ldots, A^{-1}\beta_L A\right)$.
- Alice receives a tuple from Bob, $\overline{\boldsymbol{\beta}} = \left(B^{-1}\sigma_1 B, \ldots, B^{-1}\sigma_L B\right)$.

Shared key:

The group element is the key both Bob and Alice share. $K = A^{-1}B^{-1}AB \in SAS(n)$ defined as the commutator between A and B.

- Alice uses a product to calculate K. $A^{-1} \cdot \left(B^{-1}a_{i_1}^{\varepsilon_1}B\right) \cdots \left(B^{-1}a_{i_L}^{\varepsilon_L}B\right) = A^{-1}B^{-1}AB.$
- Bob uses product to calculate $K.\left(A^{-1}b_{j_L}^{-\delta_L}A\right) \cdots \left(A^{-1}b_{j_1}^{-\delta_1}A\right) \cdot B = A^{-1}B^{-1}AB.$

5 Protocols of Data Encryption/ Decryption

This protocol explains how to use a non-commutative group SAS(n) to encrypt and subsequently decrypt a secret message M. Allow Alice to wish to communicate with Bob in secret.

1. Let be a non-commutative group SAS(n). Given A and B as public subgroups of SAS(n), let's say that for every σ in A and every β in B, $\sigma\beta \neq \beta\sigma$.
2. A selected element (x) from SAS(n) is published.
3. Bob selects secret key κ from A and makes it public. $\omega = x^{\kappa}$ as his public key.
4. Alice selects a random number r from B and computes $t = y^{r}$.
5. The encrypted message m is

   $$C = (x^{\kappa}, H(t) \oplus m),$$

6. where the symbol $\otimes$ stands for the XOR operation and H is some hash function. C is sent to Bob by Alice.
7. Bob decrypts C using the following method to recover t:

   $(x^{r})^{\kappa} = x^{r\kappa} = x^{\kappa r} = (x^{\kappa})^{r} = z^{r} = t.$

   The plain text message send by Alice is

   $m = (H(t) \oplus m) \oplus H(t).$

6 Conclusions

When expressed as matrices, the non-commutative SAS(n) group provides a great deal of versatility for use in cryptographic applications. These results highlight the close connection that exists between non-commutative structures, matrix representations, and cryptographic protocols. They also show how flexible and adaptive the non-commutative SAS(n) group is when represented as a matrix.

References

1. Wang, M.: Group Theory in Number Theory. Theor. Natural Sci. **5**, 9–13 (2023)
2. Widodo, S.A., Turmudi, T., Dahlan, J.A., Watcharapunyawong, S., Robiasih, H., Mustadin, M.: The sociograph: Friendship-based group learning in the mathematics class. Infinity J. **12**(1), 27–40 (2023)
3. Goethals, P.L., Scala, N.M., Bennett, D.T. (eds.) Mathematics in Cyber Research. CRC Press (2022)
4. De Concini, C., Gille, P., Littelmann, P.: Algebraic Groups. Oberwolfach Reports **18**(2), 1087–1148 (2022)
5. Fujita, T., Doney, J., Flanagan, R., Wegerif, R.: Collaborative group work in mathematics in the UK and Japan: use of group thinking measure tests. Education 3–13, **49**(2), 119–133 (2021)
6. Non-Commutative Key Exchange Protocol. undefined (2022)
7. Lizama-Pérez, L.A., Romero, J.M.L.: Non-Commutative Key Exchange Protocol (2021)
8. Skuratovskii, R., Osadchyy, V.: An application of Miller Moreno groups to establishment protocol Non commutative cryptography. In: 2020 Fourth World Conference on Smart Trends in Systems, Security and Sustainability (WorldS4), pp. 126–130. IEEE, July 2020
9. Van den Bergh, M.: Non-commutative crepant resolutions, an overview. arXiv e-prints, arXiv-2207 (2022)
10. Kanwal, S., Inam, S., Ali, R., Cheikhrouhou, O., Koubaa, A.: Lightweight noncommutative key exchange protocol for IoT environments. Front. Environ. Sci. **10**, 996296 (2022)
11. Lin, J., von Willich, M., Lo, H.K.: Composable Security of Distributed Symmetric Key Exchange Protocol (2023). arXiv preprint arXiv:2304.13789
12. Pointcheval, D.: Asymmetric Cryptography: Primitives and Protocols\Key Exchange. 30 Nov 2022, pp. 187–211, WILEY online (2022)
13. Furtak, J.: Data Exchange Protocol for Cryptographic Key Distribution System Using MQTT Service. In: 2022 17th Conference on Computer Science and Intelligence Systems (FedCSIS), pp. 611–615. IEEE, September 2022
14. Roman'kov, V.: An improved version of the AAG cryptographic protocol. Groups Complexity Cryptology **11**(1), 35–41 (2019)
15. Isobe, S., Iwazaki, J.Y., Koizumi, E., Suda, A.: A Construction of Cramer-Shoup Like Encryption Protocols Based on the Anshel-Anshel-Goldfeld Key Exchange Protocol. In: 2018 Sixth International Symposium on Computing and Networking Workshops (CANDARW), pp. 410–416. IEEE, November 2018
16. Grigorchuk, R., Grigoriev, D.: Key agreement based on automaton groups. Groups Complexity Cryptol. **11**(2), 77–81 (2019)
17. Muanalifah, A., Sergeev, S.: Modifying the tropical version of Stickel's key exchange protocol. Appl. Math. **65**(6), 727–753 (2020)
18. Akiyama, K., Nakamura, S., Ito, M., Hirata-Kohno, N.: A key exchange protocol relying on polynomial maps. Int. J. Math. Ind. **11**(01), 1950003 (2019)
19. Lloyd, C.: The Ko-Lee Key Exchange Protocol with Generalized Dihedral Groups (2016)
20. Oraei, H., Pourpouneh, M., Ramezanian, R.: An Improvement Over Lee et al.'s Key Agreement Protocol. J. Comput. Secur. **5**(1), 3–12 (2018)
21. Skuratovskii, R., Williams, A.: Some approach to key exchange protocol based on non-commutative groups. Int. J. Math. Models Methods Appl. Sci. **14**(2), 5–7 (2020)
22. Gupta, I., Pandey, A., Dubey, M.K.: A key exchange protocol using matrices over group ring. Asian-European J. Math. **12**(05), 1950075 (2019)
23. Eftekhari, M. (2012). A Diffie–Hellman key exchange protocol using matrices over noncommutative rings

Studying the Stability for Second Kind Mixed Volterra – Fredholm Integral Equation

Talaat I. Hassan[1(✉)] and Chiman I. Hussein[2]

[1] Automotive Technology Engineering Department, Erbil Technology College, Erbil Polytechnic University, Erbil, Iraq
talhat.hassan@epu.edu.iq

[2] Department of General Sciences, College of Basic Education, Salahaddin University, Erbil, Iraq
Chiman.hussen@su.edu.krd

Abstract. In this paper, Researchers study the stability and convergence for second kind mixed Volterra fredholm integral equation (SKMVFIE). Researcher introduce and prove several new theorems regarding the convergence of the iteration approach, driving the sufficient conditions for iterative technique this condition is very important, because it helps to understand that the numerical solutions convergent to the exact solution of the problems and to find the convergence interval for the main problems in the integral equation. Also, by this sufficient condition we can decide that the iterative technic is converging or diverging to the exact solution or not. This iterative process was supported by solving several numerical examples but in this work we wrote only two examples. In these examples, illustrated bounded and convergence intervals. Also, if the mapping Z satisfies the convergence condition, then it has a fixed point. Finally, supporting the new approach with examples that demonstrate its ability and efficiency.

Keywords: Fixed point · iterative approach · contractive mapping · stable · SKMVFIE

1 Introduction

Numerical approaches in integral equations play an important role in applied sciences, as various modules in these fields can be transformed into integral equations [1, 2]. The area of integral equations is one of the most useful in pure and applied mathematics [3, 4]. Because there are many problems in applied sciences that have been reformulated into integral equations [5, 9, 13], other problems have direct representation in terms of integral equations. [6, 7, 10]. Different types of differential equation can be reduced to different types of integral equation [14, 16]. The TDFIE-2nd expression can represent the cross section of dielectric cylinders with varying properties [13, 20]. Numerical applications have emerged recently for such problems, including the investigation of flexural waves under a uniform magnetic field. Integral equations involving functions of multiple independent variables are common in the study of many partial differential

A. J. Obaid et al. (Eds.): ICONEST 2024, ISEM 40, pp. 257–265, 2025.
https://doi.org/10.1007/978-3-031-89960-7_19

equation problems originating from diverse dynamic models [12, 20]. These equations in the case of two-dimensional integral equations have numerous compelling applications in Mechanical engineering, Physical sciences, and other applied sciences.

Many researchers have solved different kinds of integral equation problems using numerical methods to find approximations (Saadati 2009). Saadati (2009) studied integral equations, while Saeed and Berdawood (2015) solved two-dimensional linear Volterra-Fredholm integral equations. Banifatewmi et al. (2007) solved mixed Volterra-Fredholm integral equations (VFIE), and Hassan et al. (2017) found numerical solutions for VFIE-2. Sulaiman (2018) studied the stable interval (SI) and convergence interval (CI) of VFIE-2. Talhat (2019) used an iterative kernel technique to solve mixed VFIE. Abdoul and Ahmed (2020) studied the stability of two-dimensional integral equations. Momenzade et al. (2023) solved stochastic Volterra-Fredholm integral equations. To extend these works, studying the stability and convergence for stochastic mixed Volterra-Fredholm integral equations (SKMVFIE) is essential. Talaat and Chiman (2024) solved the second kind of two-dimensional Volterra integral equation.

2 Definitions and Theorems

For the next section, we will describe and discuss several definitions and theorems related to the main work.

Definition 1. The two dimensions as follows [9, 18].

$$w(r) = t(r) + \lambda \int_a^r \int_c^d H(s, q) w(q) dq ds \tag{1}$$

named TDMVFIE-2, such that $t(r)$ in $E = \{r; a \leq r \leq h_1\}$ and $H(s, q)$ in $H = \{(s, q); (s, q) \in [c, d] \times [0, T]\}$ are continuous known mappings on the connected regions respectively, while $w(r)$ is an unknown function.

Definition 2. The SKVFIE has an iterative approach of the form [2, 18].

$$w^0(r) = t(r)$$

$$w^{n+1}(r) = Z(w^n(r)) = t(r) + \int_a^r \int_c^d H(s, q) w^n(q) dq ds \tag{2}$$

for $n \geq 0$, generating the sequence $\{w^n(r)\}_{n=0}^{\infty}$ of approximate results.

Definition 3. [4, 8].

Assume (B, M) be a metric space for $D \subseteq B$ and the mapping $Z : F \to Y$ be a mapping $Z(u_1) = v_1$, if $Z(p) = p$ then Z has a fixed point (F.P) $p \in D$.

Definition 4. Assume that $(Y, \|.\|)$ be a Banach space, Z is a mapping defined by $Z : Y \to Y$ and iteration $w^{n+1}(r) = Z(y, w^n(r))$, $e_n = \|w^{n+1} - Z(y, w^n)\|$ if $\lim_{r\to\infty} e_r = 0$ then $w^n(r)$ is convergent to F.P, and $w^{n+1}(r) = Z(y, w^n(r))$ is $Z-$ stable. [8, 11].

Definition 5. Z mapping is contractive if $\exists$, δ such that $0 < \delta < 1$, and elements $e, f \in A$ we have [6, 8].

$$\boldsymbol{M(Z(e), Z(f)) \leq \delta M(e, f)}$$

Theorem 1. Let Z satisfied

$$\|Z(w) - Z(k))\| \leq L\|w - Z(w)\| + \beta\|w - k\|, \tag{3}$$

$\forall w, k \in X$ such that $L \geq 0$, $0 < \beta < 1$ then Z has F.P and its z stable.

Proof. Assume $m, n \in \mathbb{N}$, then

$$\begin{aligned}
\|Z(w^m(r)) - Z(w^n(r))\| &= \|w^{m+1}(r) - w^{n+1}(r)\| \\
&= \|t(r) + \lambda \int_a^r \int_c^d H(s, q) w^m(q) dq ds \\
&- t(r) - \lambda \int_a^r \int_c^d H(s, q) w^n(q) dq ds\| \\
= \|\lambda \int_a^r \int_c^d H(s, q) w^m(q) dq ds &- \lambda \int_a^r \int_c^d H(s, q) w^n(q) dq ds\| \\
&= \|\lambda \int_a^r \int_c^d H(s, q)[w^m(q) - w^n(q)] dq ds\| \\
\leq \|\lambda \int_a^r \int_c^d \int_a^r \int_c^d &H^2(s, q) dq ds dq ds [w^m(q) - w^n(q)]\| \\
\leq |\lambda| [\int_a^r \int_c^d \int_a^r \int_c^d &H^2(s, q) dq ds dq ds]^{\frac{1}{2}} [w^m(q) - w^n(q)]
\end{aligned}$$

Hence, if

$$|\lambda| < \int_a^r \int_c^d \int_a^r \int_c^d H^2(s, q) dq ds dq ds]^{\frac{1}{2}} \tag{4}$$

Since $a \leq r \leq d$.
By using Density theorem. $\exists$, $t_1 \in \mathbb{R}$ Where $s \leq t_1 \leq b$,
$\int_a^r \int_\Omega \int_a^r \int_\Omega [H^2(s, q, h) dq ds dq ds]^{\frac{1}{2}} \leq M$ where $M \in R^+$,

$$\beta = |\lambda| \int_a^f \int_\Omega \int_a^f \int_\Omega [H^2(s, q) dq ds dq ds]^{\frac{1}{2}} = |\lambda| M,$$

we obtain

$$\begin{aligned}
&\|Z(w^m(r)) - Z(w^n(r))\| \leqslant \beta\|w^m(r)) - w^n(r))\| \text{ and } L\|w - Z(w)\| = 0 \\
&\|Z(p_m(r)) - Z(p_n(r))\| \leqslant \beta\|p_m(r) - p_n(r)\| \text{ and } L\|p - G(p)\| = 0
\end{aligned} \tag{5}$$

The adequate condition for convergence is Eq. (4).
put $L = 0$, and $\beta = |\lambda| \int_a^r \int_\Omega \int_a^r \int_\Omega [H^2(s, q) dq ds dq ds]^{\frac{1}{2}}$,

in inequality (3), we get $L\|w - Z(w)\| = 0$ then $w = T(w)$, then we obtain Z has FP w.

$$\|Z(w^m(r)) - Z(w^n(r))\| \leq \beta\|w^m(r) - w^n(r)\|$$

Then the nonlinear mapping Z is $Z-$ stable.

Theorem 2. Let the iterative technique satisfied,

$$\begin{aligned} w^0(r) &= t(r) \\ w^{n+1}(r) &= Z\big(w^n(r)\big) \\ &= y(r) + \lambda \int_a^r \int_\Omega H(s,q) w^n(q) dqds \end{aligned} \tag{6}$$

for $n \geq 0$, generating the sequence of approximate results $\{w^n(r)\}_{n=0}^{\infty}$. Assume $|\lambda| < [(\int_a^r \int_\Omega \int_a^r \int_\Omega [H^2(s,q)dqdsdqds]^{\frac{1}{2}}]^{-1}$. If $L = 0$ and $\beta = |\lambda| \int_a^r \int_\Omega \int_a^r \int_\Omega [H^2(s,q)dqdsdqds]^{\frac{1}{2}}$ in the norm space $N^2(r)$ then Z is Z -stable.

Proof. As Theorem 1.

Theorem 3. The iterative of the form

$$\begin{aligned} w^0(r) &= t(r) \\ w^{n+1}(r,f) &= Z(w^n(r,f)) \\ &= y(r) + \lambda \int_a^r \int_\Omega H(s,q) w^n(q) dqds \end{aligned}$$

Generating the sequence $\{w^n(r)\}_{n=0}^{\infty}$, for $n = 0,1,2,3,\ldots, max$, and $\int_a^r \int_\Omega \int_a^r \int_\Omega [H^2(s,q)dqdsdqds = M^2 < \infty$, then $w(r) \in N^2(r)$ convergent to the actual solution of SKMVFIE.

Proof. By using $w^{n+1}(r) = Z(w^n(r)) = t(r) + \lambda \int_a^r \int_c^d H(s,q) w^n(q) dqds$.

Then $w^{n+1}(r) = t(r) + \lambda \int_a^r \int_c^d H(s,q) w^n(q) dqds$.

Now $\|w^{m+1}(r) - w^{n+1}(r)\| = \|Z(w^m(r)) - Z(w^n(r))\|$

$$\begin{aligned} &= \|t(r) + \lambda \int_a^r \int_c^d K(s,q) w^m(q) dqds - t(r) + \lambda \int_a^r \int_c^d K(s,q) w^n(q,h) dqds\| \\ &= \|\lambda \int_a^r \int_c^d K(s,q) w^m(q) dqds - \lambda \int_a^r \int_c^d K(s,q) w^n(q) dqds\| \\ &= \|\lambda \int_a^r \int_c^d H(s,q)[w^m(q) - w^n(q)] dqds\| \\ &\leq \|\lambda \int_a^r \int_c^d \int_a^r \int_c^d H^2(s,q) dqdsdqds[w^m(q) - w^n(q)]\| \\ &\leq |\lambda| [\int_a^r \int_c^d \int_a^r \int_c^d H^2(s,q) dqdsdqds]^{\frac{1}{2}}] \|w^m(q) - w^n(q)\| \end{aligned}$$

Since $\int_a^r\int_c^d\int_a^r\int_c^d H^2(s,q)dqdsdqds = M^2 < \infty$
getting $|\lambda| < \int_a^r\int_c^d\int_a^r\int_c^d H^2(s,q)dqdsdqds = (M^2)^{-1}$.
Therefore, obtaining

$$\|w^{m+1}(r) - w^{n+1}(r)\| = \|H(w^m(r)) - H(w^n(r))\| \le |\lambda|\|w^m(r) - w^n(r)\|$$

Then $\{w^n(r)\}_{n=0}^{\infty}$ is convergence.

Theorem 4. Let (B, M) be complete matrix space and $Z : B \to B$ is a contractive mapping then Z has unique F.P.

Proof: - Let j and k be two different F.P of mapping Z.
Since $j \neq k$, then $Z(j) = j$ and $Z(k) = k$,
Then obtaining, $M(j,k) \neq 0$
We have assumed x and y are two different F.P, then we get

$$M(Z(j), Z(k)) = M(j,k) \neq 0 \tag{7}$$

Then

$$M(Z(j), Z(k)) \le \alpha M(j,k)$$
$$\text{(By definition of contractive mapping)} \tag{8}$$

where $0 < \alpha < 1$.

From Eqs. (7) and (8), getting
$M(j,k) = M(Z(j), Z(k)) \le \alpha M(j,k) \neq 0$ then $(1-\alpha)M(j,k) \le 0$.
By suppose we have $M(j,k) > 0$, $(1-\alpha) \le 0$
Then we obtain $1 \le \alpha$, which is a contradiction because $0 < \alpha < 1$.
Therefore, $M(j,k) = 0$, hence $j = k$.
Then Z has exactly one F.P.

Theorem 5. Assume that the iterative approach holds,

$$\begin{aligned} w^0(r) &= t(r) \\ w^{n+1}(r,f) &= Z(w^n(r,f)) \\ &= y(r) + \lambda \int_a^r \int_\Omega H(s,q) w^n(q) dqds \end{aligned}$$

for $n = 0,1,2,3,\ldots,m$. and $|\lambda| < [\int_a^r\int_c^d\int_a^r\int_c^d [H^2(s,q)dqdsdqds]^{\frac{1}{2}}]^{-1}$
$L = 0$ and $\beta = |\lambda|[\int_a^r\int_c^d\int_a^r\int_c^d [H^2(s,q)dqdsdqds]^{\frac{1}{2}}]$.
Then Z has sufficient condition and the requirement of stability is sufficient for Z.

Proof. As theorem 3.

3 Numerical Examples

Display numerical examples to illustrate the SC, and satisfying all conditions for theorem (1).

Example 1. Assume SKVFIE $w(r) = r - \frac{r^2(e-1)}{4} + \int_0^r \int_0^1 (se^{q^2}) w(q) dqds.$

and $w^{n+1}(r) = Z(w^n(r))$ be the iteration which generated. The new sequence $\{Z^n(r)\}_{n=0}^{\infty}$ generating by $w^{n+1}(r) = Z(w^n(r))$, and it is convergent to SKMVFIE's result, the iteration form is $Z-$ stable. And the exact solution is $w(r) = r$.

Solution: - we have $\lambda = 1, t(r) = r - \frac{r^2(e-1)}{4}$ and $H(s, q) = se^{q^2}$,

since $r \leq d = 1$, taking $r = 1$ then

$$\int_0^1\int_0^1\int_0^1\int_0^1 H^2(s,q)dqdsdqds]^{\frac{1}{2}} = \int_0^1\int_0^1\int_0^1\int_0^1 (se^{q^2})^2 dqdsdqds]^{\frac{1}{2}}$$
$$=(1)[(0.1845)]^{\frac{1}{2}} = 0.42950$$

now $|\lambda|(\int_0^1\int_0^1\int_0^1\int_0^1 H^2(s,q)dqdsdqds)^{\frac{1}{2}}] < 1$ then $|\lambda|(0.42950) < 1$

$|\lambda| < \frac{1}{0.42950} = 2.3283$ Then the convergence interval is $-2.32830 < \lambda < 2.32830$

$$\beta = \int_0^1\int_0^1\int_0^1\int_0^1 H^2(s,q)dqdsdqds]^{\frac{1}{2}} = \int_0^1\int_0^1\int_0^1\int_0^1 (se^{q^2})^2 dqdsdqds]^{\frac{1}{2}} = (1)[(0.1845)]^{\frac{1}{2}} = 0.4295$$

Then $0 < \beta < 1$ is stable interval.

Therefore, SI is proper subset of CI. The iteration form

$$w^{n+1}(r) = \left(w^n(r)\right)$$
$$= r - \frac{r^2(e-1)}{4} + \int_0^r\int_0^1 (se^{q^2}) w^n(q) dqds$$

Generating $\{w^n(r)\}_{n=0}^{\infty}$ and it is convergence to the solution $w(r) = r$, if $\beta = 0.42950$, we obtain

$$\|Z(w^m(r)) - Z(w^n(r))\|$$
$$r - \frac{r^2(e-1)}{4} + \int_0^r\int_0^1 (se^{q^2}) w^m(q)dqds - \left[r - \frac{r^2(e-1)}{4} + \int_0^r\int_0^1 (se^{q^2}\right) w^n(q)dqds]$$
$$< \|\beta w^m(r) - w^n(r)\|,$$

putting $L = 0$ and $\beta = 0.42950$, then it has FP and its Z stable by theorem (1).

Example 2. Consider

$$w(r) = r^2 + \frac{sin(r)(cos(1) - 1)}{3} + \int_0^r \int_0^1 (cos(s) sin(q^3) w(q) dqds$$

Be SKVFIE and $w^{n+1}(r) = Z(w^n(r))$ be the iteration which generated the sequence $\{Z^n(r)\}_{n=0}^{\infty}$ such that Z is a nonlinear mapping then it converges to (SKMVFIE)'s solution, the iteration form is $Z-$ stable. Such that $w(r) = r^2$ is actual solution.

Solution. From the integral equation $\lambda = 1$, $t(r) = r^2 + \frac{sin(r)(cos(1)-1)}{3}$ and$H(s,q) = cos(s)sin(q^3)$,

since $r \le b = 1$ we take $r = 1$ then

$$\int_0^1\int_0^1\int_0^1\int_0^1 H^2(s,q)dqdsdqds]^{\frac{1}{2}} = \int_0^1\int_0^1\int_0^1\int_0^1 (\cos(s)\sin(q^3))^2 dqdsdqds]^{\frac{1}{2}}$$
$$= [0.0166]^{\frac{1}{2}} = 0.1288$$

now $|\lambda|(\int_0^1\int_0^1\int_0^1\int_0^1 H^2(s,q)dqdsdqds)^{\frac{1}{2}}] < 1$ then $|\lambda|(0.1288) < 1$

$|\lambda| < \frac{1}{0.12880} = 7.7640$ Then the converges interval is $-7.7640 < \lambda < 7.7640$

$$\beta = |\lambda| \int_0^1\int_0^1\int_0^1\int_0^1 H^2(s,q)dqdsdqds]^{\frac{1}{2}}$$
$$= |\lambda| \int_0^1\int_0^1\int_0^1\int_0^1 [\cos(s)\sin(q^3)]^2 dqdsdqds]^{\frac{1}{2}}$$
$$= (1)[(0.0166)]^{\frac{1}{2}} = 0.1288$$

Then $0 < \beta < 1$ is a stable interval.

Therefore, SI is a proper subset of CI.

Then $\{v^n(r)\}_{n=0}^{\infty}$ generated by the iteration form,

$$w^{n+1}(r) = Z(w^n(r))$$
$$= r^2 + \frac{\sin(r)(\cos(1)-1)}{3} + \int_0^r\int_0^1 (\cos(s)\sin(q^3)w^n(q)dqds$$

Convergent to actual solution $w(r) = r$ if $\beta = 0.1288$, we obtain,

$$\|Z(w^m(r)) - Z(w^n(r))\|$$
$$\le \left\| r^2 + \frac{\sin(r)(\cos(1)-1)}{3} + \int_0^r\int_0^1 (\cos(s)\sin(q^3)w^m(q)dqds - \left[r^2 + \frac{\sin(r)(\cos(1)-1)}{3} + \int_0^r\int_0^1 (\cos(s)\sin(q)^3 \right) w^n(q)dqds] \right\|$$
$$< \|\beta w^m(r) - w^n(r)\|,$$

putting $L = 0$ and $\beta = 0.1288$, then it has FP and its Z stable by theorem (1).

4 Conclusion

The SKMVFIE module is discussed for studying the stability and convergence for this problem. Proposing the sufficiently condition of convergence for SKMVFIE. Finally, the following results are concluded: -

1. The Mapping is $Z-$ stable, and it has unique F.P.
2. The new approach displays validity and applicability for solving these types of problems.
3. In general, the stable interval is proper subset of convergent interval.

5 Recommendations

For the future works I recommend these two points: -

1. Studying the stability for integro Volterra Fredholm integral equations.
2. Studying the stability of two dimensional integral equation.

Acknowledgment. The author would like to extend their sincere thanks to the referees for their helpful remarks and suggestions. Many thanks are due to the Erbil Polytechnic University for providing the financial support for this research.

References

1. Abdoul, M.A., Ahmed, S.: The stability of the solution of nonlinear integral equation in two-dimensional problems. J. Eng. Sci. **4**(8), 1–12 (2020)
2. Attah, F., et al.: On the solution of mixed Volterra- fredholm integral equations using the new Iterative method. J. Appl. Math.s **6**(6), 1–5 (2016)
3. Banifatewm, E., et al.: Two-Dimensional Legendre Wavelets Method to Mixed Volterra-Fredholm Integral Equations. J. Vibrat. Control **13**(11), 20–28 (2007)
4. Chen, Z., Jaing, W.: An approximate solutions for a mixed linear volterra fredholm integral equations. J. Appl. Math. Let. **8**(25), 20–28 (2012)
5. Hafez, R., et al.: Numerical solutions of two dimension mixed volterra- fredholm integral equation of the second kind via bernoulli collocation method. J. Roman. Phys. **3**(111), 1–11 (2017)
6. Hassan, T.I., et al.: Using aitken method to solve single volterra-fredholm integral equations of the second kind with homotopy perturbation method. J. Pure Appl. Sci. **s4**, 142–152 (2017)
7. Jafari, M., Aminatei, A.: Method of successive approximations for solving the multi pantograph delay equations. J. Gen. Math. Notes **8**(1), 23–28 (2012)
8. Lafferriere, B., et al.: Introduction to mathematical analysis. Portland State University Library (2015)
9. Mohammed, K.: Numerical solution for mixed volterra-fredholm integral equations of the second kink by using bernstein polynomials method. J. Math. Theory Model. **5**(10), 154–162 (2015)
10. Saeed, R.K., Berdawood, K.A.: Solving two-dimensional linear volterra-fredholm integral equations of the second kind by using series solution methods. J. Zankoy Sulaimani-Part A **17**(4), 253–270 (2015)
11. Pachpatte, B.G.: On a general mixed volterra- fredholm integral equation. J. Matematica Tomul **LVI**, 17–24 (2010)
12. Saadati, R.: T- Stability approach to variation iteration method for solving integral equations. J. Hindawi Publish. Corporat. **209**, 1–9 (2009)
13. Salleh, S., et al.: Fixed point and its improvement for the system of Volterra-Fredholm Integral equations of the second kind. J. Matematika **33**(2), 191–206 (2017)
14. Saleh, M.H., et al.: Approximate solution of two-dimensional Volterra integral equation by Chebyshev polynomial method and Adomian decomposition method. J. Math. Theory Model. **2**(4), 110–119 (2016)
15. Sulaiman, N.A., et al.: On the stability and convergence of linear volterra-fredholm integral equations of the second kind. J. Pure Appl. Sci. **30**(2), 25–32 (2018)

16. Talhat, I.H.: An approximate solutions of two dimension linear mixed volterra- fredholm integral equation of the second kind via iterative kernel method. J. Raparin Univ. **6**(2), 101–110 (2019)
17. Tari, A., Shahmorad, S.: A computational method for solving two-dimensional linear Fredholm integral equations of the second kind. J. The ANZIAM **49**(4), 543–549 (2008)
18. Wazwaz, A.M.: A Reliable treatment for mixed Volterra-Fredholm integral equations. J. Appl. Math. Comput. **127**(2), 405–414 (2002)
19. Momenzade, N., et al.: A numerical method for solving stochastic volterra-fredholm integral equation. J. Math. Sci. Inform. **18**(1), 145–164 (2023)
20. John, V.L., Kubilay, S.: Integral equation methods for electromagnetics" SciTech publishing, Raleigh (2012)
21. Talaat, H.I., Chiman, H.I.: Modification of Adomian decomposition method to solve two dimensions Volterra integral equation second kind. Cihan J. Sci. CUESJ **8**(1), 41–45 (2024)

Building the Foundation of Critical Thinking with Online Learning Through Writing Scientific Papers Based on Information Literacy

Imas Cintamulya(✉), Lilik Mawartiningsih, and Warli

Universitas PGRI Ronggolawe, Tuban, Indonesia
{imascintamulya,lilikmawar,warli}@unirow.ac.id

Abstract. This study examines the differences in critical thinking skills based on indicators adapted from Bloom's Taxonomy (application, analysis, inference, evaluation, and synthesis) between students with reflective and impulsive cognitive styles in writing scientific papers based on information literacy in online learning. The purpose of this study is based on the following research problem: how can reflective and impulsive students develop critical thinking skills in writing scientific papers based on information literacy through online learning? A total of 10 students became research subjects who were categorized based on the scores of cognitive style measurement results using the Matching Familiar Figures Test. Critical thinking skills were assessed using an instrument consisting of 12 sub-indicators. The results of the Mann-Whitney test showed that students with impulsive cognitive styles were significantly better at 1) formulating the conclusions developed, 2) examining various references relevant to the topic being made, 3) assessing statements and arguments of articles that will be used in the topic description; 4) the ability to use old knowledge or ideas into something new; and 5) using knowledge in new or different contexts. The study's findings indicate significant differences in critical thinking skills between reflective and impulsive students who write scientific papers based on information literacy in online learning. Further research is needed to identify specific learning interventions to support impulsive students in developing critical thinking skills. In addition, this study can contribute to developing more effective online learning models to improve students' overall critical thinking skills.

Keywords: Building Critical Thinking · Scientific Papers · Information Literacy · Reflective Cognitive Style · Impulsive Cognitive Style

1 Introduction

Advances in technology have brought education into rapid change, one of which is marked by the existence of online learning. This learning model is conducted remotely through various digital platforms [1]. Online learning relies on the Internet, a major trend in modern education [2, 3]. This development is responded to by Permendikbudristek No. 53 of 2023, which provides flexibility for educational institutions in Indonesia to learn

A. J. Obaid et al. (Eds.): ICONEST 2024, ISEM 40, pp. 266–278, 2025.
https://doi.org/10.1007/978-3-031-89960-7_20

online, face-to-face, or a combination of both. The flexibility offered by online learning allows wider access to information. However, the flood of unfiltered information is a challenge in itself. Therefore, critical thinking skills are crucial in filtering information effectively [4–6]. In addition, students are not only required to memorize information but also to develop the ability to analyze, evaluate, and synthesize various sources of information [2, 7]. In other words, critical thinking skills and information literacy are now very important core competencies in preparing individuals to face the complexity of the modern world [5, 8, 9].

Information literacy is the ability to search, evaluate and process information effectively and efficiently, as well as the skill of thinking critically about the quality of the information we encounter [10–13]. Furthermore, the importance of information literacy in online learning is further emphasized by the strong correlation between online information-seeking activities and an individual's level of information literacy [10]. However, the understanding and application of information literacy among students is still diverse and not optimal [14, 15]. Efforts to optimize information literacy skills in biology learning can be made by giving tasks to write scientific papers. This is in line with the spirit of the Minister of Research, Technology and Higher Education Regulation No. 50/2018, which encourages the improvement of the quality of students' scientific papers. Making scientific papers requires critical thinking skills and information literacy skills because the ideas presented in scientific papers must be based on facts taken from reliable sources [5, 16, 17]. Shao & Purpur's research findings evidence this [5], which shows a positive correlation between information literacy skills, students' writing scores, and final grades. This finding is in line with Shannon et al. [15] and Hollis [18], as well as the study by Buselik [19], show that information literacy is the ability to think critically and evaluate information, which is very important in today's digital era. The ability to think critically and evaluate information effectively can be honed and developed significantly through scientific article-writing activities.

One aspect often overlooked in developing critical thinking and information literacy skills is the differences in students' cognitive styles. Cognitive style includes how a person thinks, processes information, and solves problems, greatly affecting how a person learns effectively. Research by Simuth & Sarmany-Schuller [20], supports the view that each individual has a unique cognitive style. In line with this, McLoughlin [21], found that a person's learning success is greatly influenced by the suitability of how the material is delivered and his learning style. Therefore, lecturers need to understand that students have diverse cognitive styles to achieve optimal development of critical thinking skills and information literacy. Research conducted by Yuniasari [22], has identified reflective and impulsive cognitive styles that affect how students process information. Students with reflective cognitive styles tend to be more careful in making decisions, while impulsive students tend to be faster [23, 24]. Therefore, effective learning must accommodate students' cognitive styles to make learning more meaningful.

Several studies have examined the relationship between reflective and impulsive cognitive styles and students' critical thinking skills, especially in learning mathematics and biology. These studies include Aprilia et al. [25], Rofi'ah [26], and Fridanianti et al. [27], have described how these cognitive styles affect students' thought processes when solving various math problems. Meanwhile, in the field of biology, studies by

Arniyawati & Cintamulya [28], Cintamulya et al. [29], Muryani and Cintamulya [30], and Rochika & Cintamulya [31], have shown that cognitive style also plays an important role in developing students' critical thinking skills through various learning models. However, existing research has not explicitly combined reflective and impulsive cognitive styles with information literacy and critical thinking skills. This situation provides an opportunity to conduct research on information literacy integrated into scientific article writing tasks to improve critical thinking based on reflective and impulsive cognitive styles through online learning. Meanwhile, the research problem formulation is: How do students develop critical thinking skills in writing scientific papers based on information literacy that accommodates reflective and impulsive cognitive styles through online learning? The objectives to be achieved from this study are 1) to build a foundation for students' critical thinking skills in writing scientific papers based on information literacy that accommodates reflective and impulsive cognitive styles through online learning and 2) to describe the differences in students' critical thinking skills in writing scientific papers based on information literacy that accommodates reflective and impulsive cognitive styles through online learning.

2 Methods

2.1 Research Design

Critical thinking of students in making scientific papers based on information literacy by accommodating reflective and impulsive cognitive styles was investigated through online learning in Plant Anatomy using the Google Classroom application. The research method used was quantitative research with the Pre Experiment with One-Shoot Case Study design [32], in this research, there was no control class. Learning of Plant Anatomy was given to biology education students in 2018 as research subjects whose cognitive style has been measured. Student cognitive styles were measured using the MFFT (Matching Familiar Figure Test) instrument developed by Warli [33]. There are thirteen questions that students have to answer. Each question is provided with 8 different pictures and 1 standard picture. How to do it: students choose one picture according to the standard image from 8 pictures. The parameters observed included the time it took for the first time students answered (t) and the frequency with which students answered until they got the correct answer (f). The time obtained is added and divided by the number of questions and the frequency. Furthermore, to see the meantime and frequency sorted from smallest to largest. Then, a parallel line is drawn through the axis (t) and the axis (f) to classify the reflective and impulsive cognitive styles. If the time is greater than the mean and the frequency is smaller than the mean, it is considered a reflective cognitive style. Meanwhile, if time is less than the mean and the frequency is greater than the mean, it is considered an impulsive cognitive style [34].

2.2 Research Settings

Students' critical thinking skills were developed by giving them the task of writing scientific Papers about plant anatomy. At the beginning of the lesson, each student was given one topic: plant anatomy. In this research, 11 topics were provided, namely: a) cell walls; b) epidermis and stomata tissue; c) meristem and parenchyma tissue; d) secretory tissue; e) cambium vessels; f) xylem tissue; g) phloem tissue; h) root structure; i) stem structure; j) leaf and flower structure, and k) microsporogenesis and megasporogenesis. Each of these topics was used as the title of the chapter and then compiled into a textbook created by students with formats referring to Rifai [35] and Ariani [36], which were modified.

Students' information literacy skills were developed through the task of accessing plant anatomy material based on e-journal data such as Proquest, Science Direct, Google Advance, and Library Genesis [11]. Students completed their scientific papers task in one month. Furthermore, students presented their scientific papers task online through the Google Classroom application. The lecturer conducted assessments, and other students submitted questions and suggestions regarding the material presented. The material discussed and assessed was organized into textbook forms of student work according to the format provided. Students' critical thinking skills were measured using instruments of critical thinking skills that refer to Bloom's Taxonomy in Zakiyah [37] and Facione [38]. The aspects of critical thinking that were measured include a) application, b) analysis, c) inference, d) evaluation, and e) synthesis. Furthermore, the five aspects were developed into 12 indicators, namely: a) the ability to implement procedures; b) framing ideas as seen from titles and sub-headings; c) detail of the part seen from the description of subtitles; d) looking for relationships seen from the relationship between subtitles made; e) determine the parts needed to be seen from the subtitles made in support of the title of the Chapter; f) formulating the conclusions developed seen from the conclusions for each chapter); g) the ability to check multiple references relevant to the topic made; h) the ability to criticize Papers that are relevant to the topic created; i) assess the statements and arguments of the article that will be used in the description of the topic; j) the ability to use old knowledge or ideas into new things (the ability to use pictures) to clarify the description of a topic; k) combining ideas into new things (selection of images that include shapes, colours, and descriptions in support of the description of a topic); l) use knowledge in new or different contexts (seen from the evaluation of each topic.

2.3 Data Analysis

The critical thinking skills scores for each indicator (12 indicators) for reflective and impulsive students were averaged and then tabulated. The essential skills of thinking scores used were on a scale of 4. The critical thinking skill score for each indicator was analyzed using non-parametric statistics using the Mann-Whitney test. Mann-Whitney test results determined differences in students' critical thinking skills in making scientific papers based on information literacy between reflective and impulsive cognitive styles. The critical thinking conclusion criteria are presented in Table 1.

Table 1. Conclusion criteria for critical Thinking

Range of scores	Criteria for critical thinking
Score ≥ 3,90	Very Good
3,00 ≤ Score ≤ 3,90	Good
2,00 ≤ Score ≤ 3,00	Enough
1,00 ≤ Score ≤ 2,00	Less

3 Result Analysis and Discussion

3.1 Findings

Ten students measured their critical thinking skills in writing scientific papers based on information literacy: five reflective and five impulsive students. Table 2 presents the average reflective and impulsive students' critical thinking scores covering five aspects of critical thinking with 12 indicators. Next, Table 3 presents the Mann-Whitney test results for each indicator of critical thinking skills.

Table 2. Average Scores of Critical Thinking Skills of Reflective and İmpulsive Students in Writing Scientific Papers Based on İnformation Literacy

	Aspect	Indicator	Average Critical Thinking Score	
			Reflective	Impulsive
A	Application (the ability to classify material into parts, track and identify the relationships of each part, and how parts were formed)	1. The ability to implement procedures. Made Scientific papers	3.60 (good)	3.80 (good)
B	Analysis (the ability to classify material into parts, tracking and identifying the relationship of each part, and how the part was formed	2. Framing Ideas (judging by titles and sub-headings)	3,40 (good)	3.20 (good)
		3. Detailing Sections (seen from the description in sub-headings)	3.40 (good)	3.20 (good)
		4. Look for Relationships (seen from the relationship between subtitles made)	3,40 (good)	3.40 (good)

(*continued*)

Table 2. (*continued*)

	Aspect	Indicator	Average Critical Thinking Score	
			Reflective	Impulsive
C	The inference is the ability to determine an important part of a thing to form a logical conclusion	5. Determine the parts needed (seen from the subtitles made in support of the chapter title)	3.60 (good)	3.80 (good)
		6. Formulate the conclusions developed (seen from the conclusions for each chapter)	2.80 (enough)	3.20 (good)
D	Evaluation is assessing a statement's credibility and the power of logic and deciding on values for several purposes	7. Ability to check multiple references relevant to the topic made	3.00 (good)	3.40 (good)
		8. The ability to criticize Papers that are relevant to the topic created	3.40 (good)	3.60 (good)
		9. Assess the statements and arguments of the article that will be used in the description of the topic (cite people's opinions)	3.40 (good)	4.00 (good)
E	Synthesis is the ability to unite parts to form something new	10. The ability to use old knowledge or ideas into new things (the ability to use pictures) to clarify the description of a topic	3.60 (good)	4.00 (good)
		11. Combine ideas into new things (selection of images that include shapes, colours, and descriptions in support of the description of a topic)	3.60 (good)	3.40 (good)
		12. Using knowledge in new or Different contexts (seen from the Evaluation of each topic)	2.80 (good)	3.20 (good)

Based on Table 3, there is no difference in critical thinking skills between reflective and impulsive students in indicators 1, 2, 3, 4, 5, 8, and 11. There is a difference in critical thinking skills between reflective and impulsive students in indicators 6, 7, 9, 10, and 12.

Table 3. Results of the Mann-Whitney test on critical thinking skills of reflective and impulsive students in writing scientific papers based on information literacy

	Ind 1	Ind 2	Ind 3	Ind 4	Ind 5	Ind 6	Ind 7	Ind 8	Ind 9	Ind 10	Ind 11	Ind 12
Mann-Whitney U	10.000	10.000	10.000	12.500	10.000	8.000	8.500	10.500	5.000	7.500	10.000	8.000
Wilcoxon W	25.000	25.000	25.000	27.500	25.000	23.000	23.500	25.500	20.000	22.500	25.000	23.000
Z	−.655	−.655	−.655	.000	−.655	−1.342	−.949	−.516	−1.964	−1.500	−.600	−1.342
Asymp. Sig. (2-tailed)	.513	.513	.513	1.000	.513	.180	.343	.606	.050	.134	.549	.180
Exact Sig. [2* (1-tailed Sig.)]	.690[a]	.690[a]	.690[a]	1.000[a]	.690[a]	.421[a]	.421[a]	.690[a]	.151[a]	.310[a]	.690[a]	.421[a]

3.2 Discussion

Based on the Mann-Whitney test (Table 3), there were 7 indicators, or more than 50%, whose Asympton value was greater than 0.500. The seven indicators include 1) indicator 1 (the ability to implement procedures for making scientific papers); 2) indicator 2 (framing ideas as seen from the headline and sub-headings) 3) indicator 3 (detailing the section seen from the description in the sub-headline); 4) indicator 4 (looking for relationships seen from the relationships between sub-headings made); 5) indicator 5 (determine the required parts seen from the subtitles made in support of the title of the Chapter); 6) indicator 8 (ability to criticize Papers that were relevant to the topic being made); and 7) indicator 11 (combining ideas into new things seen from the selection of images that include shapes, colours, and descriptions in support of the description of a topic). This shows that the seven indicators of critical thinking skills in writing scientific papers based on information literacy between students with reflective and impulsive cognitive styles were no different.

Based on the assessment results, all indicators of students' critical thinking skills are in a good category. Analysis of the seven indicators, referring to Bloom's Taxonomy [37, 38], shows that the dominant critical thinking skills are application, analysis, and inference. This finding indicates a positive correlation between scientific writing and critical thinking skills, which is in line with previous research [16, 39]. Scientific writing is not simply putting ideas into words but is a cognitively complex task [40–42]. This process involves cognitive skills and problem-solving that stimulate and train critical thinking skills [43, 44].

Furthermore, the results of this study show that information literacy skills facilitate scientific writing in terms of accessing information, assessing sources, and managing information [10, 11, 17]. Of course, this condition aligns with several previous research results that show a positive relationship between scientific writing and information literacy [5, 13, 45]. Information literacy is a very important skill in scientific writing. Searching, assessing, and managing information effectively can help students produce quality scientific papers. Information literacy is especially necessary for students to search for credible online information about plant anatomy, such as from Proquest, Science Direct, Google Advance, and Genesis Library [10–12]. Online information search stimulates one to read. A lot of reading has an impact on critical thinking skills [16].

Based on Table 2, it is clear that students have successfully mastered the five indicators of critical thinking well, ranging from basic to complex levels. Their ability to apply, analyze, infer, synthesize, and evaluate information shows that the empowerment process through scientific writing based on information literacy has encouraged students to reach higher cognitive domains. This is in line with the findings of Reece [46] and the Association of College & Research Libraries [47] who assert that the ability to analyze, synthesize, and evaluate is key to critical thinking. Furthermore, as defined by Rofi'ah [26], critical thinking involves not only understanding information but also the ability to process that information deeply and distinguish relevant information. In other words, students are not only able to remember facts but also think more critically and analytically. The results indicate that the intervention has significantly improved the quality of students' critical thinking.

Although reflective and impulsive students have cognitively different characteristics [48]This study showed that the two groups did not show significant differences in critical thinking skills in application, analysis, and inference in writing scientific papers (Table 3). Both groups equally achieved the good category in these skills. This is interesting because reflective individuals tend to be more analytical and have better self-control [49, 50], are expected to show better performance in tasks that require critical thinking. However, this finding indicates that other factors, such as quality of guidance, availability of information, or previous writing experience, may play a greater role in determining success in this task.

Thus, the differences in the characteristics of thought processes and cognitive aspects possessed by students who have reflective and impulsive cognitive styles do not cause differences in students' critical thinking skills in the aspects of the application, analysis, and inference (indicated by 7 indicators) in writing scientific papers based on information literacy. Students can use the information obtained through information literacy skills to compile scientific papers because they can think critically even though their thinking processes and cognitive maturity are different. The results of this study indicate that the method of giving assignments in the form of scientific work based on information literacy has brought learning to the demands of the 21st century by involving students in higher-order thinking [51]. Educators develop students' critical thinking behaviour and attitudes by allowing them to express their thinking results openly through scientific papers based on information literacy. Therefore, the developed students' critical thinking skills show good categories. This is supported by the opinion of McLoughlin [21] that individuals learn best when teachers present information according to their style.

Further analysis using the Mann-Whitney test (Table 3) revealed interesting differences between reflective and impulsive students. Although both groups performed well overall in critical thinking skills in application, analysis, and inference, there were significant differences in specific indicators related to inference, evaluation, and synthesis. Impulsive students consistently performed better on these five indicators: formulating conclusions, checking references, assessing arguments, using knowledge creatively, and applying knowledge in new contexts. These findings contradict the initial expectation that reflective students, with their tendency to think analytically and carefully, would excel in tasks requiring higher-order thinking. This discrepancy may be explained by differences in thought processes and cognitive maturity between the two groups, as suggested by Cheval et al. [48] and Rozencwajg & Corroyer [49]. With their more holistic and quick-thinking style, Impulsive students may be better able to see broader connections between ideas and thus be more effective in synthesizing and evaluating. However, these findings also highlight the importance of considering task context and individual characteristics in understanding critical thinking performance.

The holistic thinking process used by impulsive students [49] has an impact on formulating conclusions, checking references, assessing arguments, using knowledge creatively, and applying knowledge in new contexts better than reflective students. The five indicators of critical thinking in preparing scientific papers based on information literacy can be achieved in the good category by these impulsive students because they respond quickly to a stimulus, do not need time to reflect on the situation and take action spontaneously [52]. In other words, the activity of impulsive students in preparing scientific papers based on information literacy is better than that of reflective students. The results of this study are in accordance with the findings of Yuniasari and Zainuddin [22] using participants, 30 impulsive students and 30 reflective students, who showed that students with impulsive cognitive styles tend to excel in descriptive writing tasks and, in general, have different learning activities compared to students with reflective cognitive styles. The difference in their cognitive strategies explains the difference in critical thinking skills performance between impulsive and reflective students. Impulsive students who prefer holistic information processing are more capable of making leaps of logic and understanding new ideas. In contrast, reflective students who tend to analyze information in depth may take longer to reach the same conclusion [38, 49, 53].

The difference in critical thinking performance between reflective and impulsive students is influenced by various factors, including differences in the thinking process and cognitive aspects they use in responding to a problem [49]. Reflective students use analytic and cognitive thinking processes, while impulsive students use holistic thinking and immature cognitive aspects. In addition, differences in attitudes also play an important role. Reflective students are more acute because they have a calm demeanour, are calmly controlled, and prefer to think before making decisions and answering questions. Still, they take a long time to respond to a problem. Meanwhile, impulsive students have high neuroticism [54, 55]. They usually make quick decisions in responding to problems and completing tasks quickly but with low accuracy [56]. Another factor influencing critical thinking is the different methods used in the learning process [57].

We recognize that this study has limitations regarding the number of participants. Although this study's results align with the findings of Yuniasari and Zainuddin [22], who used large participants. The small sample size limits our ability to conclude that the same results would apply to a broader population. Therefore, it is recommended that other researchers conduct further research involving more diverse participants and on a larger scale. Future research could also consider other factors such as task difficulty, prior learning experience and individual preferences to gain a more comprehensive understanding of how cognitive style and feedback type interact in an online learning context.

4 Conclusion

The results of this study show interesting findings, namely that students with impulsive cognitive styles are better in some aspects of critical thinking skills than reflective students. This finding contradicts previous studies and requires further analysis to reveal the underlying factors. Overall, both reflective and impulsive students showed significant improvement in critical thinking skills after participating in online learning through information literacy-based scientific writing. This indicates that this learning model is effective for both groups. However, developing specific learning strategies to optimize each group's learning is necessary. It is important to remember that this study has limitations, such as limited sample size and specific learning context. Therefore, further research with a more robust design is needed to confirm these findings and expand our understanding of the relationship between cognitive style, online learning, information literacy and critical thinking skills.

Acknowledgements. Acknowledgements to the Directorate of Research and Community Service of the Directorate General of Strengthening Research and Development at the Ministry of Research, Technology and Higher Education (Indonesia), which has provided grants for Penelitian Terapan Unggulan Perguruan Tinggi (PTUPT) for the implementation of the research entitled: Development of Group Investigation Learning Models Based-on Information Literacy to Improve HOTs that Accommodate Style Cognitive Reflective and Impulsive. In addition, the authors are also grateful to the lecturers and students who have helped carry out this research.

References

1. Banson, J.: Co-regulated learning and online learning: a systematic review. Soc. Sci. Humanit. Open **6**, 100376 (2022)
2. van Laar, E., van Deursen, A.J.A.M., van Dijk, J.A.G.M., de Haan, J.: The relation between 21st-century skills and digital skills: a systematic literature review. Comput. Human. Behav. **72**, 577–588 (2017)
3. Wong, J., Baars, M., Davis, D., Van Der Zee, T., Houben, G.-J., Paas, F.: Supporting self-regulated learning in online learning environments and MOOCs: a systematic review. Int. J. Hum.-Comput. Interact. **35**, 356–373 (2019)
4. Kirsanova, N.A.: Challenges and opportunities of online learning. Вестник Кыргызского Национального Университета имени Жусупа Баласагына 29–34 (2021)
5. Shao, X., Purpur, G.: Effects of information literacy skills on student writing and course performance. J. Acad. Librariansh. **42**, 670–678 (2016)
6. Mushtaha, E., Dabous, S.A., Alsyouf, I., Ahmed, A., Abdraboh, N.R.: The challenges and opportunities of online learning and teaching at engineering and theoretical colleges during the pandemic. Ain Shams Eng. J. **13**, 101770 (2022)
7. Scheibenzuber, C., Hofer, S., Nistor, N.: Designing for fake news literacy training: a problem-based undergraduate online-course. Comput. Human Behav. **121**, 106796 (2021)
8. Saglam, A.Ç., Çankaya, İ, Üçer, H., Çetin, M.: The effect of information literacy on teachers' critical thinking disposition. J. Educ. Learn. **6**, 31–40 (2017)
9. Goodsett, M.: Best practices for teaching and assessing critical thinking in information literacy online learning objects. J. Acad. Librariansh. **46**, 102163 (2020)

10. Çoklar, A.N., Yaman, N.D., Yurdakul, I.K.: Information literacy and digital nativity as determinants of online information search strategies. Comput. Hum. Behav. **70**, 1–9 (2017)
11. Kong, S.C.: Developing information literacy and critical thinking skills through domain knowledge learning in digital classrooms: an experience of practising flipped classroom strategy. Comput. Educ. **78**, 160–173 (2014)
12. Grafstein, A.: Information literacy and critical thinking: context and practice. In: Pathways Into Information Literacy Communities Practice Elsevier, pp. 3–28 (2017)
13. Yu, C.: Integrating information literacy and academic writing: developing a self-assessment scale of information-based academic writing. J. Acad. Librariansh. **49**, 102804 (2023)
14. Smith, J.K.: Secondary teachers and information literacy (IL): teacher understanding and perceptions of IL in the classroom. Libr. Inf. Sci. Res. **35**, 216–222 (2013)
15. Shannon, C., Reilly, J., Bates, J.: Teachers and information literacy: understandings and perceptions of the concept. J. Inf. Lit. **13**, 41–72 (2019)
16. Suhartono. Pengaruh kebiasaan Membaca, Kemampuan Berpikir Kritis, dan Penguasaan Struktur Sintaksis terhadap Keterampilan Menulis Ilmiah [The Effect of Reading Habits, Critical Thinking Ability, and Mastery of Syntax Structures on Scientific Writing Skills]. Lentera Pendidik **17**, 43–65 (2014)
17. Kocatepe, M.: Reconceptualising the notion of finding information: how undergraduate students construct information as they read-to-write in an academic writing class. J. English Acad. Purp. **54**, 101042 (2021)
18. Hollis, H.: Information literacy and critical thinking: different concepts, shared conceptions. Inf. Res. **24**(4), 1–50 (2019)
19. Buselic, V.: Information literacy and critical thinking freshman course experience. In: 2019 42nd International Convention on Information, Communication and Technology Electron Microelectron MIPRO 2019 - Proc 800–805 (2019)
20. Simuth, J., Sarmany-Schuller, I.: Cognitive style variable in e-learning. Procedia-Soc. Behav. Sci. **116**, 1464–1467 (2014)
21. McLoughlin, C.: The implications of the research literature on learning styles for the design of instructional material. Australas. J. Educ. Technol. **15**, 222–241 (1999)
22. Yuniasari, A., Zainuddin: The impact of cognitive (reflectivity/impulsivity) on tertiary EFL learners' syntactic complexity in descriptive writing. Adv. Lang Lit. Stud. **10**, 86–90 (2019)
23. Haghighi, M., Ghanavati, M., Rahimi, A.: The role of gender differences in the cognitive style of impulsivity/reflectivity and EFL success. Procedia-Soc. Behav. Sci. **192**, 467–474 (2015)
24. Aini, N., Junıatı, D., Siswono, T.: Exploring the cómbinatorial reasoning of high school students with reflective and impulsive cognitive style in solving problems. J. Educ. Gift Young Sci. **8**, 1113–1124 (2020)
25. Aprilia, N.C., Sunardi, S., Trapsilasiwi, D.: The thinking process of students with reflective and impulsive cognitive styles in solving math problems in class VII SMPN 11 Jember. J. Edukasi **2**, 31–37 (2017)
26. Rofi'ah K & M.: Identification of students' critical thinking ability in open-ended mathematical problem solving in terms of reflective and impulsive cognitive style. Mathedunesa **7**, 550–556 (2018)
27. Fridanianti, A., Purwati, H., Murtianto, Y.H.: Analysis of critical thinking ability in solving algebra problems, class VII SMP N 2 pangkah viewed from reflective cognitive style and impulsive cognitive. AKSIOMA J. Mat. dan Pendidik Mat. **9**, 11–20 (2018)
28. Arniyawati, A., Cintamulya, I.: Analysis of critical thinking ability of students with reflective and impulsive cognitive styles in biology learning through the SQ3R model (Survey, Question, Read, Review, Recite) with picture card media. Bio-Pedagogi **6**, 13–18 (2017)
29. Cintamulya, I., Warli, M.L.: Critical thinking ability after project-based learning: a comparative study on students who have different cognitive styles. In: Journal of Physics: Conference Series, IOP Publishing, p Article 12039 (2019)

30. Muryani, U., Cintamulya, I.: Analysis of critical thinking of students with reflective and impulsive cognitive style in biology learning through problem based learning learning model. Biog. (Jurnal Pendidik Sains dan Biol. **14**, 67–72(2018)
31. Rochika, N.D., Cintamulya I.: Analysis of critical thinking of students with reflective and impulsive cognitive style in biology lessons through the Means Ends Analysis (MEA) model using visual media. In: Proceeding Biology Education Conference: Biology, Science, Environmental, and Learning, pp. 562–566 (2017)
32. Leedy, P.D., Ormrod, J.E.: Practical research: planning and design, Global edition (2015)
33. Warli: Profil kreativitas siswa yang bergaya kognitif reflektif dan siswa yang bergaya kognitif impulsif dan reflektif dalam memecahkan masalah geometri (2010)
34. Muhtarom, S., Utami, R., Indriyana, K.: Metacognitive ability of male students: difference impulsive-reflective cognitive style. In: Journal of Physics: Conference Series, IOP Publishing, p Article 12118 (2018)
35. Rifai, M.A.: (2008) Tips for Packaging Research Results into Textbooks. Coll Textb Writ Publ Work AUP, UNAIR Surabaya 21–23 (2008)
36. Ariani, D.: Analysis of the validity of textbooks for the e-learning lecture system on basic chemistry courses at FKIP UMMY Solok. J. Eksakta Pendidik **1**, 104–111 (2017)
37. Zakiyah, M.: Development of Critical Thinking Teaching Materials through Learning to Write Scientific Papers of Class XI SMA/Equals (2017)
38. Facione, P.A.: Critical thinking: what it is and why it counts. Insight Assess **2007**, 1–23 (2015)
39. Wu, T.-T., Silitonga, L.M., Murti, A.T.: Enhancing English writing and higher-order thinking skills through computational thinking. Comput. Educ. **213**, 105012 (2024)
40. MacArthur, C.A., Graham, S.: Writing research from a cognitive perspective (2016)
41. Doolan, S.M.: Source-based L1 student writing development: Analyzing the relationships among functional dimensions of source use and the quality of source use. Writ. Commun. **40**, 720–753 (2023)
42. Cooney, A., Darcy, E., Casey, D.: Integrating reading and writing: supporting students' writing from source. J. Univ. Teach. Learn. Pract. **15**, 3 (2018)
43. Leibovitch, Y.M., Brown, D.J., Pearson, S., Mazzola, C., Ellerton, P.J., Watt, G.: Investigating the impact of critical thinking instruction on writing performance: a multilevel modelling analysis of relative gain data in the Australian national assessment program. Think Ski Creat **53**, 101546 (2024)
44. Myhill, D., Newman, R., Watson, A.: Going meta: dialogic talk in the writing classroom. Aust. J. Lang. Lit. **43**, 5–16 (2020)
45. Liyanage, I., Walker, T., Shokouhi, H.: Are we thinking critically about critical thinking? Uncovering uncertainties in internationalised higher education. Think Ski. Creat. **39**, 100762 (2021)
46. Reece, G.J.: Critical thinking and cognitive transfer: Implications for the development of online information literacy tutorials. Res. Strateg. **20**, 482–493 (2005)
47. ACRL: Framework for information literacy for higher education. Association of College and Research Libraries, pp. 1–18 (2015)
48. Cheval, B., Sarrazin, P., Isoard-Gautheur, S., Radel, R., Friese, M.: How impulsivity shapes the interplay of impulsive and reflective processes involved in objective physical activity. Pers. Individ. Dif. **96**, 132–137 (2016)
49. Rozencwajg, P., Corroyer, D.: Cognitive processes in the reflective-impulsive cognitive style. J. Genet. Psychol. **166**, 451–463 (2010)
50. Masfingatin, T., Suprapto, E.: Student's statistical literacy skills based on the reflective and impulsive cognitive styles. Al-Jabar J. Pendidik Mat. **11**, 273–286 (2020)
51. ŽivkoviL·, S.: A model of critical thinking as an important attribute for success in the 21st century. Procedia-Soc. Behav. Sci. **232**, 102–108 (2016)

52. Nietfeld, J., Bosma, A.: Examining the self-regulation of impulsive and reflective response styles on academic tasks. J. Res. Pers. **37**, 118–140 (2003)
53. Masduki, M., Mufidah, Z.: Writing scientific papers training for elementary school teachers in pademawu pamekasan. J. Commun. Dev. Soc. **1**, 57–65 (2019)
54. Estaji, M., Safari, F.: Learning-oriented assessment and its effects on the perceptions and argumentative writing performance of impulsive vs. reflective learners. Lang Test Asia **13**, 31 (2023)
55. Davoudi, M., Heydarnejad, T.: The interplay between reflective thinking and language achievement: a case of Iranian EFL learners. Lang. Teach. Res. Q **18**, 70–82 (2020)
56. Salehi, S., Nosratinia, M.: The interplay between oral communication strategies and willingness to communicate in impulsive and reflective EFL learners. Int. J. Foreign Lang. Teach. Res. **10**, 85–101 (2022)
57. Hajimohammadi, R., Makundan, J.: Impact of self-correction on extrovert and introvert students in EFL writing progress. English Lang. Teach. **4**, 161–168 (2011)

Computing Sunspot Parameters for Short-Period Eclipsing Binary Star CN Andromeda

Chiman I. Hussen[1(✉)] and Talaat I. Hassan[2]

[1] Department of General Sciences, College of Basic Education, Salahaddin University, Erbil, Iraq
chiman.hussen@su.edu.krd
[2] Department of Automotive Technology Engineering, Erbil Technology College, Erbil Polytechnic University, Kurdistan Region, Iraq
talhat.hassan@epu.edu.iq

Abstract. This study employs B and R filter observations to investigate and evaluate sunspot properties for CN Andromeda at lower altitudes. The analysis revealed asymmetry and irregularities in the measured light curves. The numerical values of parameters such as age, sunspot properties, bolometric magnitude, lifespan, spot parameters, age–mass connection, radius, and temperature factor of CN Andromeda were acquired and compared with those found in previous studies.

Keywords: Sunspot · Age life · Parameters · Dark spot · CN Andromeda

1 Introduction

Binary stars (BS) hold significant importance for several reasons. First, they are not the same. Second, despite variations in kinematic characteristics and chemical composition, the spectroscopic double identified in the galaxy's corona is not markedly different from those discovered in the disk. Although the measured doubling rate stands at around 20%, the actual doubling rate is likely higher due to many binaries remaining undiscovered. In the solar vicinity, the doubling rate exceeds 50%, with some stars being fake. The wealth of essential information provided by double stars regarding the basic characteristics of stars underscores their importance [1, 3]. The primary significance of double stars is in their nature. Despite differences in kinematic properties or chemical composition, spectral similarities between the disk and the galactic corona are not mutually exclusive [2, 6].

Most star frameworks consist of multiple or paired stars. In some cases, parallel BS systems have two stars that are so far apart that they can each be considered singular stars. These systems comprise two stars whose orbits cause them to eclipse each other from Earth's point of view [5, 8]. Changes in the distribution of mass between the stars occur over time as the system gains or loses precise speed. This results in an increase (if the vas versa appears) or decrease in the orbital period [6, 7, 10]. The development

A. J. Obaid et al. (Eds.): ICONEST 2024, ISEM 40, pp. 279–288, 2025.
https://doi.org/10.1007/978-3-031-89960-7_21

of the obscuring parallel light curve (LC), or size versus time examination, provides more accurate information due to increasing improvements in observational procedures [9, 10]. Since the mid-20th century, critical endeavors have been made to study and understand these BS systems. This has been invigorated by the discovery of exoplanets and assessments of movement, mass loss, and rakish force misfortune. These systems, also known as W Uma-type EB, have orbital periods typically less than two days long and mass ratios ranging from 0.1 to 0.5 [8, 11, 13]. The study of these BS systems, particularly the analysis of their LCs, is vital for furthering our understanding of the CN Andromeda.

Diagrams are created by analyzing and counting the photons that a source emits over a period [14, 17]. The evolution of a star is primarily determined by its mass and substance [4, 18]. The light and outspread speed bends are examined simultaneously, utilizing the Wilson Devinney (1971) code, to infer the optimal and actual number of BS components [15, 16]. The discovery of exoplanets prior to the 21st century promoted extensive endeavors to produce exceptionally great LCs for planetary analysis and exploration. This involved the use of instruments, perception strategies, and data analysis methods [12, 19].

In 1949, Hoffmeister discovered CN Andromeda. Oosterhoff computed variations in the period of AB Andromeda in 1950, and in 1956, Tsesevich computed the periodic orbit of CN to be 2,599 days. Bozkurt noted small changes in V and B LCs in 1976. In 2001, Van Hamme analyzed the LC of CN Andromeda, and Rucinski found changes in CN's periodic orbit in 2010. In 2016, Maxted introduced the flexible LC model, and in 2020, Menon studied contact BS. In 2023, Hammad conducted a photometric analysis of CN Andromeda, and Hussein studied LC analysis for it. As a further extension of these works, sunspot parameters for CN Andromeda with B and R filters should be examined and evaluated.

2 CN Andromeda

From extensive reviews, we obtained the following information. The mass is 0.1100 M $\odot$ at 38 $\pm$ 4 years' orbit. The variability for CN Andromeda is as follows: BD = + 39059, GSC = 2787–1815, PPM = 42831, TYC = 2787–1815-1, RA = 00 h 20 m 30.60 s, and Dec = +40 13 m 33.8 s. In 1949, Hoffmeister observed spectral type F5V near the contact binary [10, 18, 21]. The period of CN Andromeda is 0.46279 days. Figure 1 presents the LC of CN Andromeda [15, 18]. Its overshadow profundities vary by a few 0.300 mages in V. Characteristics of its first star are as follows: $M_1 = 1.4330$, $R_1 = 1.4800$, $L_1 = 3.400$, and $T_1 = 6450$; the same for the second star are as follows: $M_2 = 0.5520$, $R_2 = 0.9500$, $L_2 = 0.4000$, $T_2 = 4726$. The radius of the sun is the following: $R_0 = 695700000$ m. In this study, we also used the following information regarding CN Andromeda: periodic day = 0.462796, BH = +3959, eclipse depth = 0.3000, binary orbit = 1.500×10^{-7}, and spectral type = F (Fig. 2).

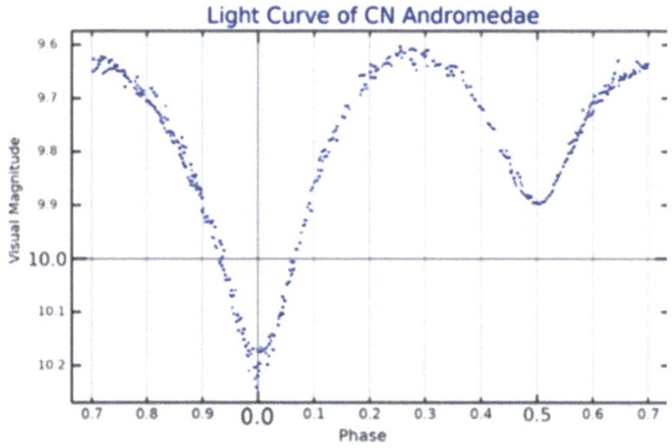

Fig. 1. Illustration of LC

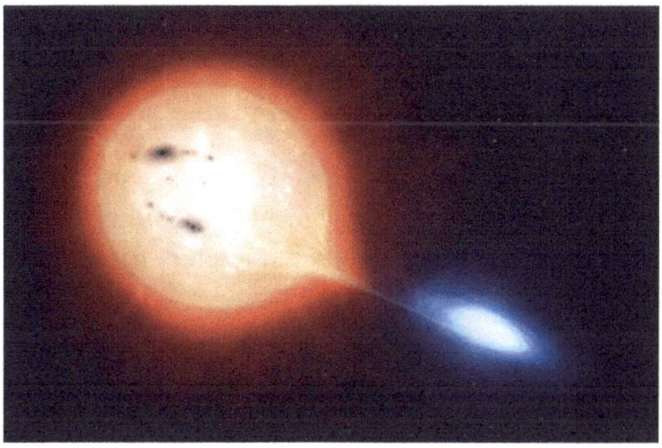

Fig. 2. The cataclysmic BS photograph of CN Andromeda

3 Photometric Analyzing for CN Andromeda

After numerous tests, we determined a set of parameters that accurately represent the observed LC. These parameters include a starting temperature of 6150 K for the primary star; LD coefficients ($x_1 = x_2$) of 0.4860 for V, 0.589 for B, 0.401 for R, and 0.475 for I groups; and gravity-darkening coefficients ($g_1 = g_2$) of 0.320, and the regard of the Albedo [2, 7].

The tuning parameters utilized are orbital incline (i), mass extent (q), ordinary temperature of the minute star (T_1), and the conceivable outcomes of the components Ω_1 and Ω_2, along with the radiance of the primary star (L_1). These parameters are presented in Table 1. However, the LC does not align well with the observed data near the 0.2500 stage, indicating a discrepancy in our analysis. To address this issue, we identified the brightest point as the moment of greatest intensity, suggesting that the primary half of the curve may be affected by a cooler region. By adjusting the parameters within the 0.500–1.000 stage, be unaffected by the point, at the same time as it were 0.5 of the confinement bend is balanced to deliver double parameters.

Table 1. The parameters for CN Andromeda

Number	First star	Second star
1	Lagrangian L_1 = 0.59581	Lagrangian L_2= 1.54060
2	C inner = 3.90190	C outer = 3.55972
3	C_1 = 3.90111	C_2= 3.90112
4	surface area one = 2.73372	surface area two = 1.14565
5	Ω_1 = = 2.6540	Ω_2 = 2.6540
6	Ω_{inner} Ω_{inner}= 2.65454	Ω_{outer} = 2.41698
7	r_1 (back) = 0.49141	r_2 (back) = 0.32506
8	r_1 (side) = 0.46473	r_2 (side) = 0.29231
9	r_1 (pole) = 0.43520	r_2 (pole) = 0.28043
10	mean radius one = 0.46378	mean radius two = 0.29926
11	f_1= 0.00231	f_1 = 0.00231

These parameters were found in [14], where Ω_1 and Ω_2 are surface potential components, C_1 and C_2 are potential components, f_1 and f_2 are fill out of the major and minor stars components, and L_1 and L_2 are absolute parameters for components in unit the of solar luminosity.

4 Age–Mass Relationship for CN Binaries

For studying the age-mass relationship of CN Andromeda, we used Eq. (1), as defined in [22]. The primary and secondary log ages are illustrated in Table 2.

5 Lifetime Calculation

The calculation of lifetime (LT) serves a significant role in research, providing valuable insights into BS. LT can be calculated using Eq. (1).

$$\frac{t_1}{t_0} = \frac{1}{2.5(\frac{M_1}{M_0})} \text{ and } \frac{t_2}{t_0} = \frac{1}{2.5(\frac{M_2}{M_0})} \tag{1}$$

6 Dark Spot Analysis

Dark spots (DS) are a key indicator for the light-bending curve. Notably, some information could have been compromised due to poor examination or weather conditions. However, in the evenings, the display was quite feasible. Two dark spots were observed in each light-bending curve [20, 22], with their respective areas outlined in Table 4. Although a different treatment routine altered some of these boundaries, the resulting lightweight bends were aesthetically pleasing. It is worth noting that not all blessings during this time were attributed to these particular spots. These lightweight bends manifested in numerous ways. Spot scope presented a substantial drawback, with mid-scope spots appearing more prominently than bright spots. The surfaces of the stars' regions were deemed perfect. However, in cooler stars, these regions were chiefly influenced by the stars' rotation rather than their transition. Star spots had the potential to impact any segment of the light curve, resembling sunspots arranged on other stars. The sizes of sunspots on certain stars are challenging to discern, as they are often too small to cause notable fluctuations in brightness. Larger star spots, covering up to 30% of the stellar surface, can be perhaps multiple times greater than those on the Sun.

Demonstrating the scotoma was fundamental for fitting the light bend accurately. Unfortunately, the demonstration was not possible due to limitations concerning examinations or weather conditions. Around evening, while the point demonstrated was conceivable, two dull spots were reproduced on each light bend. Table 4 shows the unit region of the fixed patch. The differential rectification routine significantly altered these boundaries. It is critical to note that although these slight bends appear to fit well, this does not imply that these spots are an endowment of this age. There were numerous ways of adjusting this slight bend, but a specific approach was chosen. The scope might be a significant character issue, as similarity with northern setups produces reliable outcomes as southern areas. However, there are patches in the mid-scopes, where both bright and dark patches were absent in the passband. Big-name region unit surfaces are supposed to be spotless. Nonetheless, cool stars (like the Sun) have spots, and the unit of region is made by the star's attractive transition, which is consistently achieved by the star's rotation. These star spots impact all aspects of the solar radiation curve. The unit region of a sunspot corresponds to that of a sunspot on another star. The size of sunspots is generally too small to cause observable variations in brightness, making them extremely challenging to detect in other stars. Certain spots are significantly larger than sun-oriented spots, covering up to around 30% of the star's surface, and star spots are sometimes multiple times bigger than sunspots.

7 First Contact Angle for CN Andromeda

By using the following equations, the first contact angle can be determined.

$$\theta_1 = \sqrt{(r_1 + r_2)^2 (csc(i))^2 + (cot(i))^2}, \tag{2}$$

where r_1 and r_2 are the radii of the first and second stars respectively.

$$a = \frac{+r_1}{r_2 + r_1} \tag{3}$$

$$C = \frac{cos(i)}{r_2 + r_1} \tag{4}$$

8 Empirical Relation Between *CI* and M_{bol}

In this section, the following formulae are used to find these parameters.

$$M_{bol} = M_{sun} + BC(T) \tag{5}$$

$BC_{sun} = -0.0700, M_{sun} = 4.8300$

$$B - V = -0.344Log(T) + 14.5510, \tag{6}$$

when $log(T) < 3.9610$.

$$B - v = 0.3440 \times (log(T))^2 - 3.4020 \times log(T) + 8.0370, \tag{7}$$

when $log(T) > 3.9610$

$$B - v = -8.4990 \times (\log(T) - 4)^4 + 13.4210 \times (\log(T) - 4)^3$$

$$-8.1310 \times (log(T) - 4)^2 - 3.9010 \times log(T) - 3.5620 \tag{8}$$

The parameterized equipotential surfaces can be found using these equations [5].

$$C_1 = 2\Omega B_1 + 2B_2^2 \text{ and } C_2 = 2\Omega B_1 - 2B_2^2, \tag{9}$$

where

$$B_1 = \frac{1}{1+q}, B_2 = \frac{q}{1+q} \tag{10}$$

The degree of geometrical contact (fill out factor) has the following form:

$$F = \frac{\Omega_{in} - \Omega_{1,2}}{\Omega_{in} - \Omega_{out}} \tag{11}$$

This is very important to find the empirical relationship between BC and CI.

$$BC = M_{bol} - M_y \tag{12}$$

$$CI = B - V \tag{13}$$

9 Results

The results obtained from the established models yield parameters for CN And, as illustrated. A suitable MATLAB program has been proposed for determining these parameters. The LT of CN was calculated using Eq. (2), and the results are displayed in Table 3. The spot parameters for CN are shown in Table 4. By using Eqs. (3–5), the first contact angle was calculated.

To facilitate this process, the program was processed using the MATLAB program. Table 5 shows the current first contact angle and the type of solar eclipse for CN Andromeda. Equations (6–12) were used to calculate parameters such as bolometric magnitudes, equipotential surfaces, and the degree of geometrical contact. Through Eqs. (13) and (14), the empirical relationship between BC and CI was determined, resulting in values of 82.27770 and 83.27510 for the primary star and 82.10280 and 83.1017 for the secondary star. The results are shown in Table 6. The LC fittings are displayed in Fig. 3, with B and R filters. Table 7 depicts the output spot parameters for CN and compares them with findings from other works.

10 Conclusion and Discussion

In this study, we evaluated the physical parameters of the sunspot of CN Andromeda and obtained the photometric arrangements of CN using several equations. The convective envelope f was found to be 24.90%. Novel perceptions of BS were measured using a 1.0 m telescope. Tables 2 and 3 present a photometric arrangement indicating parallels that may suggest that BS is a type of over-contact paired system. The mass ratio q was calculated to be 0.3060, which is close to the spectroscopic mass proportion q of 0.3250 acquired by a pointer. Therefore, the following conclusions were made:

1. From the assessment of the circle slant, it appears that framework CN Andromeda could be a common halfway overshadow.
2. The SP was assessed for the outline framework at a lower set examination.
3. The solution shows that the CN Andromeda is an A-type system because of its larger size and hotter temperature.
4. The results of this work, presented in Table 7, were found to align with results from other works.

(B filter) (R filter)

Fig. 3. The LC drawing of CN Andromeda

Table 2. The age-mass for CN And BS

Star	Primary	Secondary
CN And	109.036700	110.254200

Table 3. Lifetime for CN Andromeda

Star	t1/tsun*10^{83}	t2/tsun*10^{83}
CN And	4.013300	4.013300

Table 4. SP

Parameters	DK gyan
colatitude	84.0000
longitude	16.9000
radius	32.9000
Temperature factor	0.65000

Table 5. Type of eclipsing and first contact angle

Star	θ_1	a	C_0	Type
CN And	0.227110	2.517510	0.745410	F5V

Table 6. The present work of empirical numeric

Star	Component	B.C	CI
CN And	Primary	0.17490	25.61690
	Secondary	0.17341	25.03490

Table 7. Output spot parameters for CN And

Parameters	Hammad 2023 (B) filter	Soydugan 2019 (B) filter	Present work	Hammad 2023 (R) filter	Soydugan 2019 (R) filter	Present work
T_1	6500	6500	6501	6500	6350	6501
T_2	5947	5947	5948	5100	5732	5105
q	0.560	0.3935	0.565	0.560	0.387	0.572
i	0.56	0.3935	0.571	0.56	0.387	0.571
Ω_1	2.504	2.510	–	2.483	–	2.488

11 Recommendations

For future works, I recommend focusing on the following two points:

1. Studying sunspot parameters for AB BS systems
2. Conducting LC fitting for another short period BS

Acknowledgment. The author would like to extend their sincere thanks to the referees for their helpful remarks and suggestions. Many thanks are extended to the Salahaddin University and Erbil Polytechnic University for providing the financial support for this research.

References

1. Allen, C.W.: Astrophysical quantities, 3rd edn. The Athlon Press, London (1976)
2. Al-Naimiy, H.: Linearized limb-darkening coefficients for use in analysis of ES light curves. Astrophys. Space Sci. **53**(2), 181–192 (1978). https://link.springer.com/article/10.1007/BF00645913
3. Barrado, D., Fernandez, M.J., Montesinos, B.: The age-mass relation for chromo spherically active binaries. I. The evolutionary status. Astron. Astrophys. **290**, 143–157 (1994). https://adsabs.harvard.edu/full/1994A%26A...290..137B
4. Bozkurt, S., Ibanoglu, C., Gulmen, O., Gudur, N.: Information bulletin on variable stars 1087 (1976)
5. Bradstreet, D.H.: Catalog and atlas of eclipsing binaries. Department of Physics Science – Eastern University (2004)
6. Binnendijk, L.: The light variation and orbital elements of DK Cygni. Astron. J. **69**, 157–164 (1964). https://ui.adsabs.harvard.edu/abs/1964AJ.....69..157B/abstract
7. Claret, A.: A new non-linear limb-darkening law for LTE stellar atmosphere models III. Astron. Astrophys. **401**(4), 567–578 (2003). https://www.aanda.org/articles/aa/abs/2004/48/aa1673/aa1673.html
8. Claret A., Jimenez, A.: Limb-darkening for cooler stars: Standard and irradiated models. Astron. Astrophys. **230**(5), 412–428 (1990a). https://link.springer.com/article/10.1007/BF00640718

9. Chun-Hwey, K., Mi-Hwa, S., Jo-Na, Y., Wonyong, H., Min-Ji, J.: BD Andromedae: A new short-period RS CVn eclipsing binary star with a distant tertiary body in a highly eccentric orbit. Astrophys. J. **788**(134), 22–29 (2014). https://cir.nii.ac.jp/crid/1363388844142029696?lang=en
10. Daemgen, S., et al.: Banality of transit host stars. Astron. Astrophys. **498**(6), 567–579 (2009). https://www.researchgate.net/publication/41198117_Binarity_of_Transit_Host_Stars_Implications_on_Planetary_Parameters
11. Donald, R.S., Jerome, A.O., Gur, W., William, F.W.: Accurate computation of light curves and the Rossiter–McLaughlin effect in multi-body eclipsing systems. Astron. J. **156**(297), 48–63 (2018). https://arxiv.org/abs/1810.09565
12. Edward, G.: New low-mass eclipsing binary systems in Praesepe discovered by K2. Astrophys. J. **11**(1), 25–38 (2017). https://arxiv.org/abs/1706.03084
13. Essam, A., Nakhlawy, A., Rassem, M.A.: The first photometric analyses and classification of the W-UMa eclipsing binary systems GSC 1283–53 and GSC 702-1892. NRIAG J. Astron. Geophys. **2**(4), 1–7 (2013). https://www.academia.edu/53025356/The_first_photometric_analyses_and_classification_of_the_W_UMa_eclipsing_binary_systems_GSC_1283_53_and_GSC_702_1892
14. Hammad, H.Z., Murad, I.M.: Photometric analysis of CN Andromeda short period binary star. Zanco J. Pure Appl. Sci. **35**, 51–62 (2023). https://www.iasj.net/iasj/article/297536
15. Hoffmeister, C., Die, R.W.: Aurigae'Sterne und ihre nebenformen. Astronomische Nachrichten **278**(1), 24–40 (2006). https://onlinelibrary.wiley.com/doi/abs/10.1002/asna.19492780102
16. Hussein, C.I.: Investigation light curve analysis for short period eclipsing binary CN Andromeda. Cihan Univ.–Erbil Sci. J. **7**(2), 42–48 (2023). https://journals.cihanuniversity.edu.iq/index.php/cuesj/article/view/910
17. Hussein, C.I.: Investigation limb darkening for short period AB Andromeda eclipsing binary star. Polytech. J. **13**(1), 147–156 (2023). https://polytechnic-journal.epu.edu.iq/home/vol13/iss1/17/
18. Maxted, P.F.L.: ELCC - a fast, flexible light curve model for detached eclipsing binary stars and transiting exoplanets. Astron. Astrophys. **12**(5), 123–145 (2016). https://arxiv.org/abs/1603.08484
19. Menon, A., et al.: Detailed evolutionary models of massive contact binaries – I. model grids and synthetic populations for the Magellanic Clouds. Monthly Notices Royal Astron. Soc. **4**(6), 1–21 (2020). https://academic.oup.com/mnras/article/507/4/5013/6343036
20. Oosterhoff, P.T.: On the variation in the period of AB Andromeda. J. Bull. Astron Inst. **11**(2), 217–223 (1950). https://en.wikipedia.org/wiki/AB_Andromedae
21. Van Hamme, W., Samec R.G., Gothard, N.W., Wilson, R.E., Faulkner D.R., Branly, R.M.: CN Andromeda: A broken-contact binary? Astron. J. **122**(2), 3436–3446 (2001). https://iopscience.iop.org/article/10.1086/324110
22. Rucinski, S.M., Baran, A., Zola, S., Kreiner, J.M., Siwak, M., Drozdz, M.: Physical parameters of components in close binary systems: II. Acta Astronomica **54**(5), 195–206 (2004). https://acta.astrouw.edu.pl/Vol54/n2/pap_54_2_5.pdf
23. Skelton, P.L., Smits, D.P.: Modelling of W UMa-type variable stars. South African J. Sci. **105**(3), 120–127 (2009). https://sajs.co.za/article/view/10324
24. Soydugan, F., Yildirim, M.F., Alicavus, F.: CN Andromeda: A shallow contact binary with a possible tertiary component. Res. Astron. Astrophys. **19**, 10–21 (2019). https://iopscience.iop.org/article/10.1088/1674-4527/19/1/10/meta

Designing an Enhanced AI-Driven Automated Grading & Feedback Systems (AI-AGFS) in Online LMS Tools – A User Perspective

Pravitha Vijaykumar(✉), Madhumita Das, Chok Nyen Vui, Mamata Bhandar, and Dalwinder Kaur

Manipal GlobalNxt University, Kuala Lumpur, Malaysia
{PV18764,MD18687}@campus.globalnxt.edu.my, {chok.nyen, mamata.bhandar,dalwinder.kaur}@globalnxt.edu.my

Abstract. The research aims to study the perceptions of users, i.e., learners and faculties, regarding the usability and effectiveness of AI-driven automated grading and feedback systems (AI-AGFS) within online learning management systems (LMS). The study targets adult learners engaged in online learning for personal or career advancement alongside their professional work. The primary objective is to establish the relationship between the components of AI-AGFS and different user experience factors, perceived usefulness, and ease of use per TAM, to understand user preferences and expectations from an application usability perspective. The components of AI-AGFS include automated grading, real-time feedback, personalized recommendations, and customizations. The study results provide an overview of the factors that the learners would like to have in the AI system and from an overall perspective ease of use is something that the users perceive to be significantly predominant. The major findings based on the regression analysis conducted is that while the questionnaire does capture the essence of the need of AI-AGFS and its factors, there might be some factors which are either duplicating or missed out and can be further researched upon. The results can help universities or LMS vendors work towards optimizing the experiences of users and engage them on a continuous basis thereby promoting the usage of online learning environments.

Keywords: Online LMS tools · AI driven · Automated grading & feedback systems · User experience · TAM model

1 Introduction

1.1 Background

Today with technological advances and innovations, organizations and institutes are working towards enabling Artificial intelligence (AI) in their applications. AI in higher education is revolutionizing how one can experience learning, assessment, and feedback systems. AI provides learners with personalized learning experiences and can potentially

A. J. Obaid et al. (Eds.): ICONEST 2024, ISEM 40, pp. 289–302, 2025.
https://doi.org/10.1007/978-3-031-89960-7_22

improve their learning and the optimal way teachers can operate. Educational institutes using learning management systems (LMS) platforms for higher education leverages the ever-evolving AI technology. To leverage the potential of AI and enhance *administrative duties, teaching, learning, assessments*, which are the four major domains in the education field. Evaluation in the traditional manner requires teachers to go through each and every student's exam/assignments making this process very time-consuming and administratively heavy. A functioning LMS can potentially reduce the time invested by academicians in terms of developing and managing the course contents in a systematic and organized manner online. Academicians should prioritize enhancing the value and quality of the content they deliver, rather than spending excessive time to manage the tools (de Kock et al. 2016). Regarding assessments/grades and feedback, learning analytics can improve the student's performance, an aspect that AI can transform (Inria 2023). The emergence of AI-powered automated grading systems, which can be easily integrated into LMS, is one of the revolutionary changes in exam evaluation.

1.2 Problem Statement

AI-powered automated grading and feedback systems have the potential to grade exams and assignments faster with better accuracy and consistency, without any human bias, and further provide learners with personalized feedback for improvements. Today, if an institute wants to implement an automated grading system as part of their LMS platform or utilize the automated grading system in their choice of LMS platform, there is a need to understand more about the different features required in the AI-powered automated grading and feedback tool. Not all features may be required based on the courses offered or the institute's LMS objectives, which the user will find helpful from their perspective (Maslov 2020). There is a limited understanding of the essential and desirable features of the AI-enabled automated grading and feedback systems, as well as any specific tailoring requirements that instructors might need, which the users perceive to be beneficial for their use, in terms of the positive user experience provided by the tool to determine utilizing the tool.

Research Objectives (RO) and Research Questions (RQ)

Research investigates the users of online learning management systems, primarily the students and instructors, and their preferences, expectations, and perceived effectiveness of AI-AGFS. The study seeks to explore various features that can enhance the user experience of AI-AGFS within LMS. It aims to identify the critical components that contribute to optimizing user interaction, engagement, and learning outcomes based on automated grading, real-time feedback, personalized recommendations, and personalized customizations. User experience factors include accuracy of grades, ease of understanding and operation.

1. To establish a relationship between AI-AGFS into LMS and the perception of users in terms of their utility and usability.

Research Questions

- How do the users' perceptions of the usefulness and easy use of the AI-AGFS LMS correlate with users' intent to continue using the platform for their forthcoming courses?
- What are the different features that influence users perception of the usefulness and perception of the ease of use of the overall AI-AGFS in LMS?

2 Literature Review

2.1 Automated Grading and Feedback Tools

AES or Automated essay scoring, a computer-based assessment system, as part of the AES research, started in 1966 and further evolved to be an intelligent system that grades students' responses, focusing on grammar, syntactic and semantic analysis, pattern matching, etc. (Ramesh & Sanampudi 2022). US states like Ohio, Utah, and others use AES systems in school education, which provide developmental, cumulative assessments and feedback to students on their essays with the evaluation rubrics viz. idea development, style, choice of words, fluent sentences, conventions, content structure, or organization. The literature review identified that these systems lacked relevance, completeness, consistency, rationality of content, domain knowledge in the context of the post, and any additional features. Conversely, feedback should be impactful, timely, constructive, encourage positive motivational beliefs, and be relevant to the student's work (Ricci et al. 2023). Gradescope and LMS tools like Canvas utilize automated graded tools for assignment assessments with multi choice (MCQ), select-all, and short answer questions. Rubic-based grading mechanism enables transparency and consistency to make grading more straightforward and faster, though have challenges with other functionalities requiring computations and scripts (Ricci et al. 2023). Tools such as 'autograders' for grading programming languages (Messer et al. 2024) and 'gradetools' used in data science classes also assist instructors in pedagogical and administrative tasks associated with grading and providing personalized unique/repeated feedback to students (Ricci et al. 2023).

2.2 AI in Grading and Assessment/Feedback

AI in education uses algorithms that can analyze data, identify patterns/themes, and predict the learning abilities of every student and determine their progress. This enables educators to personalize learning for them. This can result in enhanced learning outcomes, greater effciency, and higher levels of student engagement. It can provide insights on the performance of learners, in addition to various administrative tasks required to manage the LMS platform. With the use of AI assessment sections can be automated, enabling students and teachers to focus on meaningful tasks (Harry 2023). Learner/student engagement can be enhanced with the usage of interaction based learning techniques and experiences. Technologies and tools utilizing adaptive learning methods, virtual assistants, and chatbots provides this capability. AI powered data analysis within LMS can offers valuable insights into student performance, and allows teachers to personalize their instructions and teaching approaches to meet the learner's individual needs. A combination of the auto-grading and feedback tools incorporated in AI-driven systems in an LMS can have features that can automatically grade the learner assessments/assignments and provide real-time feedback, which students can utilize to improve their learning. AI has its own learning ability to adapts and trains itself with the data received based on the learner's progress and interaction with the tool to support personalization.

2.3 User Experience

User experience (UX) refers to the general feelings and experiences of individuals when interacting with a product, incorporating aspects such as usability, features, content, aesthetics, and more. For any software application, the key goal is to ensure customer satisfaction by optimizing its effectiveness and efficiency, either via utility, design, or usability. A well-designed user experience contributes to improved performance and efficiency, as tools with a strong focus on UX/usability enable smoother and more productive interactions with the software (de Kock et al. 2016). The usability goals encompass effectiveness, ease of learning (learnability), efficiency, safety, proficient utility, memorability, thereby improving user engagement (Maslov 2020; Sahid et al. 2016). LMS with good usability means that the learner can focus on the educational content rather than understanding the tool operations (Nakamura et al. 2017). In the context of an LMS tool with an AI-driven automated grading and feedback system (AI-AGFS), a positive UX can be experienced by the user when the tool is efficient and effective such as, the accuracy of how well the grades are assigned and if the learners receive prompt real-time feedback on their assignments while submission as well because of their assessment. Technology usage and its adoption/ acceptance, though positively correlated with the perceived quality of user experience, is entirely subjective and depends on the perception of usefulness and perceived ease of use of technology to meet their expectations (Maslov 2020).

2.4 Literature Gap

Research exists on user adoption/acceptance of LMS tools. In the context of usability, using a holistic UX approach by establishing a correlation of the user behavior while accessing the LMS and the various UX properties, i.e., usefulness, reliability, novelty, efficiency, etc., and analysis further based on gender differences, there is lack of research availability related to the features of LMS tools (Maslov 2020). Not much research is found that directly correlates features like components of AI-driven grading and feedback systems such as automated grading, real-time feedback, and personalized customizations to the factors that contribute to a positive user experience.

2.5 Theoretical and Conceptual Framework

The theoretical framework proposed for this research is the Technology Acceptance Model (TAM) formed in 1989 by Davis which explains the acceptance and technology usage of the AI-AGFS in LMSs. Per the theory, perception of usefulness and ease of use to understand an individual's intention to use any specific technology. The study will use the TAM theory and research model based on the literature review to establish the relationship between the user experience and components of LMS related to AI-AGFS. Other theories reviewed include the Diffusion of Innovation Theory (DOI), the Theory of Planned Behaviour (TPB), and The Unified Theory of Acceptance and Use of Technology (UTAUT). DOI focuses on the initial adoption and spread of innovations within a social system with factors like relative advantage, compatibility, complexity, and observability, whereas TAM addresses long-term adoption and continuance behaviors. TPB considers attitude toward behavior, subjective norm, and perceived behavioral control, which is wider in scope and not always specific to technology adoption. UTAUT integrates elements from various theories, including TAM, TPB, and others, with performance expectancy, effort expectancy, social influence, and facilitating conditions as key determinants (Chug et al. 2013).

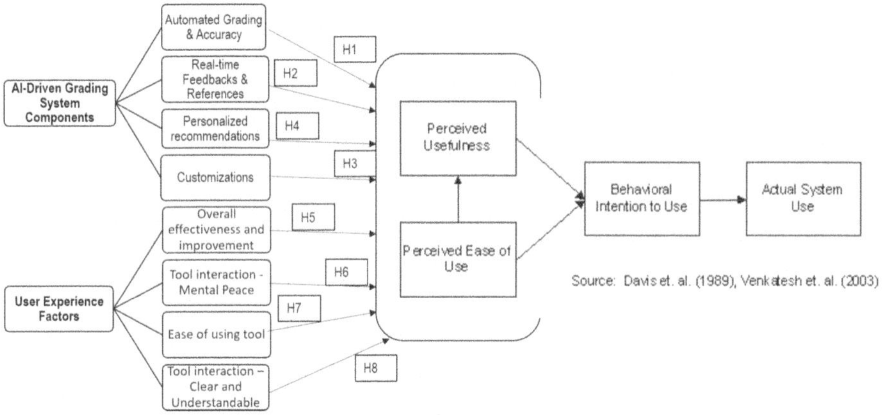

Fig. 1. Conceptual model framework.

The choice of TAM is considered since the study is interested in understanding an individual's perspective on accepting the new AI technology that can be enabled for automated grading and feedback in LMS. With this, the conceptual research model formed, including the AI-AGFS and User experience components as the independent variables and the TAM constructs as the dependent variables, can be represented as seen in Fig. 1.

2.6 Research Hypothesis

While the study would reveal the features that the users would prefer, for an improved user experience of AI-AGFS. The hypothesis determined to establish the relationship between the independent and dependent variables is listed in Table 1.

Table 1. Hypothesis

For AI-driven grading system components	
	Automated Grading and Accuracy: (PU5)
H1	Automated grading significantly influences PU and PEOU
	Real-time Feedback & References (PU6)
H2	Real-time feedback significantly influences PU and PEOU
	Customizations: (PU7)
H3	Customizing grading criteria significantly influence PU and PEOU
	Personalized recommendations: (PU3 & PU4)
H4	Personalized recommendation significantly influences PU and PEOU
For User Experience Factors	
	Overall effectiveness and improvement: (PU1 & PU2)
H5	Overall effectiveness and improvement significantly influence PU and PEOU
	Tool interaction - Mental Peace: (PEOU1)
H6	Interacting with the tool leading to mental peace significantly influences PU and PEOU
	Ease of using tool: (PEOU2)
H7	Easy access to the tool significantly influences PU and PEOU
	Tool interaction – Clear and Understandable: (PEOU3)
H8	Clear and understandable interaction with the tool significantly influences PU and PEOU

Table 2. Descriptive Statistics for Perceived Usefulness (PU) and Perceived Ease of Use (PEOU)

	PU							**PEOU**		
	1	**2**	**3**	**4**	**5**	**6**	**7**	**1**	**2**	**3**
Count	*47*	*47*	*47*	*47*	*47*	*47*	*47*	*47*	*47*	*47*
Mean	*3.9*	*3.8*	*4*	*3.9*	*3.8*	*3.9*	*3.8*	*3.5*	*3.7*	*3.7*
SD	*0.95*	*0.88*	*0.91*	*0.86*	*1*	*0.95*	*0.94*	*1.18*	*0.93*	*0.99*
Min	*1*	*1*	*1*	*1*	*1*	*1*	*1*	*1*	*1*	*1*
25%	*3*	*3*	*3*	*3*	*3*	*3*	*3*	*3*	*3*	*3*
50%	*4*	*4*	*4*	*4*	*4*	*4*	*4*	*4*	*4*	*4*
75%	*5*	*4*	*5*	*4.5*	*4.5*	*5*	*5*	*4*	*4*	*4*
Max	*5*	*5*	*5*	*5*	*5*	*5*	*5*	*5*	*5*	*5*
Mode	*4*	*4*	*4*	*4*	*4*	*4*	*3*	*4*	*4*	*4*
Median	*4*	*4*	*4*	*4*	*4*	*4*	*4*	*4*	*4*	*4*

3 Research Methodology

Quantitative research is suited for studying the user perspectives on AI-enabled automated grading and feedback systems as it allows the establishment of relationships based on the user demographics/ preferences with the features potentially to be utilized, please see the Appendix for more information about the Features perceived.

3.1 Research Design

The researcher uses a Cross-sectional survey to capture the participant's thoughts, feelings, perceptions, etc., as the data will be collected once (Cincala et al. 2023). Descriptive research is conducted to understand the impact of the identified attributes/variables without the intent to establish a cause-and-effect relationship (Baker 2017). In addition to the mean, mode, and standard deviation, Pearson correlation is utilized to determine the relationship between the variables, viz., the user perceptions/behaviors and the various features of the AI-driven AGFS (Cincala et al. 2023). Senior academicians reviewed the questionnaire before sharing it with the learner community.

The study population includes students and faculty members who have enrolled or conducted online education programs or courses utilizing LMS platforms. The academic levels of students can vary from undergraduate, graduate, postgraduate, or PhD/DBA students, and there are no specific criteria regarding the courses/disciplines they have enrolled in as long as they are utilizing the online learning platforms/LMS tools as part of their course curriculum. The unit of analysis will be the individual users within this study sample who have volunteered to participate in the data collection.

The researcher opted for convenience sampling, one of the non-probability sampling methods, as it was easier for the researcher to select individuals who were more accessible to reach, thereby providing higher chances of collecting data i.e. students and faculties of Manipal GlobalNxt University (MGNU). The sample may not represent the population, and as part of the analysis, the researcher classified individuals based on students/faculties and their age, education, and sex to understand the relationship between these variables (Rahman et al. 2022). Based on the G*Power version 3.1.9.7 software for a multiple linear regression f-test at a 95% confidence level (alpha, $\alpha = 0.05$), f of 0.18, i.e., moderate variance, power of $(1-\beta) = .80$, the sample size required is 57 (Fig. 2).

The research instrument for data collection is developed and adopted from the questionnaire TAM survey (Davis 1989), to measure the intended behavior of users to accept any technology based on the results from users as to how they perceive the technology's usefulness and ease of use. This scale item has been investigated for content and construct validity after reliability testing in studies and used widely for acceptance of information systems situations and environments (Sandema-Sombe 2019).

F tests - Linear multiple regression: Fixed model, R^2 deviation from zero			
Analysis:	A priori: Compute required sample size		
Input:	Effect size f^2	=	0.18
	α err prob	=	0.05
	Power (1-β err prob)	=	0.8
	Number of predictors	=	2
Output:	Noncentrality parameter λ	=	10.26
	Critical F	=	3.168246
	Numerator df	=	2
	Denominator df	=	54
	Total sample size	=	57
	Actual power	=	0.802877

Fig. 2. G*Power Analysis

3.2 Data Collection, Analysis, Findings and Discussions

Data was collected through a survey questionnaire shared with participants via online Google forms by sharing the survey URL through WhatsApp messages. The survey was kept accessible to users for two weeks, from 18th May to 1st June 2024, and over the weekend on 8th and 9th June 2024. The survey was rolled out to students and faculty of Manipal GNU. The survey questions include close-ended Likert items and single/multiple-choice questions that capture the users' age and demographics and their perspectives/behaviours towards using AI-AGFS in the LMS tool. The research instrument used for data collection was adopted from the TAM survey questionnaire to understand the PEOU and PU of AI-AGFS. Data analysis is conducted using the descriptive and inferential statistics of MS Excel and parallelly correlated with statistical and exploratory data analysis developed in Python.

Forty-seven participants, aged 20 to 62, responded to the survey, including 72.3% males, 23.4% females, and 4.3% others who did not prefer to specify their gender. The participants' qualifications included 53% postgraduates, 36% graduates, 7% doctorates, and 4% undergraduates. 17% of the participants have used AI-enabled LMS tools, 23% believe that they might have used or encountered such tools, and 60% have not used them.

The reliability of PEOU and PU questions was verified using Python code by Cronbach's alpha reliability coefficient, which equated to 0.955, indicating high reliability and likely redundancy of the variables in the construct (Fig. 3).

PU and PEOU Components

The results of descriptive statistics show that the median, mode, and mean for both PU and PEOU are approximately 4, indicating a positive perception by the participants of the system's usefulness and ease of use. Based on the interquartile range between 25% to 75%, since the responses are between 3 and 5 for all the questions, it can be said that there is a skewness towards higher ratings, with fewer outliers and moderate variability since the standard deviation is ~1. Overall, the strong consistency in the responses indicates that the participants perceive AI-AGFS in LMS as beneficial.

```
Overall Cronbach's Alpha: 0.9558705769120133
Cronbach's Alpha for each column (treated as a scale with remaining items):
PU1: 0.9492278895648105
PU2: 0.9498971420799401
PU3: 0.9506323827705525
PU4: 0.952412639343168
PU5: 0.9512211009014955
PU6: 0.9491851928984224
PU7: 0.9511603429686297
PEOU1: 0.9576431613819214
PEOU2: 0.9526898493380004
PEOU3: 0.9479633024288786
```

Fig. 3. Cronbach's Alpha Reliability for PU and PEOU

Figure 4 shows a positive correlation between the variables, indicating relationships, though the correlation strength varies amongst the variables. A strong correlation is seen between the PU values in the range of 0.66 to 0.86, and for PEOU, it's between 0.66 and 0.8 (Table 2).

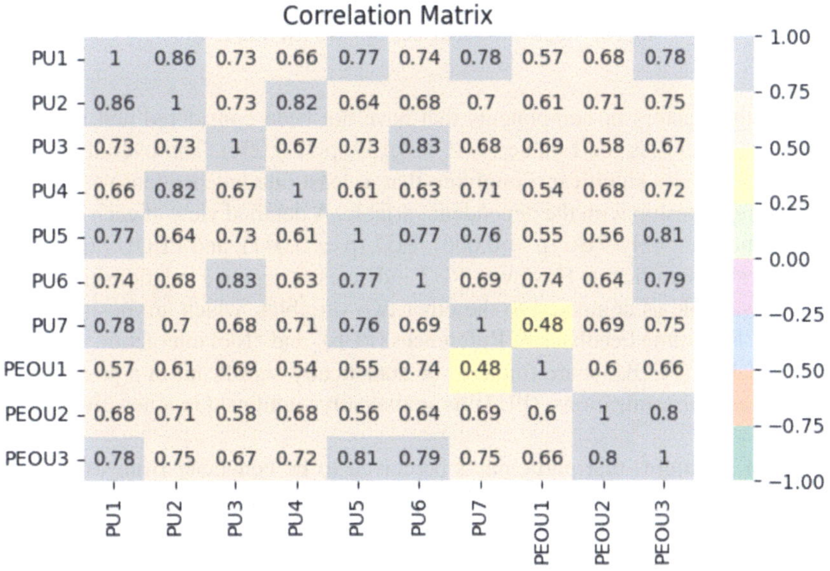

Fig. 4. Correlation Matrix

The multiple regression analysis conducted on the 47 observations indicates a moderate to strong correlation between the dependent and independent variables, as the multiple correlation coefficient is 0.731. The R2 value of 0.535 indicates that the model can explain 53.5% variability in the independent variables. However, Adjusted R2 is relatively low, with only 40.6% of the variability explanation, and this would mean that while some of the independent variables significantly contribute to the model, there

Regression Statistics	
Multiple R	0.731
R Square	0.535
Adjusted R Squ	0.406
Standard Error	0.191
Observations	47

ANOVA

	df	*SS*	*MS*	*F*	*Significance F*
Regression	10	1.502	0.150	4.139	0.001
Residual	36	1.306	0.036		
Total	46	2.809			

	Coefficients	*Standard Error*	*t Stat*	*P-value*	*Lower 95%*	*Upper 95%*	*Lower 95.0%*	*Upper 95.0%*
Intercept	0.659	0.146	4.506	0.000	0.363	0.956	0.363	0.956
PU1	0.092	0.082	1.127	0.267	-0.074	0.258	-0.074	0.258
PU2	-0.067	0.093	-0.718	0.477	-0.255	0.122	-0.255	0.122
PU3	-0.085	0.066	-1.281	0.208	-0.220	0.050	-0.220	0.050
PU4	0.002	0.070	0.026	0.979	-0.140	0.143	-0.140	0.143
PU5	-0.034	0.065	-0.527	0.602	-0.167	0.098	-0.167	0.098
PU6	0.198	0.068	2.924	0.006	0.061	0.335	0.061	0.335
PU7	-0.126	0.061	-2.053	0.047	-0.250	-0.002	-0.250	-0.002
PEOU1	0.017	0.040	0.420	0.677	-0.064	0.097	-0.064	0.097
PEOU2	-0.093	0.059	-1.573	0.124	-0.213	0.027	-0.213	0.027
PEOU3	0.168	0.078	2.153	0.038	0.010	0.327	0.010	0.327

Fig. 5. Regression statistics for AI-AGFS in LMS

might be other factors or components that have not been considered and have scope for future research. As the f-value is ($F = 4.139$, $p < 0.001$), The overall regression model is statistically significant, implying that at least one independent variable has a significant relationship with the dependent variable. A detailed view of each of the variables specifies that only PU6 ($p = 0.006$), PU7 ($p = 0.047$), and PEOU3 ($p = 0.038$) are statistically significant. However, PU7 indicates a negative relationship with the dependent variable as compared to the other two variables, which are positive. Of the hypothesis 'Real-time Feedback & References (PU6)' and 'Tool interaction – Clear and Understandable: (PEOU3)' are found to be statistically significant in a positive direction where-as 'Customizations: (PU7)' is statistically significant in a negative direction (Fig. 5).

The top 3 features that participants perceived to be beneficial from usability and ease of use standpoint are 'Real-time feedbacks', 'Personalized recommendations and adaptive learning' and 'Dynamic results with auto grading'. The faculties did find the 'Grading/Rubric criteria', and 'Customizations' to be additionally beneficial considering reduction in their workload please see the Appendix for more information about the Features perceived.

4 Conclusion and Research/Business Contributions

The research intended to address the gap in today's literature where even though AI-enabled automated tools for grading, assessment feedback etc., exists, there is no study which identifies what are the usability factors/features that the any usable AI tool intended to be integrated with LMS should have. The researcher defined AI-AGFS components which have the potential to be usable and increase the usability of the tool. The research outcome based on survey results with the university using convenient sampling, reveal that despite the Cronbach Alpha reliability is very high, depicting that the questionnaire is reliable, the variables under consideration are repetitive and needs to be revisited. While the users have rated features based on their preferences considering usability and ease of use that an AI-AGFS integrated LMS should have, not all factors are considered statistically significant except for the factors 'Real-time Feedback & References (PU6)', 'Tool interaction – Clear and Understandable: (PEOU3)', and 'Customizations: (PU7)', irrespective of the direction and strength. The study is a starting point which gives direction in terms of what at a minimum the learners expect from the AI integrated LMS tools.

The study can be further enhanced to take it across other universities or working professionals and data with a larger sample size and additional variables can provide more factors that can help contribute to the enhancing the AI-AGFS in LMS.

4.1 Contribution to Research and Business

The study is significant for institutes that intend to implement an LMS platform with an AI-enabled set of features and integrations. Assessments/evaluations being one of the significant elements in any LMS platform, a better understanding of the expectation from the learners as well as the faculty instructors on the features required for automated grading and feedback system would lead to a smooth application implementation as well as enhance the experience of the usage of the LMS platform. The outcome of this study provides information to internal and external stakeholders in identifying what contributes to providing an enhanced experience to users within online LMS platforms and what the key parameters are, which can be used to model the AI algorithms to improvise on the learning outcomes.

Internal Stakeholders

- Students/Learners: Preferences of the learners who use LMS tools to access course materials, submit assignments, and receive feedback for evaluating the usability and utility of AI-AGF is understood.
- Faculty Members: Capturing the requirements of faculty teaching the courses offered via LMS are vital members who are involved in developing/implementing AI-AGFS to obtain a hassle-free, usable, and effective product with optimized evaluation, feedback, and tailoring capabilities.
- University/Institute Administration: Department heads, Deans, Administrators, Decision makers, etc., who oversee academic programs and policies need the right information to make decisions about technology adoption and development support.

- IT Department: To manage the technical infrastructure and support services required for the AI-AGFS LMS deployment, maintenance, and validation.

External Stakeholders

- Educational Technology Companies/Industry Partners: LMS with AI-integrated solution providers for education would benefit from the research findings to prioritize the product features and develop their development strategy and plans.
- Professional Associations: Associations representing educators, educational technologists, or designers who promote best practices, innovations, and advocate for policy changes, etc., in online learning.
- Government Agencies: Research will benefit Government bodies by informing funding priorities, grant programs, education policy and regulation, and promoting the adoption of technology-enhanced LMS.
- Nonprofit Organizations: NGOs focused on education, technology, or student advocacy can use the research findings to promote policies/initiatives to address digital equity and inclusion and provide quality learning experiences.

The significance of this research lies in its potential to understand the features of AI-AGFS LMS that users perceive to be beneficial. By investigating the perceived usefulness and ease of use among users, including DBA students, faculty, and other online learners, the study addresses critical gaps in current understanding and practice within educational technology.

- Enhancing User Experience: This study's findings have the potential to identify the features and user experience factors most valued by users. Educators and developers can prioritize these features to improve, innovate, and optimize the usability and utility of AI-AGFS.
- Improving Learning Outcomes: The research can provide insights into students' needs or perceptions of improving learning outcomes/feedback/personalized recommendations. This information can then be utilized to train the AI model and design the system so that more effective feedback mechanisms promoting deeper and adaptive learning are established.
- Informing Institutional Decision-Making: Based on research data, institutions, and educational organizations can make informed decisions about technology adoption, resource allocation, and strategic planning.

The study is not limited to a particular institute for LMS platform implementation but can be further extended as an advancement to online education. This can help realize the full potential of AI technologies, thereby improving learning and assessment.

4.2 Limitation of the Study

The current study has limitations considering that all the variables are not considered. Qualitative research to understand the needs of the stakeholders in-depth can add-on to further refine this study and identifying the variables that can enhance the AI-AGFS integration in LMS for future.

Appendix

Features perceived to be usable and easy to use from user perspective.

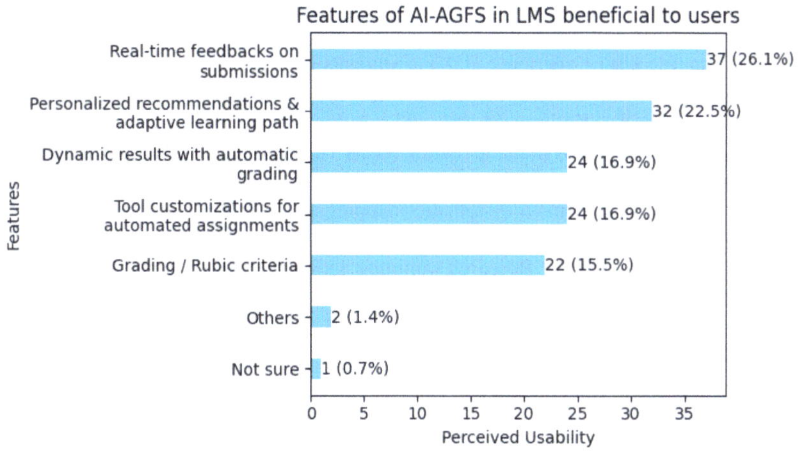

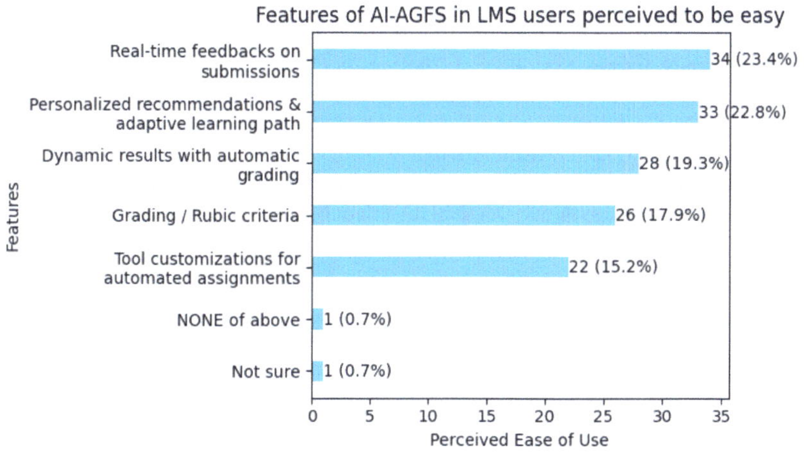

References

Baker, C.: Quantitative research designs: experimental, quasi-experimental, and descriptive. Evid.-Based Pract.: Integrat. Approach Res. Administ. Pract. **2**, 155–183 (2017)

Chung, C.H., Pasquini, L.A., Koh, C.E.: Web-based learning management system considerations for higher education. Learn. Perform. Quar. **1**(4), 24–37 (2013)

Cincala, P., Penno, D., Zubkov, P., Wa-Mbaleka, S.: Quantitative research for practical theology (2023). http://books.google.ie/books?id=FzTZzwEACAAJ&dq=Quantitative+Research+for+Practical+Theology+(pp.103-114)Publisher:+Andrews+University&hl=&cd=1&source=gbs_api

Davis, F.D.: Perceived usefulness, perceived ease of use, and user acceptance of information technology. MIS Quar. 319–340 (1989)

de Kock, E., van Biljon, J., Botha, A.: User experience of academic staff in the use of a learning management system tool. In: Proceedings of the Annual Conference of the South African Institute of Computer Scientists and Information Technologists, pp. 1–10 (2016)

Harry, A.: Role of AI in education. Interdicipl. J. Humman. **2**(3), 260–268 (2023)

Maslov, I.: Evaluating User Experience (UX) of students using a Learning Management System Moodle in a Finnish university through a holistic UX model approach (2020)

Messer, M., Brown, N.C., Kölling, M., Shi, M.: Automated grading and feedback tools for programming education: a systematic review. ACM Trans. Comput. Educ. **24**(1), 1–43 (2024)

Nakamura, W.T., de Oliveira, E.H.T., Conte, T.: Usability and user experience evaluation of learning management systems-a systematic mapping study. In: International Conference on Enterprise Information Systems, vol. 2, pp. 97–108. Scitepress (2017)

Oribhabor, C.B., Anyanwu, C.A.: Research sampling and sample size determination: a practical application. J. Educ. Res. (Fudjer) **2**(1), 47–57 (2019)

Rahman, M.M., Tabash, M.I., Salamzadeh, A., Abduli, S., Rahaman, M.S.: Sampling techniques (probability) for quantitative social science researchers: a conceptual guidelines with examples. Seeu Rev. **17**(1), 42–51 (2022)

Ramesh, D., Sanampudi, S.K.: An automated essay scoring systems: a systematic literature review. Artif. Intell. Rev. **55**(3), 2495–2527 (2022)

Ricci, F.Z., Medina, C.M., Dogucu, M.: Designing and implementing an automated grading workflow for providing personalized feedback to open-ended data science assignments. arXiv preprint arXiv:2309.12924 (2023)

Rutner, S.M., Scott, R.A.: Use of artificial intelligence to grade student discussion boards: an exploratory study. Inform. Syst. Educ. J. **20**(4), 4–18 (2022)

Sahid, D.S.S., Santosa, P.I., Ferdiana, R., Lukito, E.N.: Evaluation and measurement of Learning Management System based on user experience. In: 2016 6th International Annual Engineering Seminar (InAES), pp. 72–77. IEEE (2016)

Sandema-Sombe, C.N.: Relationship between perceived usefulness, ease of use, and acceptance of business intelligence systems (Doctoral dissertation, Walden University) (2019)

The Effect of Scaffolding Strategies on the Ability to Construct Proofs: An Exploration of Abstract Algebra Proofs

Warli(✉), Imas Cintamulya, and Edy Nurfalah

Universitas PGRI Ronggolawe, Jl. Manunggal 61, Tuban, Indonesia
warli@unirow.ac.id

Abstract. This study aims to explore the effect of providing scaffolding on the ability to construct proofs of abstract algebra problems. This research is a mixed method, firstly providing learning treatment with a scaffolding approach then testing, secondly conducting qualitative exploration. The research subjects were 64 mathematics education students taking algebraic structures courses. Exploration in constructing proof is studied based on stages, including the entry, attack, and review stages. At the entry stage, students with high abilities fulfilled the know, want, and introduce aspects and understand the group problems well. At attack stage, the student fulfilling the try, maybe, and why aspects, prove the conjecture, and convince the truth of the proof well. At review stage, it meets the check and reflect aspects. Students with low abilities can meet the know, want, and introduce aspects well. At the attack stage, it fulfills the try, maybe, and why aspects, but less able to propose a conjecture and convince the correctness of proof. At the review stage, it fulfills the aspects of check and reflect, but is not sure about the truth. Providing scaffolding has a positive effect on the process of constructing proofs of abstract algebra problems.

Keywords: abstract algebra · construct proof · exploration · scaffolding strategies

1 Introduction

Please Advanced mathematics learning has the main goal of improving the ability to construct formal proofs [1–3]. Mathematicians and mathematics students must have the essential ability, namely constructing mathematical proofs [4–8]. In fact, the proof is considered an essential component for working, communicating, knowing, and understanding mathematics [9, 10]. As a result, in recent years there has been a lot of educational research focused on how students engage with proving and verifying. There is a great deal of literature on teacher and student proof conceptions [11], about how students construct mathematical proofs [6, 12], and how pedagogy can be designed to develop students' understanding of proof [13–15]. However, it is also recognized that despite the growing literature, studies that focus on reading activity are proving [7, 16, 17] relatively unusual, and cannot distinguish valid proofs from invalid arguments [7, 18].

A. J. Obaid et al. (Eds.): ICONEST 2024, ISEM 40, pp. 303–321, 2025.
https://doi.org/10.1007/978-3-031-89960-7_23

Many studies reveal that constructing mathematical proofs is a difficult skill for most students [1, 12, 18–24]. Research that focuses on students' particular difficulties when writing proofs, such as having a poor understanding of advanced mathematical concepts, lack of an evidentiary strategy [6], and don't know where to start to write proof. Although there has been a great deal of research on students' difficulties with proof, some researchers have noted that there is little about the processes students use when they try to construct a proof, and further research of this type is needed [6, 7, 23, 25]. For this reason, this research explores students' abilities in constructing proofs, especially on abstract algebra topics. Formal proof of mathematics and abstract algebra are two things that are interrelated with the arguments needed for reference in making a conjecture [26].

The proof is recognized as the essence of mathematical thinking [22]. According to Furinghetti and Morselli [27] proof is a special case of problem solving. Given the specificity of the mathematical proof construction, namely convincing someone logically about the truth of a mathematical statement, the stage of mathematical proof construction requires a separate thinking stage. The results of exploration when compiling proof can provide better clues to identify student difficulties. Mathematical proof lies in the formal world and is formally deductive [28]. Tall [29] states that written deductive formal proof is the final stage of mathematical thinking. Thus, the process of constructing mathematical proofs can only take place in the formal world or involve the real and symbolic worlds.

According to Stacey [30] Mathematical thinking is 1) an important goal of education, 2) a way of learning mathematics, and 3) an important means of teaching mathematics. Mathematical thinking helps 1) understand the importance of using knowledge and skills, and 2) ability achievement and implementation of independent learning [31]. Focusing on thinking skills is essential to support active cognitive processes that make students learn better. Students are given the opportunity to express their ability to construct proof, make student exploration more explicit, and reflect on the strategies students apply. The process of constructing proof carried out by students consists of two terms, namely: the term 'natural' is used to describe the process of extracting meaning, and the term 'formal' is used for the process of providing meaning that works in a standard way. In addition [32] added the term 'procedural' learning for students who only try by rote to understand formal definitions and proofs.

The process of constructing evidence can be dug through the stages of problem solving according to [33]. There are three stages that a person goes through when solving math problems, namely the entry, attack, and review stages. Aspects that include the entry stage, namely know, want, and introduce. Several indicators at the entry stage, namely students understand the questions carefully and classify and sort information. Aspects that include the attack stage, namely try, maybe, and why. Several indicators at the attack stage are proposing an assumption about solving the problem and trying to solve the problem. While the aspects that include the review stage are check, reflect, and extend. One of the indicators at the review stage is checking the accuracy of the calculation of the problem-solving process. In solving problems, apart from the entry, attack, and review stages, there are other important processes that need to be known. The four important processes are specializing, generalizing, conjecturing, and convincing [33].

In information processing theory, the process of proving mathematically necessarily involves human cognitive architecture. Human cognitive architecture is defined as a way of organizing the structures and functions needed in human cognitive processes. This cognitive structure consists of working memory and long-term memory [34]. Working memory is a place for information processing, while long-term memory is a place where information is stored in the form of schemes (schemata) [35]. The acquisition of a new schema must be consciously processed and requires a thought process [36]. The more complex schema reconstructed from the old schema contains the information in the previous schema. So that the level of complexity of the schema in the human mind continues to evolve.

Exploring in constructing the proof studied is after students have learned using the scaffolding strategy. In this research, the level of scaffolding used was that developed by [37] particularly level 2, namely explaining, reviewing, and restructuring, and level 3, namely developing conceptual thinking, and making connections. The scaffolding level is combined with five scaffolding strategies includes: invite students to contribute to mentoring; involving student participation; models appropriate behavior; clarify and verify student understanding; and explain something that is still ambiguous.

There are several reasons why using scaffolding in evidentiary learning (especially abstract algebra) including research results [38] said that to improve students' understanding of mathematical proof, scaffolding can be applied. Several studies have shown the effectiveness of scaffolds in supporting student learning [39, 40]. According to Lowe et al. [41], the right approach for students to learn more skills and solve problems independently is scaffolding. According to [42], said that the scaffolding process can be successful within reason; 1) the scaffolding process can motivate the problem solving procedure for students, 2) the scaffolding method will increase the level of students 'capacity and social relations, 3) the scaffolding process will increase students' confidence in solving difficult mathematical problems, and 4) the scaffolding education method can show students' mistakes and misconceptions in problem solving procedures.

Referring to the description above, this research will explore the effect of providing a scaffolding strategy on the process of constructing proofs in abstract algebra problems. So the problem is how to explore the effect of providing scaffolding strategies on students' process of constructing proofs of abstract algebra problems? This research aims to explore the effect of providing a scaffolding strategy on students' process of constructing proofs of abstract algebra problems.

2 Methods

This research is classified as mixed methods research. First, carry out learning for group material using a scaffolding strategy. To test the impact of the scaffolding treatment, a pretest and posttest were given with valid and reliable proof problems. Second, to explore the ability to construct proof of Group problems using a qualitative method. According to [43], several characteristics of qualitative research, namely (1) natural setting; (2) researchers as a key instrument; (3) Multiple sources of data. The researcher reviews all data, assigns it meaning, and processes it into categories or themes that cut across all data sources. The results of this research are descriptive data in the form of written words from the subjects and observed behavior.

The research subjects were 64 mathematics education study program students who were taking the Abstract Algebra (Algebraic Structure) course. To explore the process of constructing evidence, 6 subjects were selected to be interviewed (with initials: AN, TN, JK, MRS, SEL, and HK), the first three subjects had good skills in constructing proofs, and the other three had poor abilities in constructing the proof. Besides, the subject selection is based on the ability of students to express their ideas/thoughts both in writing and orally. According to [44] syntactic strategy and semantic strategy in constructing the mathematical proof. The syntactic strategy occurs when the entire proving process occurs in the proof representation system, while the semantic strategy occurs when the proving process involves aspects outside the proof representation system. Alcock & Inglis [44] defines a proof representation system as a system of generally accepted symbolic statements that can be combined into allowable configurations via accepted rules of proposition logic and proof framework. The categories of semantic strategy and syntactic strategy are more sufficient to explain the possible lines of thinking in proof.

The processes observed in this research were student activities when constructing proof of a given Group problem and interviews conducted after the verification process. The supporting instruments in this research are test instruments for the ability to prove Group problems and validated interview guidelines. Exploration indicators in research are viewed from the three stages of problem solving discussed by [33], namely the entry, attack, and review stages (Table 1).

In the entry, attack, and review stages, in problem solving there are four important processes, namely specializing, generalizing, conjecturing, and convincing [33]. For data analysis, the process of constructing evidence uses a model [45] namely data reduction, data presentation, and concluding. For data analysis on the ability to construct proofs of abstract algebra problems before (pretest) and after (posttest) receiving scaffolding treatment, using a paired sample t-test. The calculation process uses SPSS version 26 with the significance level set at $p = 0.05$.

3 Results and Discussion

The following consecutively discussed the student's explorations in constructing evidence from high-ability students and low-ability students after being given learning scaffolding techniques.

3.1 The Process of Students Who Have High Abilities in Constructing Proofs of Abstract Algebra Problems

The ability to construct proof before giving JK scaffolding when it shows a closed and associative nature using examples so that one example is correct, it is assumed to be proven. This finding is in line with [13, 46, 47] research that students tend to use real examples when they prove it. The subject of AN in constructing proof of closed and associative nature of the initial steps of proof is not appropriate, as well as the use of related concepts is not relevant. The arguments and language of evidence used by AN do not contain cause and effect. These findings are in line with Knapp [19] that students have difficulty in logic, using formal mathematics. The same thing was done by TN.

Table 1. Stages of problem solving [33].

Phases	Rubric	Activities
Entry	I know	Read the question carefully Specialize in finding out what's involved What ideas/skills/facts seem relevant? Do I have any analog or similar questions?
	I want	Classification and brief information Be aware of the ambiguity Specialize in finding out what the real questions are
	Introduce	Pictures, diagrams, symbols Representation, notation, organization
Attack	Try	conjecture, Cyclic process
	Maybe	Specialize systematically Analogy
	Why	Justification
Review	Check	Calculation Arguments to ensure that the calculation is correct The impact of the conclusion is to check whether it makes sense That the resolution is in accordance with the question
	Reflect	About key ideas and moments About the implications of conjectures and arguments Regarding your resolution: could you make it clearer?
	Extend	The results take to a wider context by generalizing By looking for new paths to resolution By changing some restrictions

When proving the existence of the unit element and the inverse of each element, both TN, AN, and JK, do not know the initial proving steps that must be taken to obtain the unit element and the inverse of each element.

The process of constructing proof for high ability students (subject: TN) after giving scaffolding when constructing evidence of problem 1 can be observed properly. When asked to explain the steps taken to understand the problem, TN can properly explain the information that is known, which will be proven and needed to construct a proof of the problem. The following is an excerpt from the interview between researchers and TN during the entry stage.

R 1103: *You have read this question, and what is being done to understand this problem?*
TN1103: *Yes sir, those who read are carefully carefully*
R 1104: *What do you know in this matter?*
TN1104: *Eeee … The set of G and its operation of the star (*)*
R 1107: *What has to be proven?*
TN1107: *Mmmhh …. have to prove that G in the star operation (*) is a Group*

R 1108: *Then what needs to be shown to prove a group?*
TN1108: *There are four sirs, mmhhh ....*
R 1109: *Anything?*
TN1109: *Closed, associative, elements of identity and inverse, sir*

The results of the interview showed that TN was able to explain the information provided on problem 1 and was able to explain the steps taken to obtain the information needed to prove the problem. Based on the entry stages, TN fulfills the know, want, and introduce aspects. The results of the interview showed that TN was able to sort the information contained in the problem given. Apart from that, TN also wrote symbols correctly, both those that were known and those that had to be shown in building proof of problem 1.

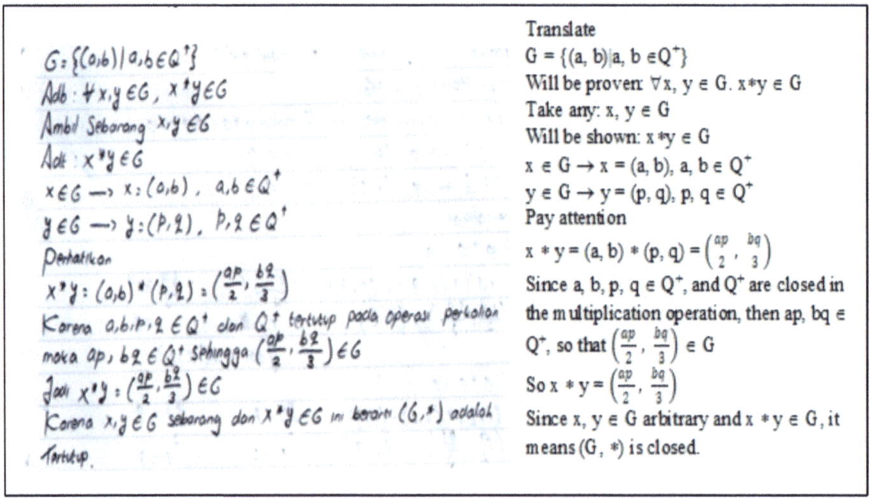
G = {(a, b)|a, b ∈ Q*}
Adb: ∀x, y ∈ G, x*y ∈ G
Ambil Sebarang x, y ∈ G
Adt: x*y ∈ G
x ∈ G → x = (a, b), a, b ∈ Q+
y ∈ G → y = (p, q), p, q ∈ Q+
Perhatikan
x*y = (a, b)*(p, q) = (ap/2, bq/3)
Karena a, b, p, q ∈ Q+ dan Q+ tertutup pada operasi perkalian
maka ap, bq ∈ Q+ sehingga (ap/2, bq/3) ∈ G
Jadi x*y = (ap/2, bq/3) ∈ G
Karena x, y ∈ G sebarang dan x*y ∈ G ini berarti (G, *) adalah tertutup.

Translate
$G = \{(a, b) | a, b \in Q^+\}$
Will be proven: $\forall x, y \in G.\ x*y \in G$
Take any: $x, y \in G$
Will be shown: $x*y \in G$
$x \in G \rightarrow x = (a, b),\ a, b \in Q^+$
$y \in G \rightarrow y = (p, q),\ p, q \in Q^+$
Pay attention
$x * y = (a, b) * (p, q) = \left(\frac{ap}{2}, \frac{bq}{3}\right)$
Since a, b, p, q $\in Q^+$, and Q^+ are closed in the multiplication operation, then ap, bq $\in Q^+$, so that $\left(\frac{ap}{2}, \frac{bq}{3}\right) \in G$
So $x * y = \left(\frac{ap}{2}, \frac{bq}{3}\right)$
Since $x, y \in G$ arbitrary and $x * y \in G$, it means (G, *) is closed.

Fig. 1. TN answers when constructing proof of closed properties after being given scaffolding

Figure 1 shows that the initial steps were taken by TN to construct a proof of problem 1 (about the Group) when proving a closed nature, then associative. The following are excerpts from interviews between researchers and TN when constructing the proof.

R 1111: *After you know what is known and what has to be proven, what do you do to construct the proof?*
TN1111: *Hmmm... (pause) I have to prove the four conditions to prove the Group.*
R 1112: *Anything, Try to explain.*
TN1112: *Hmmm ... (pause) one has to show closed properties, the second shows associative properties, the third must show the presence of neutral elements, and the fourth must show that each element has an inverse. Then ..... (pause) if one is not fulfilled, then it is not proven*
R 1115: *Then how do you do it?*
TN1115: *Yes .... it must be proven one by one*

In the review stage, the thinking activities do not appear explicitly, to explore at this stage is done through interviews. The following is an excerpt from an interview with TN at the review stage.

R 1119: Have you checked the results of your proof?
TN1119: Yes
R 1122: How?
TN1122: Hmmm ... (silence) I checked it step by step, then I also checked the results of the operation, then ... (silent) the sentence, hmmm .... I meant the argument and language
R 1125: Have the conclusions been checked and appropriate?
TN1125: ..... (Pause) Sir
R 1126: Can an idea that has to be proven can be demonstrated?
TN1126: Hmm .... Yes sir,
R 1128: How about the conjecture that must be proven by your argument
TN1128: .... (silent) The argument refers to what has been proven before

At the review stage, TN fulfills the check and reflection aspects. The checking aspect was fulfilled because based on the results of the interview it was discovered that TN re-checked the evidence that had been constructed. TN also checks whether the operation, related concepts, as well as the argumentation and proof language are correct? This can be seen from the interview excerpt below.

R 1132: Are you sure that your proof is correct?
TN1132: Hmmm ... (smile) .. sure sir.
R 1133: What makes you believe that your proof is correct?
TN1133: I've been able to show what has to be proven.
R 1134: Is it true?
TN1134: I hope it is true sir, hmmm ... who are still in doubt about the reasons, hmm ... I mean the argument
R 1137: So you are sure of the proof that has been written?
TN1137: Hmmm ... (silence) sure sir..
R 1139: Was there any difficult part of proving the Group's problems?
TN1139: Earlier sir, hmm ... The sentence was the reason hmm ... I meant the argument, the related concept was sometimes doubtful sir.

The process of constructing AN and JK proofs at the entry, attack, and review stages is similar to the process of constructing TN proofs. There is a slight difference in JK's process when constructing proof of closed and associative characteristics, especially the attack stage on the why aspect in justifying. The evidentiary language used tends to be less relevant to the related concepts, so the truth still looks less logical. Likewise, in proving question 2, exploration results were found that were relatively the same as proving question 1. Thus, students who had high abilities when constructing proofs after receiving scaffolding fulfilled the entry, attack and review stages very well.

Based on the analysis of the proof construction process for question 1 (about groups), it shows that TN can explain the proof process well. Figure 1 shows the try aspect of TN in building proofs of closed properties, with the initial steps of proving and writing down what must be shown well. This process shows that TN is carrying out a specialization stage [33] in constructing proof. In the maybe aspect, TN can show the proposed conjunction in the try aspect. Then carry out the operations from the members that have been described down to their simplest form. Starting from the try aspect to the maybe, TN carries out it systematically so that the proof flow is easy to understand. This process shows that TN carried out a generalization stage [33] in constructing proof. In the aspect of why TN tries to convince (justify) with two things, namely 1) ensuring that the results of the operation meet the specified membership requirements; 2) ensure that the proof is valid for each member of the set. To support his statement, TN uses several concepts related to propositions and evidentiary language that are presented logically. This process shows that TN is carrying out the convincing stage [33] in constructing proof. The following is the construction process to prove the closed and associative nature of high ability students after being given scaffolding (Table 2).

Table 2. The process of constructing proofs in closed and associative nature of high ability students after being given scaffolding.

Code	Description
E1	Be able to name what is known
E2	Be able to say what to prove
A1	Doing the beginning of the proof well starts "Take any x, y ∈ G"
A2	Write down what the symbolic proves
A3	Write down the members of the set according to the membership requirements of set G
A4	Perform operations on two members of a set, to the simplest form (according to the set G)
A5	uses related concepts to argue that the final result is a member of the set G
A6	conclude with arguments and good proof language
R1	reviewing written proof by checking the flow of evidence from the initial steps of proof
R2	Do a review of the proof which is carried out with a reflection step

Note: Ei = Entry.
Ai = Attack.
Ri = Review.

When constructing proof about the existence of neutral elements, TN takes different initial proof steps with closed and associative characteristics. The initial step of proof taken by TN is to suppose that G in the "*" operation has a neutral element, then suppose that $e \in G$. Furthermore, using the definition of a neutral element, TN argues that $a * e = e * a = a$ for every a member of G. To find e (neutral element), TN decomposes an into $a = (p, q)$, $p, q \in Q^+$, likewise $e = (r, s)$. Then to get e (neutral element), TN performs operations using the members of the set that have been described following the membership requirements and the defined definition of the operation. TN performed two operations to get a left neutral element and a right neutral element, and the results were the same, namely (2, 3). At the review stage, TN does the same thing as when constructing proof of closed and associative properties, fulfilling the check and reflect aspects.

In the final proving of the Group Problem, TN constructs the inverse proof of an element. This section of TN takes the initial steps of proof which is a combination of closed properties and neutral elements. The initial step of proving that TN takes is "Take any $x \in G$". To clarify what is going to be constructed, TN wrote what it will show. Then supposing that x has an inverse, then let's say $x^{-1} \in G$. The next step is to use the inverse definition, TN argues that $x * x^{-1} = x^{-1} * x = e$, with $e = (2, 3)$. To get x^{-1}, TN decomposes x into $x = (p, q)$, $p, q \in Q^{+}$, likewise $x^{-1} = (r, s)$. Then to get x^{-1}, TN operates using the set member has been described following the terms of its membership and the defined operating definition. TN performed two operations to get left inverse and right inverse, and the results were the same $x^{-1} = \left(\frac{4}{p}, \frac{9}{q}\right)$. At the review stage, TN did the same thing as when constructing the proof of closed and associative characteristics. The check aspect is fulfilled because based on the results of the interview it is known that TN re-checks the proof that has been constructed, whether the proving flow is following what must be shown/proven or not. TN also checks whether the operation is correct, then the related concepts used are appropriate, as well as the arguments and language of proof whether they are correct. Besides, TN also fulfills the reflect aspect because the subject can reflect on the conjecture in solving and which parts are difficult in the given problem (Fig. 2).

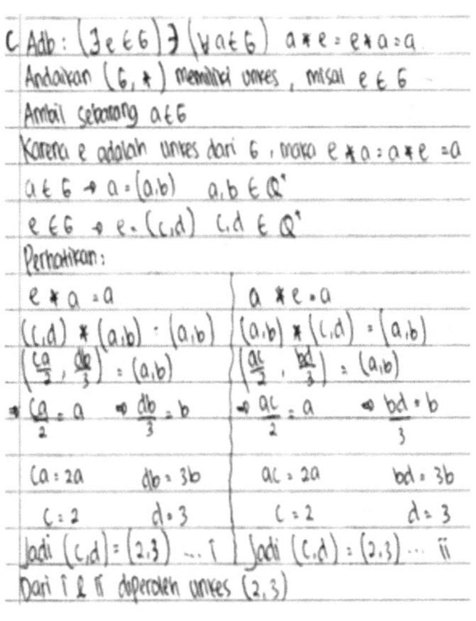

Translate
Will be proven: $(\exists e \in G) \ni (\forall a \in G)\ a * e = e * a = a$
Suppose (G, *) have unkes (element of unity), suppose $e \in G$
Take any $a \in G$
Since e is the unitary element of G, then $a * e = e * a = a$
$a \in G \rightarrow a = (a, b)$, $a, b \in Q^{+}$
$e \in G \rightarrow e = (c, d)$, $c, d \in Q^{+}$
Pay attention

$e * a = a$	$a * e = a$
$(c, d) * (a, b) = (a, b)$	$(a, b) * (c, d) = (a, b)$
$\left(\frac{ca}{2}, \frac{db}{3}\right) = (a, b)$	$\left(\frac{ac}{2}, \frac{bd}{3}\right) = (a, b)$
$\frac{ca}{2} = a$ $\frac{db}{3} = b$	$\frac{ac}{2} = a$ $\frac{bd}{3} =$
$bca = 2a$ $db = 3b$	$ac = 2a$ $bd = 3b$
$c = 2$ $d = 3$	$c = 2$ $d = 3$
So $(c, d) = (2, 3)$ i	So $(c, d) = (2, 3)$..... ii

From i & ii we get the element of unity (2, 3)

Fig. 2. AN's the answer when constructing a neutral element proof after being given scaffolding

In proving the existence of unit elements and the inverse of each element, the AN and JK construction processes, both entry, attack and review stages, are similar to the TN process. There are slight differences in the construction process of AN, JK when constructing the proof of the existence of the unit element and the inverse of each element, especially the attack stage on the why aspect. The evidentiary arguments that were constructed were not written well, so they lacked logic. However, during the interview, AN immediately realized there was something odd and explained the meaning of the argument he wrote. Arguments can be formed from the construction of students' reasoning or thinking based on problem solving procedures, concepts and logic [7, 48]. Likewise, in proving question 2, the process turned out to be relatively the same as proving question 1. The following is the process of students who have high abilities when constructing proof of neutral elements, and inverse an element after being given scaffolding based on stages (Table 3).

Table 3. The process of constructing a proof of the unit element theorem and the inverse of each element of high ability students after being given scaffolding.

Code	Description
E1	Be able to name what is known
E2	Be able to say what to prove
A1	Doing the beginning of the proof well starting "Take any $x \in G$"
A2	Write down what is indicated symbolically
A3	Suppose that x has an inverse, then let $x^{-1} \in G$
A4	use the inverse concept to argue that $x * x^{-1} = x^{-1} * x = e$
A5	Write down the members of sets x and x^{-1} according to the requirements for membership of sets G
A6	Perform operations on two members of the set until they get the value x^{-1}
A7	do the operation twice to get left inverse and right inverse
A8	compile conclusions based on the findings of left inverse and right inverse
R1	Reviewing the evidence by checking the flow of proof from the initial steps of proof
R2	Do a review of the evidence which is carried out with a reflection step

Note: Ei = Entry.
Ai = Attack.
Ri = Review.

Thus, students who have high abilities in constructing proofs after receiving scaffolding fulfill the entry, attack and review stages very well. This is in line with the results of research [38] which states that to increase students' understanding of mathematical proof, scaffolding can be applied.

3.2 The Process of Students Who Have Low Ability in Constructing Proofs of Abstract Algebra Problems

Low ability students (HK subject) after being given scaffolding when constructing proof for problem 1, the stage of constructing proof can perform well. HK can explain well the information that is known, which will be proven and needed to construct evidence of the problem. The following is an excerpt from the interview between the researcher and HK at the entry stage.

R 2102: *You have read this question, and what is being done to understand this problem?*
HK2102: *Mmhh ..... Yes sir.*
R 2104: *What do you know about that problem?*
HK2104: *Mmmhhh ... There is a set G, then there is a star operation ("*")*
R 2107: *What has to be proven?*
HK2107: *Mmmhh ..... group, ... that means G in operation * is Group*
R 2108: *What must be shown to prove G group?*
HK2108: *Mmmhh There are closed, associative, neutral elements, and mmhhh .... the last inverse, sir*

The results of the interview showed that HK was able to explain the information provided in problem 1 and was able to explain the information needed to prove the problem. Based on the entry stages, HK fulfills the know, want, and introduce aspects. Besides, HK is also able to write correctly in symbolic form, both what is known and what must be shown in constructing proof of the question.

In Fig. 3, it can be seen that the initial steps taken by HK to construct a proof of problem 1 (about the Group) by proving the closed nature, then associative, continued to show the existence of unit elements, and the last shows that each element has an inverse. The following is an excerpt from the interview between the researcher and HK when constructing proof of the Group Problem.

R 2110: *What did you do to construct the proof?*
HK2110: *Mmmhh ... (pause) I have to prove the four provisions, sir.*
R2112: *Anything, Try to explain.*
HK2112: *Mmmhh ... (pause) first closed, second associative, third mmhh ... neutral elements, and four inverse.*
R 2115: *Then how do you do it?*
HK2115: *Mmmhh .... must be one one*

Refer on the answers and the results of the interviews in the process of constructing proof of problem 1 (about the Group), it appears that HK can explain the proof process that was carried out well. Look at Fig. 3, it can be seen from the try that HK constructs proof of closed properties with the initial steps of proof it well (see: HK starts. Take any x, y $\in$ G), then write down what should be shown in closed properties. This process shows that HK undertakes the specializing stage [33] in constructing proof by compiling parts of the problem that must be proven, such as what is known and what must be demonstrated. In the maybe aspect, HK tries to be able to show according to the conjecture proposed in the trying aspect but does not describe each member who is taken in the initial steps of proof according to the requirements of association membership. This has an impact when performing the x * y operation difficulty, in the end, HK replaces x = (a, b), and y = (p, q) so that they can perform operations according to the rules. The emergence of x = (a, b), and y = (p, q) is sudden and the membership conditions a, b, p, q are not clear, but HK can perform operations. HK concludes x * y ∈ G, this is not followed by a clear reason, is $\left(\frac{ap}{2}, \frac{bq}{3}\right)$ qualify for membership G? Starting from the try to maybe aspect HK, it seems that it is not systematically carried out so that the flow of proof is difficult to understand. This process shows that HK carries out the generalizing stage [33] in constructing proof which is an activity to look for proof steps and how to test the truth steps of the proof conjunctions that have been made.

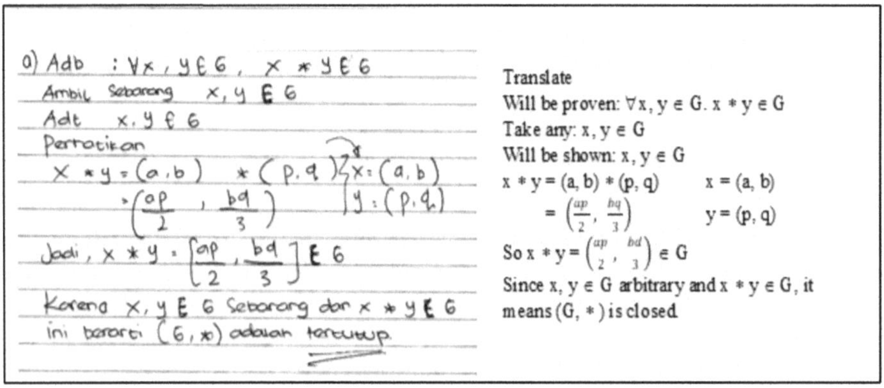

Fig. 3. HK's answer after being given scaffolding proved its closed nature

In the review stage, the thinking activities do not appear explicitly, to explore at this stage is done through interviews. Excerpts from the review stage interview with HK are presented below.

R 2118: *Have you checked the results of your evidence?*
HK2118: *Yes*
R 2120: *How?*
HK2120: *Mmmh.... (pause) I checked step by step, and... the operation.*
R 2122: *What is the argument or language of evidence?*
HK2122: *Mmmh... I'm confused that the argument wasn't made*
R 2123: *Have the conclusions been checked and appropriate?*
HK2123: *..... (Pause) Sir*
R 2125: *Can an idea that has to be proven can be demonstrated?*
HK2125: *Mmmh.... looks like it has been sir,*
R2128: *How about the conjecture that must be proven by your argument*
HK2128: *.... (silent) I'm confused hmm... didn't write down the reason.*

At the review stage, HK fulfills the check and reflect aspects. HK also checks whether the operation is correct, but is confused in making arguments. In addition, HK also fulfills the reflect aspect, but HK has difficulty presenting arguments or using related concepts in proof. This can be seen from the interview excerpt below.

R 2130: *Are you sure that your proof is correct?*
HK2130: *Hmmm .... (shaking his head) ... not sure sir.*
R 2132: *What makes you unsure of the evidence?*
HK2132: *Hmm ... the argument is sir .., still in doubt.*
R 2133: *Does that mean your proof is wrong?*
HK2133: *Hmm ... I think so because there is no argument*
R 2135: *So you are not sure about the proof that has been written?*
HK2135: *Hmmm ... (silence) yes sir, but if the operation is correct sir, but hhmm .. the example is not quite right, sir?*
R 2136: *What was the hard part of proving the Group's problems?*
HK2136: *Earlier sir, the sentence was the reason hmm ... I mean the argument, the related concept is sometimes doubtful sir.*

The process of constructing closed and associative proofs carried out by MRS and SEL after being given scaffolding, at the entry, attack and review stages, has similarities with the process of constructing HK proofs. There is a slight difference in the processes of MRS and SEL when constructing proof of closed and associative properties. The evidentiary arguments used tend not to use related concepts that are relevant, especially in terms of sets, so that the truth still looks less logical. However, during the interview, they realized there was carelessness, so the writing was wrong. Likewise, in proving problem 2, exploration results were found that is relatively the same as proving problem 1. Thus, students who have the low ability when constructing proof after receiving scaffolding fulfills the entry, attack, and review stages quite well. The following is the process of constructing proof of closed and associative properties after being given scaffolding to students with low abilities (Table 4).

Table 4. The process of constructing proof of closed and associative properties of low-ability students after being given scaffolding.

Code	Description
E1	Be able to name what is known
E2	Be able to say what to prove
A1	Doing the beginning of the proof well starts "Take any x, y $\in$ G"
A2	Write down what the symbolic (but false) proved
A3	Perform the operation on two members of the set, with x and y according to the members in G
A4	draw conclusions with good arguments and language proof
R1	reviewing evidence by examining the flow of evidence, but not knowing the truth
R2	reviewing the evidence which is carried out by reflective steps and finds it difficult to use related concepts

Note: Ei = Entry.
Ai = Attack.
Ri = Review.

The case of not describing G's membership in accordance with the requirements of its membership was also carried out by HK when constructing proof of the existence of neutral elements, taking the initial steps of proof that were different from being closed and associative. The initial step of proof by HK is to suppose that G in the "*" operation has a neutral element, then suppose with $e \in G$. Furthermore, HK uses the definition of a neutral element, HK argues that $a * e = e * a = a$ for every a member of G. As $a * e$ cannot be operated, then HK takes a to be $a = (p, q)$, $p, q \in Q^+$, likewise $e = (r, s)$. Then continue the operation until you get $e = (2, 3)$. HK only performs one operation to get the right neutral element, and then it is concluded that the neutral element is the same, namely (2, 3). At the review stage, HK does the same thing as when constructing proof of closed and associative characteristics, fulfilling the aspects of check and reflect. Note the HK proof when constructing the neutral element proof in Fig. 4 below.

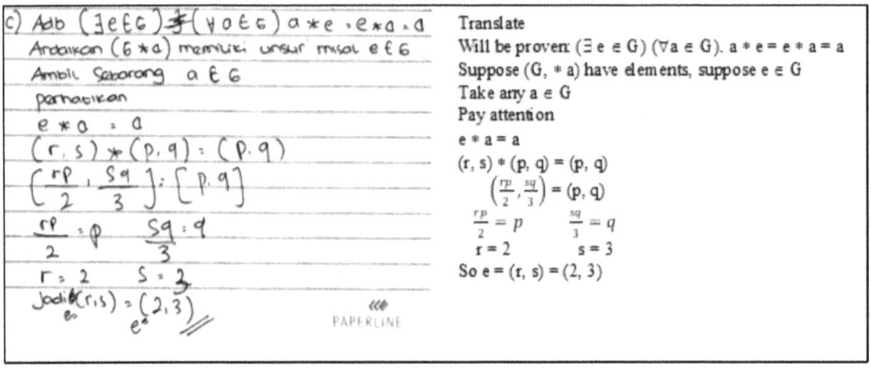

Fig. 4. HK's answer when constructing proof of neutral elements after being given scaffolding

When constructing an inverse proof of an element, HK carries out a construction process that is almost the same as when constructing a proof of a neutral element. In this section HK takes the initial steps of proof by "Take any $x \in G$". To clarify what is going to be constructed, HK writes what it will show. Then suppose that x has an inverse, then let's say $x^{-1} \in G$. The next step, using the inverse definition, HK writes that $x * x^{-1} = x^{-1} * x = e$, with $e = (2, 3)$. Just as in proving a neutral element HK immediately operates $x * x^{-1} = e$ to get x^{-1} because it cannot operate, then HK takes x to be $x = (p, q)$, $p, q \in Q^+$, so too $x^{-1} = (r, s)$, HK continues to carry out operations using members of the set that have been described following the requirements for membership and the defined definition of the operation, and the same is obtained, namely, $x^{-1} = \left(\frac{4}{p}, \frac{9}{q}\right)$. Similar to proving neutral elements, HK only gets the right inverse part, while the left inverse is not shown, but HK concludes that $x^{-1} = \left(\frac{4}{p}, \frac{9}{q}\right)$. In the review stage, HK did the same thing as when constructing evidence of closed and associative characteristics. The check aspect is fulfilled because based on the results of the interview it is known that HK re-checks the evidence that has been constructed. HK admits that it still has doubts about its correctness, including the concept of left and right neutral elements and left and right inverse, so that it only shows one side. Besides, HK also fulfills the reflect aspect because the subject can reflect on conjectures in constructing proof.

The following is the process of constructing proof of the existence of the unit element and the inverse of each element after being given scaffolding to students with low abilities (Table 5).

Table 5. The process of constructing proof of the existence of the unit element and the inverse of each element of low-ability students after being given scaffolding.

Code	Description
E1	Be able to name what is known
E2	Be able to say what to prove
A1	Doing the beginning of the proof well starts "Suppose G has a neutral element, for example, $e \in G$"
A2	Write down what the symbolic proves
A3	using the concept of neutral elements then write down $x * e = e * x = x$
A4	Operate $x * e = x$ to get e (right neutral element), with x and e corresponding to the membership of G
A5	conclude that $e = (2, 3)$ without showing the left neutral element
R1	reviewing the evidence by examining the flow of evidence
R2	reviewing the evidence which is carried out by reflective steps and finds it difficult to use related concepts

Note: Ei = Entry
Ai = Attack.
Ri = Review.

The ability to construct a proof of students who have low abilities before giving scaffolding of MRS and HK subjects when showing closed and associative characteristics has the same way as JK, namely using examples so that with one example it is correct it is assumed to have been proven. These results have similarities with the findings [13, 46, 47] that real examples are often used as proof by students. The subject of HK in constructing proof of closed and associative nature of first stage of proof is not appropriate, as well as the use of related concepts is not relevant. The arguments and evidentiary language used by HK do not contain cause and effect so that the conclusions made are not based on correct premises. These findings are similar to the research results [19] that students have difficulty in logic, using formal mathematics. Besides, HK does not know the specific mathematical concepts needed to prove it. The students didn't understand the definition of certain mathematical objects or concepts that are required in proof. When proving the unit element and the inverse of each element, both HK, MRS, and SEL, do not know the initial steps of proof that must be taken to obtain the unit element and the inverse of each element, students do not know certain mathematical concepts required for proof, do not know how to use them.

In the aspect of why HK tries to convince (justify) by only making sure that the initial step of the proof is to take whatever it is to show that the results of this operation ultimately apply to each member of the set. To justify all of his statements, HK does not appear to use related concepts, nor does it appear to express logical arguments and language of evidence, so that the evidence is less convincing. This process shows that HK is carrying out a convincing stage [33] in constructing proof, which is an activity to explain the reasons for proof not based on mathematical concepts.

In proving the existence of unit and inverse elements, each element of the MRS and SEL construction processes including the entry, attack and review stages also has similarities to the HK process. Just prove the neutral element is left, and the inverse is right. During the interview they said the reason was because they believed the result would be the same. In line with [47] that students only remember and use existing procedures or rules. The similarities are when explaining the requirements for association membership and are less systematic. The difference between MRS and SEL when constructing proof of the existence of the unit element and the inverse of each element, in the why aspect of the justification. The evidentiary arguments are not written well, so the truth is less logical.

However, when interviewing MRS and SEL, it seemed that they were still unsure about the arguments and language of evidence used, including the use of related concepts that were used. MRS and SEL are not sure about the veracity of the evidence that has been built. These findings are similar to the results of research [19] that students have difficulty in logic, using formal mathematics. Proving strategies need to be considered, according to Yeo [49] students have inadequate knowledge and strategies and are unable to use mathematics properly. The same thing was also expressed by [6] that students lack strategic knowledge, namely knowledge of how to determine the facts or theorems that will be used in constructing the proof. Likewise, in proving question 2, a process was found that was relatively the same as proving question 1. Thus, students who have high abilities when constructing proofs after receiving scaffolding fulfill the entry, attack and review stages quite well.

3.3 The Effect of Providing Scaffolding on the Ability to Construct Proofs of Abstract Algebra Problems

The first step before testing the difference in means in the ability to construct proofs of abstract algebra problems, was to test the normality of the data and homogeneity of variance. Normality test results, for pretest data obtained Shapiro-Wilk statistics 0.978 with $p = 0.294$, while posttest data obtained Shapiro-Wilk statistics 0.976 with $p = 0.246$. Based on these results, the pretest and posttest data are normally distributed. For the homogeneity of variance test, Levene's statistic was 0.642 with $p = 0.498$, so it was concluded that homogeneity of variance was met.

Then for the mean difference test, based on Table 6, the t count is -52.728 with probability $p = 0.000$. For a two-sided test the probability number is $0.000/2 = 0.000$. Because $0.000 < 0.025$, it can be concluded that there is a difference in the results of the ability to construct proofs of abstract algebra problems before and after giving scaffolding. This means that providing scaffolding has a positive impact on students' ability to construct proofs of abstract algebra problems. This is in line with [38] who

Table 6. Paired sample test results

		Paired Differences					t	df	Sig. (2-tailed)
		Mean	Std. Dev	Std. Error Mean	95% Confidence Interval of the Difference				
					Lower	Upper			
Pair-1	Pretes-Postest	−45.047	6.835	0.854	−46.754	−43.340	−52.728	63	0.000

said that to improve students' understanding of mathematical proof, scaffolding can be applied. Likewise [41] said that the right approach for students to learn more skills and solve problems independently is scaffolding.

4 Conclusion

Students with high abilities after being given scaffolding can understand the evidentiary problem given by mentioning what is known, and what must be proven in the Group's proof problem. At the time of the interview, able to explain related information as well as the information used to properly construct a proof of Group problems. Based on the aspects of the entry stage, this shows that highly capable students meet the know, want, and introduce aspects. At the attack stage, high ability students can construct a good proof. Based on the aspect of the attack stage, it shows that high-skilled students go through the try, maybe, and why stages very well. At the stage of reviewing the process of constructing proof, high-ability students fulfill the aspects of check and reflect.

Students with low abilities after being given scaffolding can understand the evidentiary problem given by mentioning what is known, and what must be proven in the Group's proof problem. Based on the aspects of the entry stage, this shows that low-ability students meet the know, want, and introduce aspects. At the attack stage, low ability students are less able to construct proof correctly. Based on the attack stage, it shows that low-ability students through aspects of try, maybe, and why, but not maximally. At the stage of reviewing the process of constructing proof, low-ability students fulfill the check and reflect aspects. At the time of the interview, he stated that he checked again about the initial steps of the proof, but the use of related concepts, language, and arguments proof did not know the truth.

References

1. Alcock, L., Weber, K.: Undergraduates' example use in proof construction: purposes and effectiveness. Investig. Math Learn. **3**, 1–22 (2010)
2. Samkoff, A., Lai, Y., Weber, K.: On the different ways that mathematicians use diagrams in proof construction. Res. Math Educ. **14**, 49–67 (2012)
3. Mejia-Ramos, J.P., Weber, K.: Why and how mathematicians read proofs: further evidence from a survey study. Educ. Stud. Math. **85**, 161–173 (2014)

4. Hanna, G.: The ongoing value of proof. In: Theorems Sch. Brill Sense, pp. 1–16 (2007)
5. Gabel, M., Dreyfus, T.: Affecting the flow of a proof by creating presence—a case study in number theory. Educ. Stud. Math. **96**, 187–205 (2017)
6. Weber, K.: Student difficulty in constructing proofs: the need for strategic knowledge. Educ. Stud. Math. **48**, 101–119 (2001)
7. Selden, A., Selden, J.: Validations of proofs considered as texts: can undergraduates tell whether an argument proves a theorem? J. Res. Math. Educ. **34**, 4–36 (2003)
8. Weber, K.: How mathematicians determine if an argument is a valid proof. J. Res. Math. Educ., 431–459 (2008)
9. Cirillo, M.: Challenges to teaching authentic mathematical proof in school mathematics. ICMI Study **19**, 130–135 (2009)
10. Köğce, D., Aydin, M., Yildiz, C.: The views of high school students about proof and their levels of proof (the case of Trabzon). Procedia Soc. Behav. Sci. **2**, 2544–2549 (2010)
11. Knuth, E.J.: Teachers' conceptions of proof in the context of secondary school mathematics. J. Math. Teach. Educ. **5**, 61–88 (2002)
12. Weber, K., Alcock, L.: Semantic and syntactic proof productions. Educ. Stud. Math. **56**, 209–234 (2004)
13. Stylianides, A.J.: Proof and proving in school mathematics. J. Res. Math. Educ., 289–321 (2007)
14. Stylianides, G.J., Stylianides, A.J.: Facilitating the transition from empirical arguments to proof. J. Res. Math. Educ., 314–352 (2009)
15. Warli, C.I., Rahayu, P.: Scaffolding process based on students diagnostic difficulties in proving group problems by using mathematics mapping. J. Phys. Conf. Ser. (2020). https://doi.org/10.1088/1742-6596/1422/1/012012
16. Hazzan, O., Zazkis, R.: Mimicry of proofs with computers: the case of linear Algebra. Int. J. Math. Educ. Sci. Technol. **34**, 385–402 (2003)
17. Mamona-Downs, J., Downs, M.: The identity of problem solving. J. Math. Behav. **24**, 385–401 (2005)
18. Weber, K.: Mathematics majors' perceptions of conviction, validity, and proof. Math. Think. Learn. **12**, 306–336 (2010)
19. Knapp, J.: Learning to prove in order to prove to learn. Acad. **23**, 1–11 (2005). Accessed Feb 2005
20. Selden, A., McKee, K., Selden, J.: Affect, behavioural schemas and the proving process. Int. J. Math. Educ. Sci. Technol. **41**, 199–215 (2010)
21. Sowder, L., Harel, G.: Case studies of mathematics majors' proof understanding, production, and appreciation. Can. J. Math. Sci. Technol. Educ. **3**, 251–267 (2003)
22. Hanna, G., de Villiers, M.: ICMI study 19: proof and proving in mathematics education. ZDM Int. J. Math. Educ. **40**, 329–336 (2008)
23. Harel, G., Sowder, L.: Toward comprehensive perspectives on the learning and teaching of proof. In: Second Handbook of Research on Mathematics Teaching and Learning, vol. 2, pp. 805–842 (2007)
24. Selden, A., Selden, J.: Overcoming students' difficulties in learning to understand and construct proofs. In: Making the Connection: Research and Teaching in Undergraduate Mathematics Education, pp. 95–110 (2008)
25. Weber, K.: Investigating and teaching the processes used to construct proofs. In: Research in Collegiate Mathematics Education VI, pp. 197–232 (2006)
26. Pedemonte, B.: Argumentation and algebraic proof. ZDM Int. J. Math. Educ. **40**, 385–400 (2008)
27. Furinghetti, F., Morselli, F.: Every unsuccessful problem solver is unsuccessful in his or her own way: affective and cognitive factors in proving. Educ. Stud. Math. **70**, 71–90 (2009)

28. Tall, D., Mejia-Ramos, J.P.: The long-term cognitive development of reasoning and proof. In: Hanna, G., Jahnke, H., Pulte, H. (eds.) Explanation and Proof in Mathematics, pp. 137–149. Springer, Boston (2010). https://doi.org/10.1007/978-1-4419-0576-5_10
29. Tall, D.O.: The transition to formal thinking in mathematics. Math. Educ. Res. J. **20**, 5–24 (2008)
30. Stacey, K.: What is mathematical thinking and why is it important (2006)
31. Katagiri, S.: Mathematical Thinking and How to Teach It. Criced, pp. 1–53 (2004)
32. Weber, K.: Traditional instruction in advanced mathematics courses: a case study of one professor's lectures and proofs in an introductory real analysis course. J. Math. Behav. **23**, 115–133 (2004)
33. Mason, J., Burton, L., Stacey, K.: Thinking Mathematically. Pearson Higher Ed. (2011)
34. Sweller, J., Ayres, P., Kalyuga, S.: Measuring cognitive load. In: Sweller, J., Ayres, P., Kalyuga, S. (eds.) Cognitive Load Theory. Explorations in the Learning Sciences, Instructional Systems and Performance Technologies, vol. 1, pp. 71–85 Springer, New York (2011). https://doi.org/10.1007/978-1-4419-8126-4_6
35. Sweller, J.: Cognitive load theory. In: Psychology of Learning and Motivation, pp. 37–76. Elsevier (2011)
36. Sweller, J., Sweller, S.: Natural information processing systems. Evol. Psychol. (2006). https://doi.org/10.1177/147470490600400135
37. Anghileri, J.: Scaffolding practices that enhance mathematics learning. J. Math. Teach. Educ. **9**, 33–52 (2006)
38. Miyazaki, M., Fujita, T., Jones, K.: Flow-chart proofs with open problems as scaffolds for learning about geometrical proofs. ZDM Math. Educ. **47**, 1211–1224 (2015)
39. Cho, K.L., Jonassen, D.H.: The effects of argumentation scaffolds on argumentation and problem solving. Educ. Technol. Res. Dev. **50**, 5–22 (2002)
40. Simons, K.D., Klein, J.D.: The impact of scaffolding and student achievement levels in a problem-based learning environment. Instr. Sci. (2007). https://doi.org/10.1007/s11251-006-9002-5
41. Lowe, J.R., Erickson, S.J., MacLean, P., Schrader, R., Fuller, J.: Association of maternal scaffolding to maternal education and cognition in toddlers born preterm and full term. Acta Paediatr. **102**, 72–77 (2013)
42. Amiripour, P., Amir-Mofidi, S., Shahvarani, A.: Scaffolding as effective method for mathematical learning. Indian J. Sci. Technol. **5**, 3328–3331 (2012)
43. Cresswell, J.W.: Planning, conducting, and evaluating quantitative and qualitative research. Educ. Res. (2012).https://doi.org/10.4135/9781483349435
44. Alcock, L., Inglis, M.: Doctoral students' use of examples in evaluating and proving conjectures. Educ. Stud. Math. **69**, 111–129 (2008)
45. Miles, B.M., Huberman, A.M., Saldana, J.: Qualitative Data Analysis. A Methods Sourcebook (2014). https://doi.org/10.1177/239700221402800402
46. Stylianides, G.J., Stylianides, A.J.: Ability to construct proofs and evaluate one's own constructions. Math. Educ. **166** (2009)
47. Zazkis, R., Liljedahl, P., Chernoff, E.J.: The role of examples in forming and refuting generalizations. ZDM Int. J. Math. Educ. **40**, 131–141 (2008)
48. Imamoglu, Y., Yontar Togrol, A.: Proof construction and evaluation practices of prospective mathematics educators. Eur. J. Sci. Math. Educ. **3**, 130–144 (2021)
49. Yeo, K.K.J.: Secondary 2 students' difficulties in solving non-routine problems. Int. J. Math. Teach. Learn. (2009)

AI-Powered Automated Detection of Cervical Cancer Using Deep Learning Techniques: A CNN and Transfer Learning Approach

Amit Sharma[1] and Abid Hussain[2(✉)]

[1] Department of Computer Science and Application, Vivekananda Global University Jaipur, Jaipur, Rajasthan, India

[2] School of Computer Application and Technology, Career Point University, Kota, Rajasthan, India

abid.hussain@cpur.edu.in

Abstract. This study proposes picture enhancement in conjunction with enhancement through a deep convolution generative adversarial network to enhance categorization of cervical cancer images. With this method (DCGAN), the [cervical cancer] images can be classified in a more precise way. In order to start the right treatment, cervical cancer, a leading cause of cancer-related death for women globally, has to be detected early. Cervical cancer may be difficult to diagnose in photos because of its complexity and wide range of symptoms. We propose to generate synthetic pictures that are comparable to each category in the current dataset using a deep convolutional generative adversarial network (DCGAN). We use three different dataset layers to train the DCGAN for each class in a different session. They are then used once again to train a deep learning model for the categorization of pictures of cervical cancer after the production of fake images. These produced pictures together with the augmented dataset are used to in the pre-training of the classification model. The next conventional practice during the transfer learning technique is to further train the classifier on the training images. The proposed method is applied using a cervical cancer images dataset, which demonstrated that the combination of transfer learning and artificial data generated through DCGAN is more accurate than using only the original dataset.

Keywords: AI ANN · Machine learning · Deep Learning · DCGAN · image segmentation · Cervical Cancer Screening · Classification Models · Transfer Learning

1 Introduction

Owing to the fact that the premalignant phase of cervical cancer takes approximately 10 years, this type of cancer can be almost completely prevented. Cervical cancer is perhaps reversible if the early warning signs are detected and dealt with immediately [1]. Among the preventative measures done to prevent cervical cancer, screening for cervical intraepithelial neoplasia is effective. In the case of both primary and secondary

A. J. Obaid et al. (Eds.): ICONEST 2024, ISEM 40, pp. 322–330, 2025.
https://doi.org/10.1007/978-3-031-89960-7_24

prevention, it is strongly advised to take the HPV vaccine and to get check and treated as early as possible [2]. The ignorance of cancers are defined to correlate with the results from various imaging examinations and clinical assessment for cervical cancer. In term of presentation, the most accurate method of detection is biopsy done under colposcopy because it provides the most sensitive results. This working method has proven to be efficient in the countries with high per capita incomes only when teamed with professional support of pathologists, cytopathologists, colposcopy experts, and when backed by a well-designed infrastructure. In recent years, there has been increasing adoption of several effective screening programmes [3]. On the other hand, it is difficult to put into practice such programmes and to see to it that they will continue to be successful because educated workers are hard to come by, and a sound health care delivery system is still nonexistent [4].

1.1 Objective to Accomplish

This initiative is primarily focused on achieving the following three goals:

To Overcoming the limitations that occur due to limited data by utilizing DCGAN based synthetic image augmentation in cervical cancer image dataset.

The following is a deeper and more detailed explanation of these:

1. **Dataset Expansion Using DCGAN:** Generate synthetic cervical cancer images with a tailored DCGAN architecture to augment the dataset and address data scarcity issues.
2. **Model Evaluation and Comparison**: Develop two CNN models—one trained on the original dataset and another incorporating synthetic images—and compare their classification metrics to assess performance improvements.
3. **Performance Optimization:** Enhance the CNN models by optimizing key performance parameters such as precision, recall, accuracy, and F1-score, ensuring more reliable and robust classification.

1.2 Cervical Cancer Stages

Role of Staging in Cancer Treatment: Classification is critical in both determining how advanced the cancer is, whether it is malignant or benign (Classification Stage 0 to Classification Stage 6).

Stage 0 and Stage 1: At stage 0 cancer cells are found only in the cervix epithelium and has not yet spread deeper into the cervix. Stage 1 as characterized by micro invasive cancer that is diagnosed by the naked eye only in a microscope.

Intermediate Stages (2–4): Stage 2 limited invasions with a depth of up to 3 mm and width of 7 mm are presented at Stage 3 with wider invasion zone (>7 mm) and depth of up to 5 mm. In Stage 4, the sizes of tumors are little bigger but they cannot be seen through a microscope.

Advanced Stages (5–6): Stage 5 tumours are microscopic and smaller than 4 cm, less than 5 mm in depth while stage 6 are gross and larger than 4 cm in size can be seen with naked eye.

2 Review of Literature

The main objective of Robert P. and Celine Kavida's 2019 research classification of microscopic pictures of the cervix is to determine the real impact of cancer and help in the patient's suitable medication. The most accurate medical test is the Pap smear test; nevertheless, issues occur when it is examined under a microscope. To fix this issue, automatic cancer detection was developed. A few image processing techniques, including segmentation and an improved SVM (Support Vector Machine) classification algorithm, are used in this detection procedure. The results of this suggested strategy are contrasted with those of other classification methods like ANN (Artificial Neural Network), KNN, and others (KNearest Neighbor). The experimental findings show that the suggested method produces decent results & performance evaluation.

2.1 Problem Formulation and Motivation

Neha Sharma and Shefali Gupta et al. The GI tract is one of the most crucial components of the human body. The WHO has shown that GI tract infections are responsible for 1.8 million fatalities annually. Radiation therapy may improve survival chances in people with GI cancer. By segmenting the gastrointestinal tract, radio oncologists may treat cancer patients more successfully and quickly. This study provides a way for doing so [1].

Mayur Virarkar and Sai Swarupa Vulasala et al. Gynecological cancers must be identified early to treat the patient and extend survival. In this field, the use of molecular imaging techniques like MRI and PET (positron emission tomography) has grown. The PET/MRI technique combines metabolic information from PET scans with structural and functional data from MRI scans [2].

Leila Allahqoli and Antonio Simone Laganà et al. With prompt identification of aberrant cervical cells, the probability of receiving cervical cancer treatment rises. Manual identification is labor-intensive, prone to mistakes, and needs pathologists with specialized training. It has been shown that the use of AI techniques is essential in cervical lesion screening programs for both malignant and precancerous lesions. The accuracy of the algorithms' cervical cancer prediction varied from 70% to 100% [3].

S.R. Sylaja Vallee Narayan and R. Jemila Rose. The cervix's tissue is where cervical cancer develops. Most cervical cancer screening procedures include invasive procedures. It is necessary to segment and analyze the collected picture using a decision tree technique. Data may be extracted from images effectively using image processing. It is used to measure the uterus's size and the size of cervical cancer [4].

Wan Azani Mustafa and NurAin Alias et al. The fourth most common reason for mortality in women first presents with no symptoms. Cells for cervical cancer steadily multiply at the cervix. For cancer to be adequately treated, early diagnosis of cervical tumors is crucial. Healthcare experts are today faced with the urgent challenge of diagnosing such cancer before it spreads rapidly [4].

Ejay Nsugbe. In women, cervical cancer is a common condition that has a high fatality rate. Even in its early stages, this cancer is still difficult to accurately detect. This study's goal was to provide a cybernetic system that illustrated how people and computers may collaborate to develop a super intelligence framework, which would then make it possible to use superior clinical care procedures [5].

Kritsasith Warin and Wasit Limprasert. Evaluating the accuracy with which oral photographic pictures of oral squamous cell carcinoma (OSCC) and other potentially malignant diseases (OPMDs) may be used to detect and classify them using deep convolutional neural network (CNN) methods. 980 oral photographic pictures were broken down into 365 photos of OSCC, 315 photos of OPMD, and 300 photos that weren't pathological. The best multiclass picture classification was achieved by DenseNet-196, with AUC values on OSCC and OPMD of 1.00 and 0.98, respectively [6].

Shan Fang and Jiahui Yang et al. The categorization of cervical precancerous lesions using ShuffleNet is suggested. The proposed CNN models have an accuracy of between 81.23 and 81.38%. Our classifier had an AUC of 0.99, demonstrating its superior performance. The colposcopy image classification network has excellent performance and model size, according to the experimental findings [7].

Derya Yeliz Coşar Soğukkuyu and Oğuz Ata. Any sort of cancer pre-diagnosis has a profound effect on a human being's existence. One of the deadliest types of cancer that mostly affects women is cervical cancer. Artificial intelligence-based systems are now utilized for decision-making, but machine learning approaches enable faster and more effective automated diagnosis of cervical cancer [8].

3 Methodology

How to Implement the DCGAN Algorithm?
The following are some examples of how image data augmentation using DCGAN may be implemented:

A) Acquire a group of photographs that you want to utilize for the augmentation process. This dataset is going to be used in the training of both the generator and discriminator networks.
B) Perform preliminary processing on the photos by resizing, standardizing, and converting them to a format that is compatible with the neural network.
C) Using deep convolutional neural networks, define both the generator networks and the discriminator networks. Both the generator and the discriminator should be made up of regular convolutional layers, however, the generator should be made up of transposed convolutional layers.
D) Begin by setting the weights for the generator network and the discriminator network.
E) Start the repeating cycle of training:
   a) Obtain a set of images from the dataset; b) create a random noise vector and feed it to the generator network so it can produce a synthetic image; c) feed the real images from the dataset and the synthetic images produced by the generator into the discriminator network; d) determine the loss for both the generator and the

discriminator networks. e) Use back propagation to make any required modifications to the weights of the generator and discriminator networks. f) Carry out steps a through e once again for the specified number of epochs.

F) When the training is finished, you should utilize the generator network to create fresh synthetic pictures by feeding it random noise vectors. These computer-generated pictures may serve as supplementary data for the job that you are working on.
G) Perform the procedure once again, this time with a new dataset or even the same dataset but with a different random seed and generator network design to produce a variety of unique enhanced photos.
H) If you want to increase the quality and variety of the produced photos, you may choose to fine-tune the generator network using a smaller dataset of images taken from the particular topic you are working on. This is an optional step.

Note that the technique described above is just a high-level overview, that there are many other ways to put a DCGAN into practice to enhance picture data. Depending on the size of your dataset, the level of complexity of your generator and discriminator networks, and the particular needs of your job, it's possible that some adjustments to the technique described above will be required.

3.1 Methods for Proposed Algorithms

Here the enhancement of picture data through DCGAN and the work are to be divided into three parts for giving a better results.

3.2 Implementation Diagram

Implementation Diagram of Generator + Discriminator. The design of the discriminator and generator in a DCGAN model. Conv stands for convolution, while FSC stands for fractionally-strided convolution (Figs. 1 and 2).

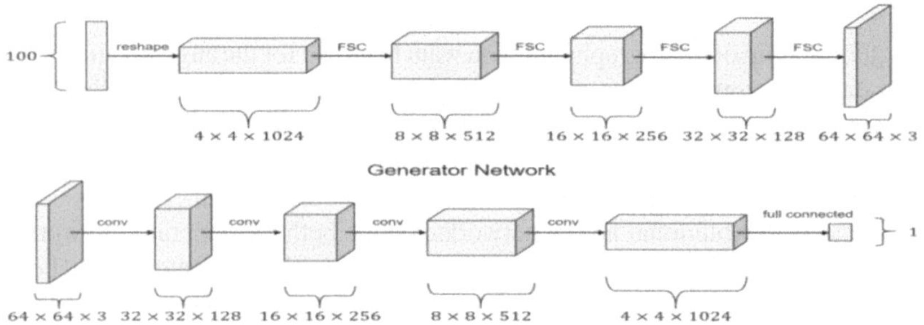

Fig. 1. Discriminator Network Ref: https://www.researchgate.net/figure/The-architecture-of-the-generator-and-the-discriminator-in-a-DCGAN-model-FSC-is-the_fig4_343597759

The generator creates the target by picture using these deconvolutional layers. The discriminator is then given this picture in order to classify it. The two networks would be randomly initialized in a normal training process. The dataset would then be iterated

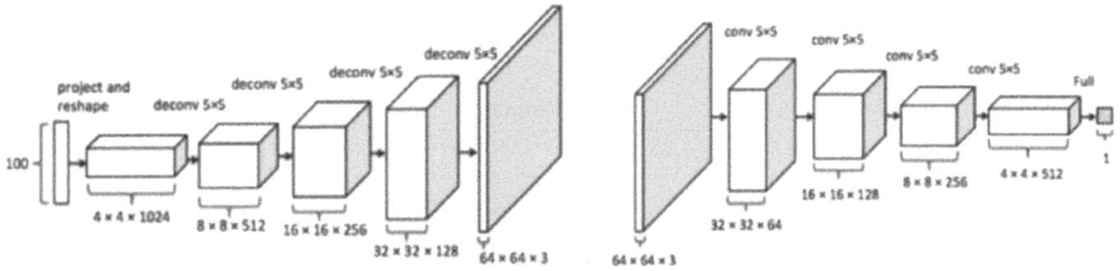

Fig. 2. The DCGAN pipeline Ref: https://tjmachinelearning.com/lectures/1718/gan/

through. In order to have an equal amount of false and genuine photographs, we would produce the batch size number of fake images from the generator at each iteration. The discriminator might then be trained using a binary cross-entropy loss on these pictures (Tables 1 and 2).

Table 1. An overview of the relevant literature on cervical cancer screening.

S.no	Methods	Dataset	Advantages	Disadvantages
1	Voting strategy and gene-assistance module [7]	Venezuela's Universitario De Caracas and a Chinese hospital	(i) More useful and scalable	(i) Limited datasets
2	KNN algorithm-based PSO	Images of cervical smears	(i) Improved accuracy (ii) Effective feature selection	(i) Time-consuming since feature selection is done in two phases
3	machine learning techniques and the Fourier transform [9]	Microscopic pictures	(i) A completely automated system (ii) Preserving the microscopist's valuable time	(i) The degree of intricacy is higher
4	DenseNet pretrained using transfer learning [2]	Kaggle, a maternal and pediatric health center in Fujian	(i) More practical and efficient	(i) Limited information

Table 2. The DCGAN architecture with output size 224 × 224.

Layers Name	Type Learnable	Properties
Generator		
Fully connected	Completely Interconnected Layer	Bias 381408 × 1; Weight 381408 × 100
Trans Conv 1	Layer of Transposed Convolution	3 × 3 × 128 × 128 in weight, 128 × 1 in bias, and 2 strides
TransConv 2		3 × 3 × 64 × 128 in weight, 64 × 1 in bias, and 2 strides
Conv 1	Layer of Convolution	3 × 3 × 64 Weight 1 × 1 Bias Step 1

(continued)

Table 2. (*continued*)

Layers Name	Type Learnable	Properties
Discriminator		
Conv 1	Layer of Convolution	Bias 32 × 1 Weight 3 × 3 × 1 × 32 Step 2
Conv 2		3 x 3 x 32 x 64 Weight 64 x 1 Bias Step 2
Conv 3		3 × 3 × 32 × 128 in weight, 128 × 1 in bias, and 2 strides
Conv 4		3 × 3 × 32 × 256 in weight, 128 × 1 in bias, and 2 strides
Fully connected	Fully Connected Layer	1 × 4096 in weight and 1 × 1 in bias

4 Result Analysis

4.1 Output

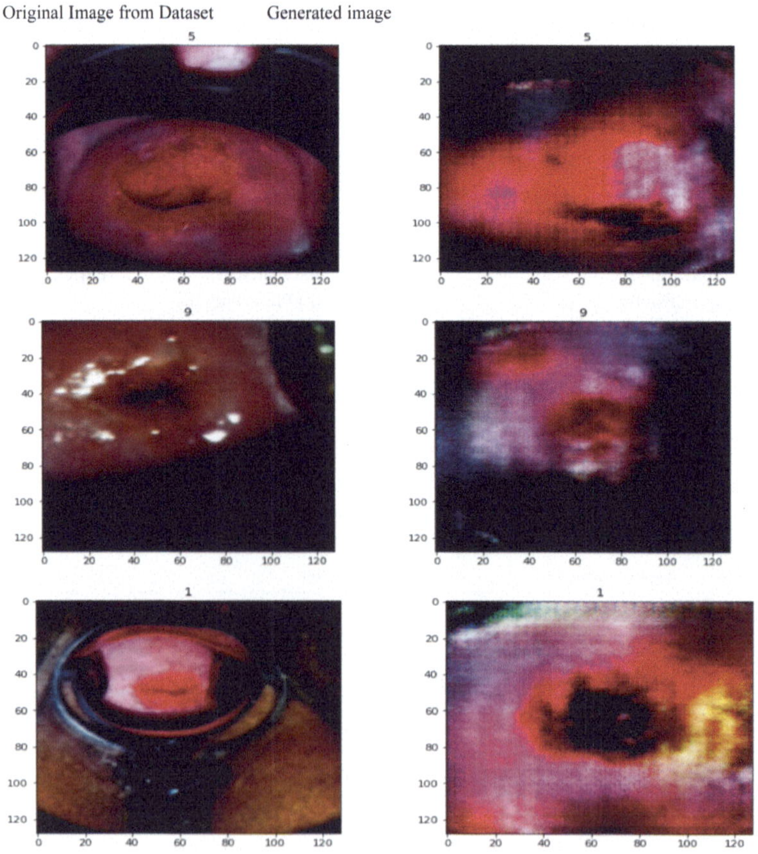

Table 3. Augmented vs. Original images

TYPES OF IMAGES	ADDITIONAL IMAGES	Quantity of Original Images	DCGAN Images
TYPE 1	778	350	3650
TYPE 2	3662	899	0
TYPE 3	899	545	1220

Benefits of Data Augmentation: By leveraging existing data to modify copies of a dataset, data augmentation allows us to artificially extend the training set (Table 3).

We supplement data in order to decrease a data model's overfitting and improve the model's accuracy. Due to the imbalance in our dataset and the duplicated nature of the photographs. We perform data augmentation as a result.

5 Conclusion

In this study, we examined a wide range of cervical cancer detection methods. The accuracy of conventional testing was increased for cervical cancer screening by using advanced mathematical methods and image processing. Among the techniques used in the discovery process to provide trustworthy results are preprocessing, splitting, highlight extraction, and arranging Middle sifting/Gaussian channel systems for preprocessing, adaptive division technique for division, shading as a component evacuation post-handling method, and a decision tree as a classifier yielded the best assessment results. It may be possible to determine the extent of the illness by combining these techniques and applying them to a typical clinical picture. To get precise forecasts on cervical cancer, a variety of data mining techniques may be used. Three sets of cervical cancer data were evaluated by the author, who then provided a prognosis.

References

1. Sharma, N., Gupta, S., et al.: U-Net model with transfer learning model as a backbone for segmentation of gastrointestinal tract. Bioengineering **10**, 119 (2023). https://doi.org/10.3390/bioengineering10010119
2. Virarkar, M., Vulasalaet, S.S., et al.: Current update on PET/MRI in gynecological malignancies—a review of the literature. Curr. Oncol. **30**, 1077–1105 (2023). https://doi.org/10.3390/curroncol30010083
3. Allahqoli, L., Laganà, A.S., et al.: Diagnosis of cervical cancer and pre-cancerous lesions by artificial intelligence: a systematic review. Diagnostics **12**, 2771 (2022). https://doi.org/10.3390/diagnostics12112771
4. Sylaja Vallee Narayan, S.R., Jemila Rose, R.: Cervical cancer detection based on novel decision tree approach. Comput. Syst. Sci. Eng. https://doi.org/10.32604/csse.2023.022564
5. Mustafa, W.A., Alias, N.A., et al.: A recent systematic review of cervical cancer diagnosis: detection and classification. J. Adv. Res. Appl. Sci. Eng. Technol. **28**(1), 81–96 (2022). ISSN: 2462-1943

6. Nsugbe, E.: Towards the use of cybernetics for an enhanced cervical cancer care strategy. ISSN: 2667-1026. https://doi.org/10.1016/j.imed.2022.02.001. Accepted 24 Feb 2022
7. Warin, K., Limprasert, W.: AI-based analysis of oral lesions using novel deep convolutional neural networks for early detection of oral cancer. PLoS ONE **17**(8), e0273508. https://doi.org/10.1371/journal.pone.0273508
8. Fang, S., Yang, J., et al.: An improved image classification method for cervical precancerous lesions based on ShuffleNet. Comput. Intell. Neurosci. **2022**, Article no. 9675628, 8 p. https://doi.org/10.1155/2022/9675628
9. Dineshkumar, P., SenthilKumar, R., Sujatha, K., Ponmagal, R., Rajavarman, V.: Big data analytics of IoT based health care monitoring system. In: Proceedings of the 2016 IEEE Uttar Pradesh Section International Conference on Electrical, Computer and Electronics Engineering (UPCON), Varanasi, India, December 2016, pp. 55–60 (2016)

A Comparative Analysis of SVM and BPNN Algorithm for Stock Price Prediction in Banking Sector

Yogita Punjabi(✉) and Mainaz Faridi

Banasthali Vidhyapith, Banasthali 304022, Rajasthan, India
yogita234@gmail.com, fmainaz@banasthali.in

Abstract. Due to the variable and non-linear structure of financial stock markets, correctly forecasting stock prices is a stressful job. For the investment in stock market, which is an extremely composite & difficult process, prediction plays a very essential role. In fact, the latest extension in financial stock market forecast technology is machine learning, which generates the forecast through training on the price of present financial stock market indicators and their history. Machine learning uses different models in order to come up with correct predictions. In this extensive research work, all the models employed for evaluating the future movement in the banking industries are accurately computed. Two of these models comprise Back Propagation Neural Network (BPNN) & Support Vector Machine (SVM). When comparing with SVM and BPNN, the later usually gains higher accuracy than that of the former in each of the evaluated criteria. In terms of Mean Absolute Error (MAE), for example, BPNN model demonstrates much lower values across all the banks that were investigated, when compared to SVM. MSE (Mean Squared Error) values for the BPNN are significantly less for several data sets and the results prove higher calibration of the forecasts. Moreover, on the analysis of the R-squared (R2) values, there is evidence for a more accurate estimate of fluctuations in the stock market movement by the BPNN. Evidently, such a view provides the stakeholders with a convincing tool that aids them more safely and economically navigate the world of issues related to the banking sector investments.

Keywords: SVM · BPNN · MSE · MAE · R^2

1 Introduction

Despite the fact that, stock market is very important determinant of business performance, the people and governments have been holding some sort of fear towards it for a long time. The extrinsic factor could make it difficult to determine how certain market conditions are likely to transform the stock price [1]. Since the discovery of information technology several decades ago, there have been necessary changes in corporate processes [2]. The notion of financial markets stands out as especially noteworthy because

A. J. Obaid et al. (Eds.): ICONEST 2024, ISEM 40, pp. 331–342, 2025.
https://doi.org/10.1007/978-3-031-89960-7_25

of the fundamental way in which they transform the landscape of corporate operations operating inside countries. The aggregate value of worldwide stocks was estimated to have surpassed 68.654 quadrillion US dollars in 2018, as published by the World Bank [3]. This provides more evidence of the huge importance of global equities. As a result of technological improvements, particularly in the present time, there has been key rise in the number of people contributing to the stock market. Extensively, investors have been looking for ways to maximize profits while also limiting risks.

Nevertheless, because the stock market is characterized by its quick volatility, inherent unpredictability, and occasional mistakes, forecasting the movements in the stock market continues to be a difficult attempt [4]. During the rapid expansion of information technology, deep learning technology has appeared as a strong instrument that can be used to navigate the large amounts of data and complicated models that are endemic to the banking business.

The role of the stock market in the banking industry cannot be emphasized. It acts as an economic barometer, impacting lending policies, investment strategies, and risk management processes at financial institutions [5]. However, the fundamental constraints of predicting stock market movements continue, prompting both scholars and practitioners to explore novel ways to improve prediction accuracy and inform strategic decision-making.

Indeed, according to researchers, Deep learning models were used to carry out a study on historical market data. Thus, the groundwork for increasing the precision of foreseeing systemic risks and volatility on the markets has been prepared [1]. This type of application of deep learning appears to yield a relatively high degree of effectiveness in identifying potential risk events in the financial markets, at least for risk management purposes that require greater precision and adaptability [6]. Another industry area within the finance sector where high value has been noted of deep learning is in portfolio optimization. This further goes to show the potential of deep learning in the future in transforming the markets, as we have already seen to some extent.

Market factors can be easily predicted using deep learning model, enabling investors to better manage their investment portfolios with a view of affecting the current market prices [7]. Also, there has been a lot of discussion on the applicability of the deep learning technology to the marketplace forecasting technique. Thus, this work proposed an SVM and BPNN model for stock market forecasting as follows.

1.1 Back-Propagation Neural Network (BPNN)

Recent trends in stock market analysis have identified backpropagation neural networks (BPNNs) as the most appropriate models for the analysis of patterns in the finance business. The computational structure of these networks is extensible and can be applied to a broad class of nonlinear difficulties [8]. In general, BPNNs are superior to traditional models of sample functions and, due to their universal approximator capabilities, provide a more accurate approximation of a number of functions. Specifically, BPNNs can analyse data at the same time and the model does not need to make any assumptions when building a model, and information extraction can be fast. In most situations, a BPNN involves three separate layers: input level, the hidden levels & the output level [9]. Deduced from its structure of multiple layers that ease the analysis of information then, applications for stock market forecasting in the banking sector can produce more precise.

Each layer of a Backpropagation Neural Network (BPNN) has many processing units that are linked together via acyclic connections. The input layer accommodates the first inputs of network. The hidden layer is then filled with several neurons or hidden units that are stacked equivalent. Every neuron makes a weighted sum of its inputs before going via a nonlinear activation function (Fig. 1).

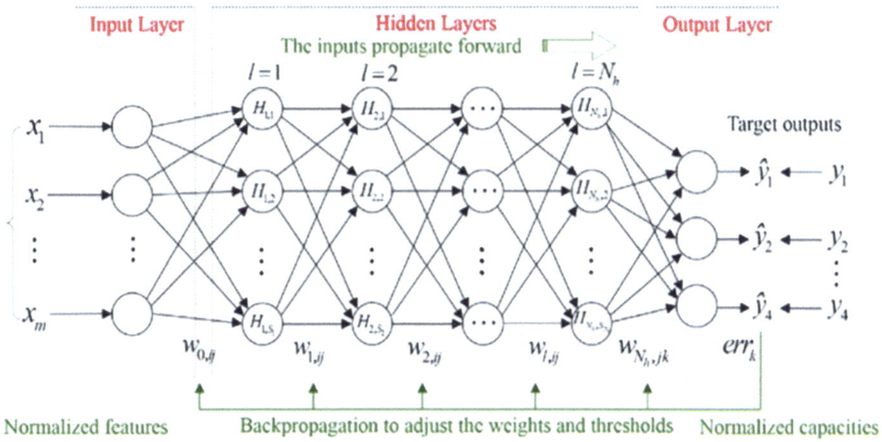

Fig. 1. Back-Propagation Neural Network (BPNN) Architecture

Algorithm 1 BPNN Algorithm

1: Import necessary libraries, such as NumPy and Tensorflow.
2: Assuming X_{train}, y_{train}, X_{test}, and y_{test} are arrays representing the training and testing data. Reshape the data for univariate time series if necessary.
3: Define the neural network architecture using Tensor Flow's Kera's API:
4: Create a sequential model.
5: Combine a dense layer with 64 units and ReLU activation function for the input layer. Specify the input shape as (1,) since the input data is univariate.
6: Insert a dense layer with 32 units and ReLU activation function for the hidden layer.
7: Include the output layer with one unit (for regression), without specifying any activation function.
8: Compile the model:
9: Use Adam optimizer.
10: Apply Mean Squared Error (MSE) as the loss function since it's a regression problem.
11: Train the model:
12: Fit the model to the training data (X_{train}, y_{train}).
13: Specify the number of epochs (e.g., 50), batch size (e.g., 32), and validation data (X_{test}, y_{test}).
14: Evaluate the model:
15: Evaluate the trained model on the test data to compute the mean squared error (MSE).
16: Print the MSE as the evaluation metric.

1.2 Support Vector Machine (SVM)

The support vector machine (SVM) is a predictive analytic data classification tool which assigns new data items to one of the previously recognized categories [10]. In most circumstances, Since SVM is a binary classifier, it thinks data might have two possible target values.

The SVM algorithm has a variation called "Multiclass SVM" that enables it to be applied as a classifier on a dataset with several classes (grouping or category). Image identification, medical diagnosis, and text analytics are just a few of the areas where SVM has proved effective.

Assume you've transformed each image into a data matrix. The support vector machine then receives two primary inputs:

- Historical (training) data: This set of matrices correlates to photos of apples and pears that have previously been seen.
- The new (unseen) data is an image that has been turned into a matrix. The goal is to automatically determine if the image depicts an apple or a pear.

To forecast the unknown label, the support vector employs a mathematical function recognized as a kernel function, which associates fresh data to the best image from the training data (apple or pear). SVM models often employ polynomial, sigmoidal, and radial basis functions (RBFs) (Fig. 2).

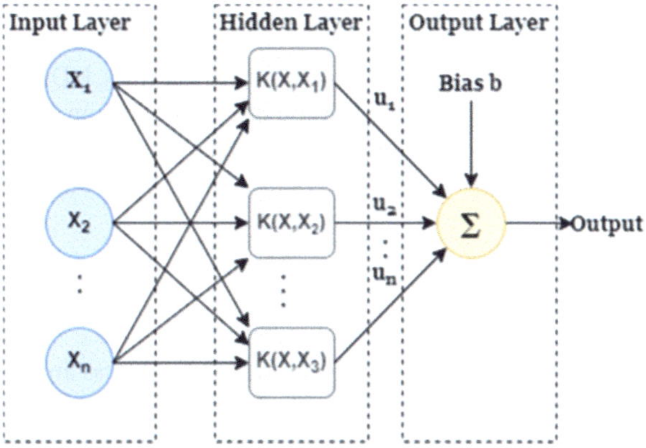

Fig. 2. Support Vector Machines (SVM) Architecture

Algorithm 2 SVM Algorithm

1: Import necessary libraries (e.g., from sklearn, svm import SVC, import NumPy as np).

2: Define or load the training data (X_{train}, y_{train}) & the test data (X_{test}).

3: Instantiate the SVM classifier with an RBF kernel.

4: svm ← SVC(kernel =' rbf')

5: Train the SVM classifier on the training data.

6: svm.fit (reshape(X_{train}, y_{train})

7: Make predictions on the test data using the trained SVM classifier.

8: ypred ← svm.predict (reshape(X_{test}))

9: Estimate the performance of the SVM classifier (e.g. precision, accuracy & recall etc.).

10: Optionally, tune hyper parameters using techniques like grid search or cross validation.

2 Methodology

This section illustrates the technique developed for stock market forecast in the banking industry, which uses a Support Vector Machine (SVM) & Backpropagation Neural Network (BPNN) approach. Pre-processing, feature selection, model architecture design, data segmentation, model training, evaluation metrics, results interpretation, and fair deployment issues are all included in the technique. By implementing this structured process, we suppose to create a robust predictive model capable of expecting stock values in the dynamic and unstable banking industry.

2.1 Data Collection

The data of major 6 financial organizations including PNB, Axis, HDFC, ICICI, KOTAK and SBI is obtained from NSE (National Stock Exchange). From these banking organizations, this vast data set consists of historical daily stock market indexes such as stock prices opening and closing along with trading volume and other critical economic parameters.

2.2 Data Pre-processing

Strict validation and cleaning procedures were applied to address any missing values or inconsistencies to preserve data integrity. The historical stock market data for all banks was pre-processed in a uniform manner. Missing data were addressed using imputation methods, outliers were managed, and features were standardized or normalized to support uniform scales. The categorical variables were encoded, and feature engineering approaches were used. These stages refined the datasets for model building and validation, resulting in reliable and precise forecasts across all banks.

2.3 Train-Test Split

The data is separated in two subsets—training and testing, after pre-processing. The goal of this split is to prepare the model on single slice of the data while calculating its performance on another, previously unknown part. The training set, that usually has 80% of the complete data, is applied to train the model by exposing it to known input-output pairings. The testing set, which holds another 20% of the data, is put aside for assessment reasons, enabling the model's generalization capacity to be examined on hidden data.

3 Model Development

3.1 Model Assessment

The working of predicted model is measured using the metrics - R-Square Score, Mean Absolute Error (MAE) and Mean Square Error (MSE), The following formulae are employed:

R-Square Score (R2): It calculates an approximation of the part that the independent variables in the model describe by looking at the dependent variable. A regression model's overall efficacy and explanatory power can be better recognized with the use of R2, a useful statistic.

$$R^2 = 1 - \frac{SSR}{SST}$$

Where,

SSR is computed as the summation of the squared differences between the real & anticipated values.

The sum of all squares.

SST computes the complete variance in the dependent variable.

Mean Absolute Error (MAE): The name "Mean Absolute Error" (MAE) indicates the average of mistakes between paired observations that reflect the same trends. Comparisons between projected and actual data, time after an event and the time immediately before it, and one measuring method and another are a few examples of Y vs X. Total absolute errors divided by sample size is how MAE is computed.

$$MAE = \frac{1}{l}\sum_{i=1}^{l}|a_i - y_i|$$

Mean Squared Error (MSE): It computes the square root of the average variation between a dataset's real and forecasted values. MSE is often used in regression analysis to measure the effectiveness of prediction models.

$$MSE = \frac{1}{n}\sum\nolimits_{i=1}^{n}(x_i - y_i)^2$$

4 Results and Discussion

4.1 Results

See Table 1 and Figs. 3, 4, 5, 6, 7, 8, 9, 10, 11, 12, 13, 14, 15.

Table 1. Evaluation Metrics

Banks	Model	MAE	MSE	R2
PNB	SVM	122.47	36880.8	0.73862
	BPNN	93.5367	20711.1	0.85322
Axis	SVM	178.293	57070.4	0.43704
	BPNN	151.223	40418.5	0.6013
HDFC	SVM	348.171	225987	0.14254

(continued)

Table 1. (*continued*)

Banks	Model	MAE	MSE	R2
	BPNN	320.831	161687	0.38652
ICICI	SVM	167.769	59309.7	0.46397
	BPNN	178.03	54264.6	0.50957
KOTAK	SVM	209.396	65414.4	0.75107
	BPNN	159.798	39658.7	0.84908
SBI	SVM	417.819	370130	0.52453
	BPNN	293.775	176955	0.77269

4.2 Discussion

When it comes to stock market forecast for several institutions, a comparison of performance metrics of different models within a sphere is very interesting. The examination of the stock market prediction models of six integrated banks—PNB, Axis, HDFC, ICICI, KOTAK and SBI—is given based on the findings.

Therefore, these models have employed Back Propagation Neural Network (BPNN) & Support Vector Machine (SVM).The forecast accuracy analyses clearly depict that the BPNN model remarkably outperforms the SVM in all the banks most of the time.

The result also shows that in majority of cases, this BPNN model predicts with much fewer errors than with standard SVM and higher R-squared (R2). This proves the flexibility and efficiency of this model to capture all the complicated characteristic of the stock markets, which exist in the banking sector. For instance, if we look at the SBI set the proposed BPNN model, has higher R-squared (R2) 0.77269 and comparatively lesser MAE which is 293.775 rather than the MAE of 417.819 of the proposed SVM model. This is evident in other banks also. As observed for Axis, HDFC, ICICI, KOTAK and PNB also the proposed BPNN outperforms SVM in all three aspects of MAE, MSE, and R2 respectively.

Furthermore, the PNB's performance reveals the highest result where the model BPNN got the maximum R2 Score at 0.8533 which provided a strong association between estimated & factual values. On the other hand, HDFC is shown to have larger prediction errors for all models indicating possible challenges in capturing the firm's market appropriately. In general, multilayer backpropagation neural network confirms its efficiency when forecasting the motions of the banking segment in the stock market and can contribute more valuable information for investors and analyst. But there might be a need for some further modifications of the model to capture more of High Frequency dynamics of banks and to increase the accuracy of its forecasts for banks, like HDFC whose position within the observed period seems to be much more variable.

Fig. 3. Evaluation Metrics MAE, MSE, R^2

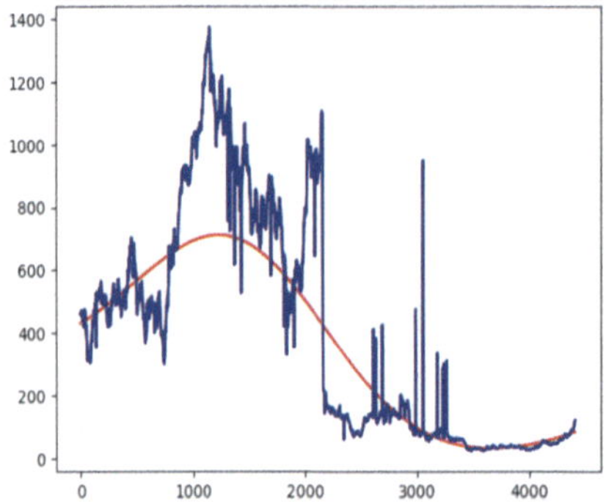

Fig. 4. The results received for Punjab Bank using SVM

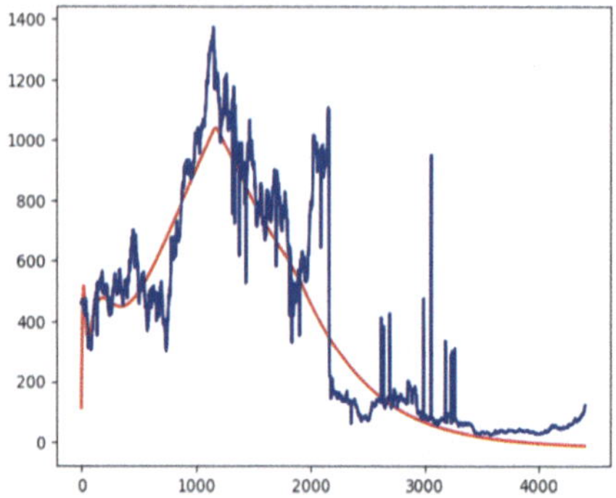

Fig. 5. The results received for Punjab Bank using BPNN

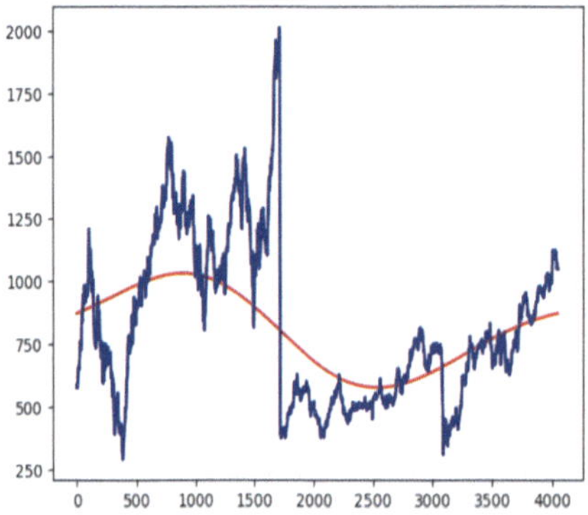

Fig. 6. The results received for Axis bank using SVM

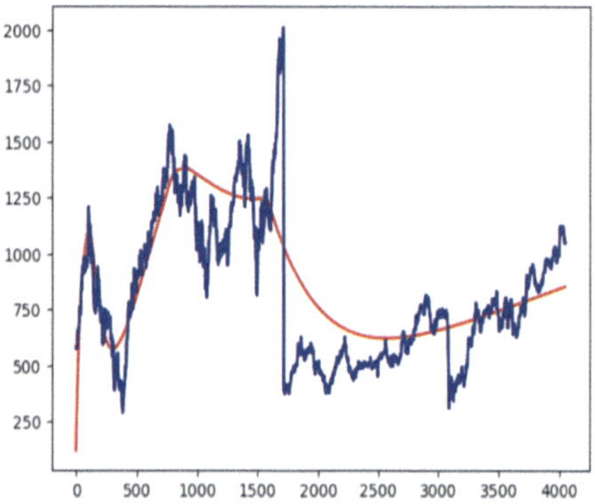

Fig. 7. The results received for Axis bank using BPNN

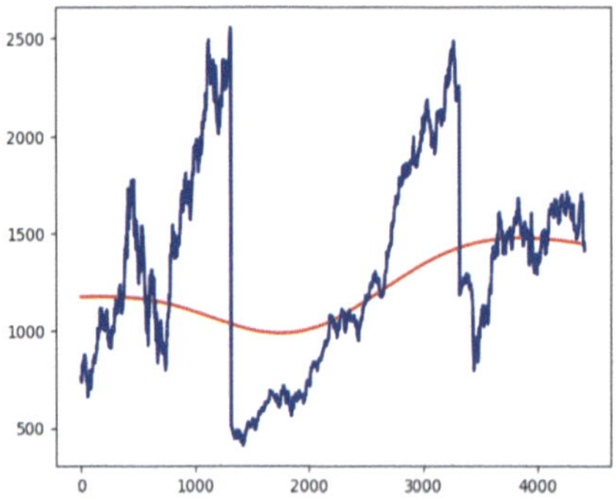

Fig. 8. The results received for HDFC using SVM

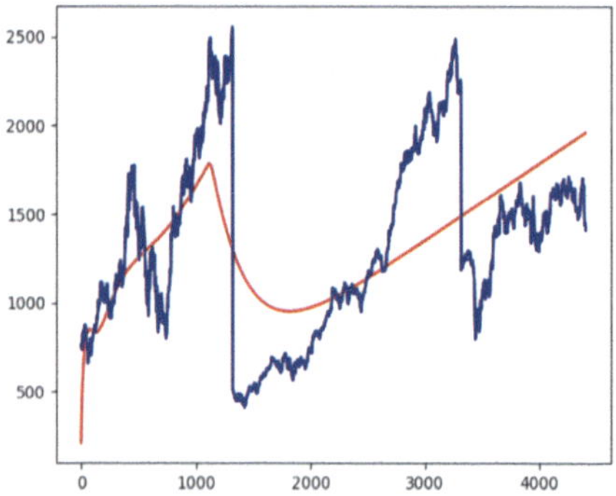

Fig. 9. The results received for HDFC using BPNN

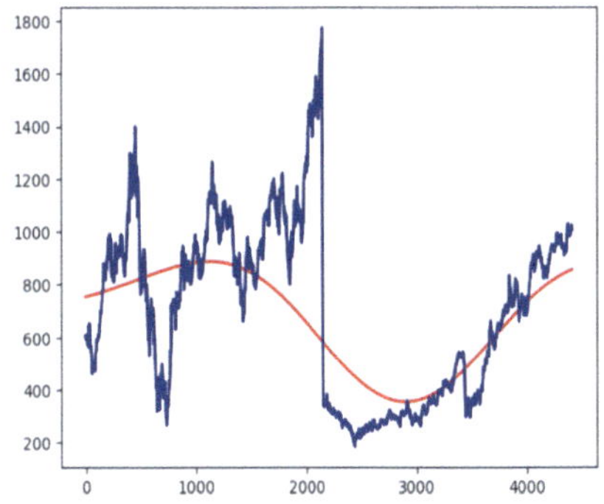

Fig. 10. The results received for ICICI using SVM

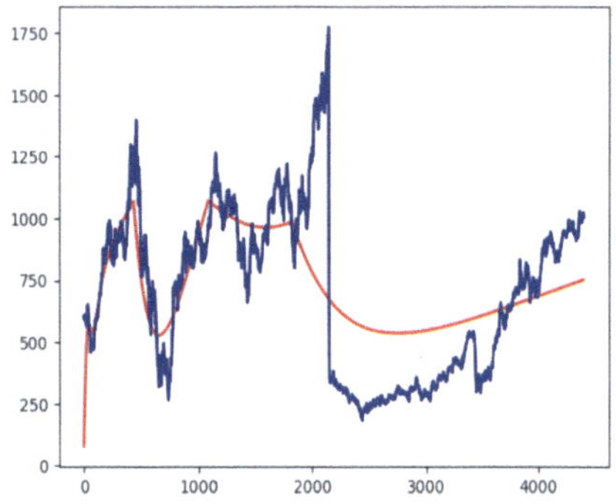

Fig. 11. The results received for ICICI using BPNN

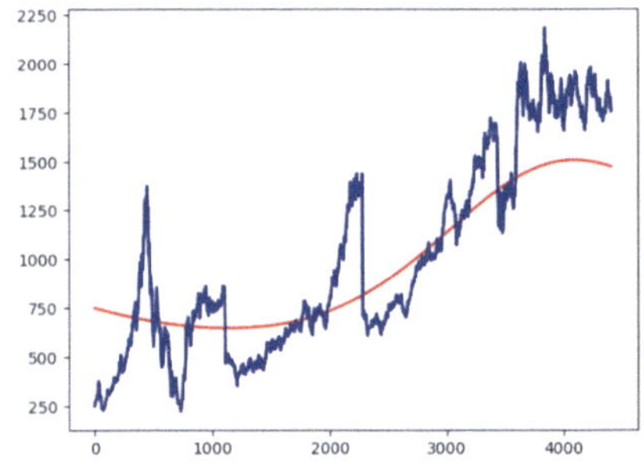

Fig. 12. The results received for KOTAK using SVM

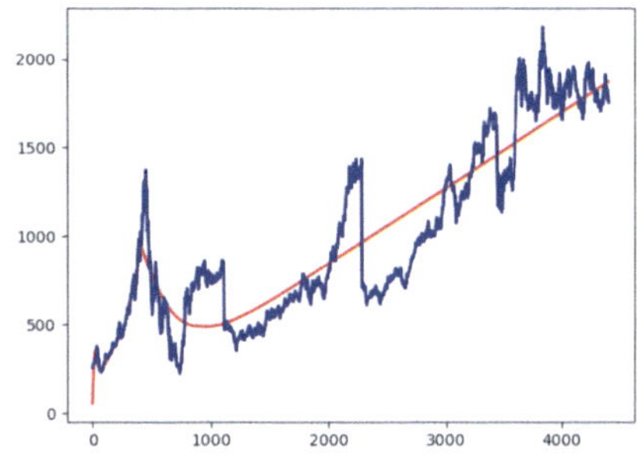

Fig. 13. The results received for KOTAK using BPNN

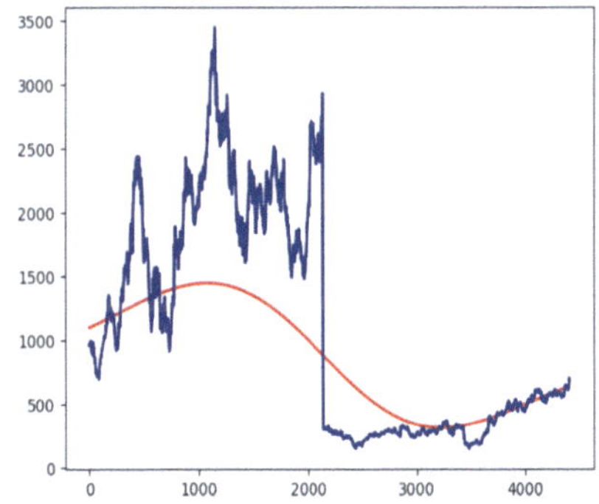

Fig. 14. The results received for SBI using SVM

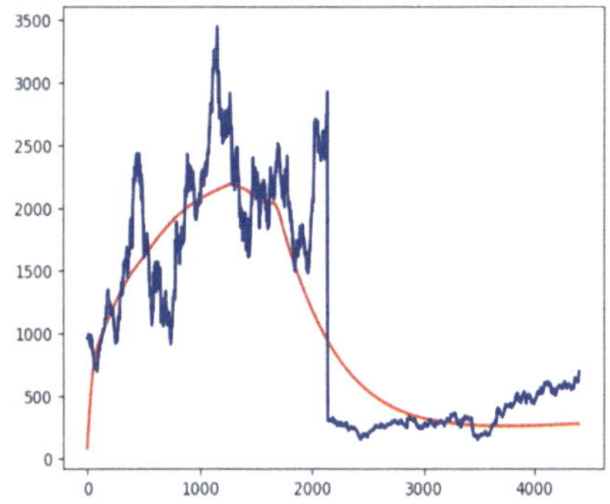

Fig. 15. The results received for SBI using BPNN

5 Conclusion

The predictive powers of PNB, Axis, HDFC, ICICI, KOTAK, and SBI are thus revealed by the methodical examination of stock market prediction models, which also clarifies the many facets of the six major banks' predictive potentiality. Throughout the whole study, the Back Propagation Neural Network (BPNN) model is the most accurate, outperforming SVM in every performance metric examined. Its consistent low MAE and MSE scores along with a strong R-Square (R2) profile further support this, indicating that it is stable in its capacity to capture the fleeting volatility that characterizes the stock market for the banking industry.

In conclusion, a particularly clear example of the BPNN model's superiority is its capacity to provide forecasts that are more accurate and exact across all banks. This approach gives stakeholders a convincing tool that helps them accurately and strategically manage the financial environment's intricacies.

References

1. Tan, T.Z., Quek, C., Ng, G.S.: Biological brain-inspired genetic complementary learning for stock market and bank failure prediction. Comput. Intell. **23**(2), 236–261 (2007). https://doi.org/10.1111/j.1467-8640.2007.00303.x
2. Chang, P.C., Di Wang, D., Le Zhou, C.: A novel model by evolving partially connected neural network for stock price trend forecasting. Expert Syst. Appl. **39**(1), 611–620 (2012). https://doi.org/10.1016/j.eswa.2011.07.051
3. Khan, W., Ghazanfar, M.A., Azam, M.A., Karami, A., Alyoubi, K.H., Alfakeeh, A.S.: Stock market prediction using machine learning classifiers and social media, news. J. Ambient. Intell. Humaniz. Comput. **13**(7), 3433–3456 (2022). https://doi.org/10.1007/s12652-020-01839-w
4. Zhang, Y.J., Wang, J.L.: Do high-frequency stock market data help forecast crude oil prices? Evidence from the MIDAS models, vol. 78. Elsevier B.V (2019). https://doi.org/10.1016/j.eneco.2018.11.015
5. Hiransha, M.E.A.G., Gopalakrishnan, E.A., Menon, V.K., Soman, K.P.: NSE stock market prediction using deep-learning models. Procedia Comput. Sci. **132**, 1351–1362 (2018). https://doi.org/10.1016/j.procs.2018.05.050
6. Rouf, N., et al.: Stock market prediction using machine learning techniques: a decade survey on methodologies, recent developments, and future directions. Electronics **10**(21) (2021). https://doi.org/10.3390/electronics10212717
7. Agrawal, M., Shukla, P.K., Nair, R., Nayyar, A., Masud, M.: Stock prediction based on technical indicators using deep learning model. Comput. Mater. Contin. **70**(1), 287–304 (2021). https://doi.org/10.32604/cmc.2022.014637
8. Selvamuthu, D., Kumar, V., Mishra, A.: Indian stock market prediction using artificial neural networks on tick data. Financ. Innov. **5**(1) (2019). https://doi.org/10.1186/s40854-019-0131-7
9. Wijesinghe, G.W.R.I., Rathnayaka, R.M.K.T.: ARIMA and ANN approach for forecasting daily stock price fluctuations of industries in Colombo stock exchange, Sri Lanka. In: Proceedings of the ICITR 2020 - 5th International Conference on Information Technology Research Towar. New Digit. Enlight. (2020). https://doi.org/10.1109/ICITR51448.2020.9310826
10. Jiang, Z., Ji, R., Chang, K.-C.: A machine learning integrated portfolio rebalance framework with risk-aversion adjustment. J. Risk Financ. Manag. **13**(7), 155 (2020). https://doi.org/10.3390/jrfm13070155

Machine Learning in Heart Failure Prediction: Enhancing Accuracy and Developing Early Detection Method

Ahmed A. Alethary[1], Rana Ghalib[2], and Luis Cardoso[3](✉)

[1] Department of Quranic Sciences, Faculty of Jurisprudence, University of Kufa, Najaf, Iraq
[2] General Directorate of Education in Najaf Governorate, Najaf, Iraq
[3] Polytechnic Institute of Portalegre and Centre for Comparative Studies of the University of Lisbon, Lisbon, Portugal
lmcardoso@ipportalegre.pt

Abstract. Heart disease is a major global health issue that has an impact on rates of morbidity and mortality everywhere. Even with advancements in diagnostic processes, traditional methods struggle with efficiency, scalability, and accuracy. To enhance the prediction of heart frailer, this study proposed a multi-voting ensemble Machine Learning (ML) model that utilizes a several ML techniques as base models such as Random Forest (RF), Logistic Regression (LR), Support Vector Machine (SVM), Naïve Bayas (NB) and Multiplayer perceptron (MLP). To maximize performance like accuracy, sensitivity, and specificity, the proposed method uses the preprocessing techniques including filling missing values, data scaling, and managing categories variables. Utilizing a heart disease dataset containing 918 patient records, the model surpassed the conventional classifiers with a perdition accuracy of 94.13%. This model proposed a useful method for enhancing prompt medication and better patient outcomes. The result shows that the ensemble learning is useful in medical decision-making and in development of diagnostic tools.

Keywords: Classification · Heart Failure · Machine Learning · Data Analytics

1 Introduction

The fast growth of medical data and advancements in data science have encouraged various organizations to develop indicators for predicting diseases. Heart disease (CVD) is the leading cause of death globally, responsible for an estimated 17.9 million deaths per year, or 31% of all deaths. One common condition caused by CVDs is heart failure. Early and accurate prediction of heart disease and high cardiovascular risk persons (those who have one or more risk factors, such as diabetes, hypertension, hyperlipidemia, or pre-existing disease) is critical for timely intervention, potentially reducing the associated morbidity and mortality rates [1].

However, traditional diagnostic methods often face difficulties in accuracy, specificity, and scalability. A machine learning model can be highly helpful in solving this

A. J. Obaid et al. (Eds.): ICONEST 2024, ISEM 40, pp. 343–353, 2025.
https://doi.org/10.1007/978-3-031-89960-7_26

challenge. Using this method, we try to automate another natural problem using AI approaches so that we can fight it and move on to the next one. In recent years, the advent of machine learning (ML) has revolutionized the field of medical diagnostics by enabling data-driven insights and improving predictive accuracy [2].

Machine learning techniques, such as DT, SVM, Neural Networks, and LR, have been extensively employed in heart disease prediction. These algorithms predicate disease with acceptable accuracy by using patterns and relationships found in medical datasets [3]. Regardless of their success, single classifier struggle with inherit complexity of medical data which contains outliers, variability, and missing information By using ensemble learning techniques that combing multiple base classifiers to handle these complexities. The ensemble techniques like hard and soft voting leverage the strength of single classifies and qualifying their limitations [4]. Recent research has showed that ensemble classifiers, such as RF and Gradient Boosting Machines, are successful in obtaining high performance metrics for the prediction of cardiac disease [5].

This study further explores the ability of ensemble learning by blending advanced techniques such as outlier removal, feature engineering, and novel voting mechanisms. The study shows the ability of these methods on heart disease datasets to evaluate their performance across metrics like accuracy, sensitivity, and specificity. By concentrating on machine learning's ability to improve diagnostic precision, this research contributes to the ongoing efforts to develop reliable, cost-effective, and scalable solutions for heart disease prediction.

The reset structure of the paper is arranged as follows: Sect. 2 is a summary of previous research on related system while Sect. 3 discuss the architecture and implementation of the system covering aspects like dataset distribution preprocessing steps and used ML algorithms and ensemble learning methods whereas Sect. 4 shows the evolution of the proposed system and compassion with previous research. Lastly, the conclusion section highlights the finding and theoretical implications.

2 Related Works

In earlier years, multiple predictive models have been developed for a wide range of diseases, but relatively few have concentrated specifically on heart failure prediction systems. Researchers have recently increasingly concentrated on heart failure prediction due to its crucial influence on global health. Different systems have been developed using different procedures, including statistical approaches and ML models. In this section, related works on heart failure prediction using ML will be introduced, covering advancements in methodologies, dataset usage, and model construction.

Patel et al. (2016) [6] used machine learning and data mining approaches to predict heart disease, focusing on Decision Tree algorithms. LM, RF, and j48 classifiers are used on key attributes extracted from the Cleveland dataset from the UCI repository. According to the result, the J48 method superior LM and RF in terms of accuracy and efficiency. The study also showed how datamining techniques can be used to discover meaningful patterns from clinical data to assist in medical diagnoses.

Gonsalves et al. (2019) [7] studied the performance of supervised learning techniques such as NV,DT, and SVM using the south African heart disease dataset containing 462 records. Their findings showed that NB achieved the highest sensitivity and accuracy, surpassing J48 and SVM in recognizing patients with CHD. However, J48 and SVM demonstrated more acceptable specificity in correctly classifying non-CHD cases.

Jawalkar et al. (2023) [8] suggested a decision tree random forest classifier optimized with stochastic gradient boosting (SGB) to improve heart failure prediction. They trained the classifier with SGB to minimize loss utilizing the Cleveland Heart Disease dataset and assessed the model performance with 86% precision, 86% recall, and an 85% F1 score, the suggested model enhanced conventional techniques. With this method, issues like outliers, missing data, and overfitting are treated while highlighting the potential of SGB-enhanced classifiers in clinical decision-making.

Sawinder Kaur et al. (2024) [9] The researcher studies the essential challenge of predicting heart failure prediction using ML techniques and presents a new approach that leverages Reinforcement Learning for Hyperparameter Optimization. The study uses a benchmark dataset holding medical records of 299 heart failure patients, featuring 13 clinical and lifestyle variables. The RLHPO+ANN approach integrates multiple metrics, including Matthews Correlation Coefficient (MCC) and binary cross-entropy loss, to overcome the limitations of traditional accuracy-focused optimizations, particularly for imbalanced datasets.

Amrit Singh et al. (2024) [10] Key studies have explored various ML models for heart disease detection. Techniques such as Decision Tree (DT), Random Forest (RF), Support Vector Machine (SVM), and Principal Component Analysis (PCA) have been widely employed. Among these, RF consistently outperforms other models, achieving up to 97% accuracy in prediction tasks on datasets like the Framingham Heart Study. This dataset, with over 4,000 records, is extensively used for risk assessment over a 10-year horizon, emphasizing key predictors like blood pressure, cholesterol levels, and smoking status.

3 Methodology

Heart disease is a major global health issue that claims many lives each year. Conventional techniques for identifying cardiac disease can be costly, time-consuming, and imprecise. Therefore, more accurate and efficient techniques for heart disease diagnosis and prediction are required [11]. The authors of this study have suggested a hybrid decision support system for heart disease prediction. Two stages make up this hybrid system: Firstly, gathering data and pre-processing data then building the model using a Multiple Voting Method, a robust ensemble technique, to improve the accuracy and reliability of heart failure prediction models A combination of five classifiers have been used so that the overall model forming. The detail of the proposed system is illustrated in Fig. 1.

3.1 Datasets Description and Preprocessing

Heart disease dataset is collected from the Kaggle. This data contains the following input characteristics and prediction attributes (sex, age, type of chest pain, resting blood pressure, cholesterol, maximal heart rate, angina brought on by exercise, resting ECG

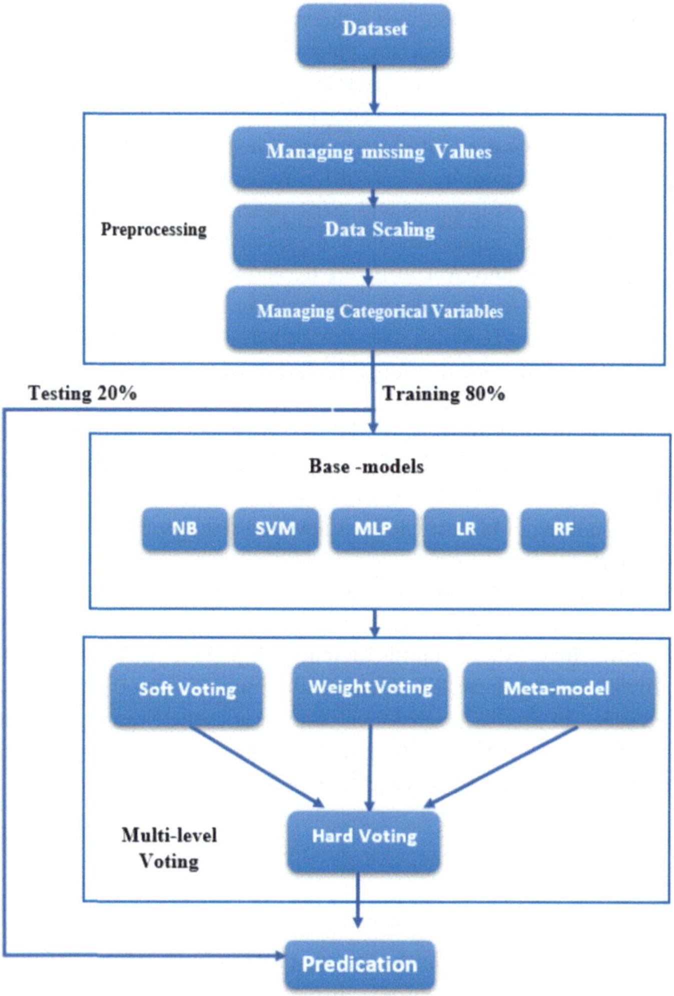

Fig. 1. The layout of the proposed methodology.

readings, and depression numerical value). The target we want to predict is the Death Event or not. Since the number of positive and negative samples is different, Fig. 2 shows a sample distribution for some attributes.

Data preprocessing such as (Data Scaling, Managing missing Values, and Managing Categorical Variables is an important stage in machine learning since the quality of the data and the insights that can be drawn from it directly impact our model's capacity to learn [12]. As a result, it is crucial to preprocess our data before incorporating it into our model. KNN imputation is used to fill in the missing values.

The machine learning model cannot understand the value of features. It doesn't know what the input means; it just treats it as a basic integer. This makes scaling the data required. Both standardization and normalization are used for data scaling. Normalization is used for features whose data does not exhibit a normal distribution, and standardization is applied to normally distributed features whose values are extremely tiny or large about other features [13]. Converting categories into numerical representations is essential since most ML models work with numerical data. Label Encoding method was used for this purpose.

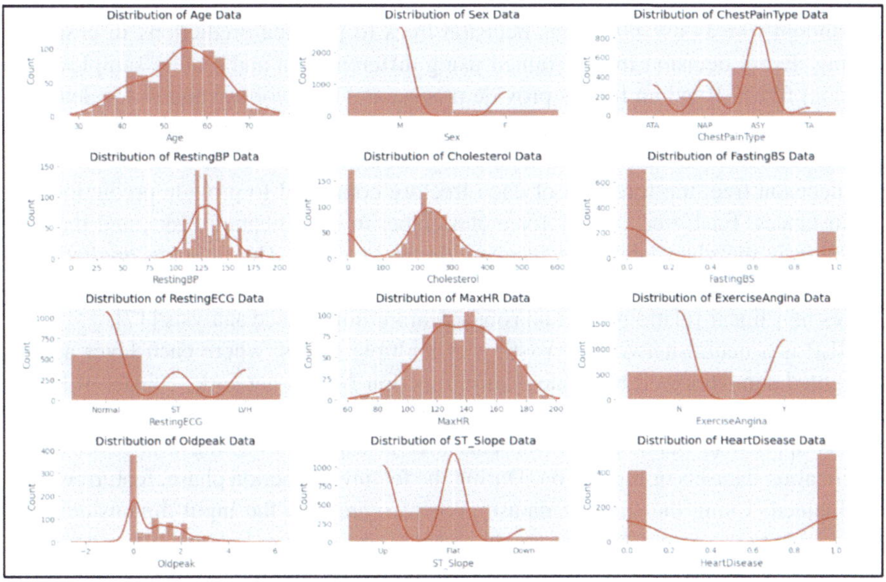

Fig. 2. A sample distribution for dataset attribute

3.2 Base Models

The proposed system for the research is based on a multi-level voting ensemble approach, which integrates preprocessing, base-model training, and advanced ensemble techniques to optimize predictions. The process can be described as follows:

3.2.1 Classification Algorithms

The proposed system integrates five classification algorithms as a base model NB, SVM, LR, RF, and MLP—into a unified voting-based framework. The system aims to leverage the strengths of individual algorithms and improve the overall classification performance through a multiple-voting mechanism. Below, we describe the workflow from model training to final decision-making.

NB is a probabilistic classifier The classifier assumes feature independence and assigns the class with the highest posterior probability Because of its efficiency and simplicity with smaller datasets that have been used [14].

SVM finds the optimal hyperplane that separates classes by maximizing the margin. The sigmoid function is used in logistic regression to model the likelihood of a binary outcome. For its strength in high-dimensional feature spaces is selected [15].

A statistical technique for binary or multi-class classification problems is logistic regression. Logistic regression forecasts the likelihood that an instance will belong to a particular class, as opposed to linear regression, which forecasts a continuous value. This makes it particularly useful for problems like binary classification (e.g., spam detection, disease prediction) [16].

Random forests use numerous decision trees to produce predictions in ensemble learning. Every decision tree is trained using different data and feature samples from a random forest. Random forests provide precise and trustworthy models for detecting heart disease by utilizing the predictions of many trees. Complex feature interactions and high-dimensional data can be handled by random forests. Following the growth of each decision tree, the forecasts of each tree are combined to provide predictions for new instances. Each tree "votes" for a class label in classification tasks, and the final prediction is the class label that receives the most votes. The ultimate prediction in regression problems is determined by averaging the predictions made by each tree. It handles non-linear relationships and robust feature importance analysis [17].

MLP is a neural network consisting of multiple layers, where each layer applies a weighted sum followed by an activation function Neural networks consist of highly interconnected neurons designed to address specific problems in parallel. In this study, MLP is applied to our dataset. This classifier is capable of being trained on either classification datasets or regression. During the feature extraction phase, feature vectors are obtained. Using the training dataset Where represents the input dimensions, and corresponds to the output dimensions. In an MLP, one or more hidden layers (nonlinear layers) may be inserted between the input and output layers. The input layer is made up of neurons, where each neuron acts as the input feature which is fed into the hidden layer. Then, the hidden layer calculates the weighted summation. Lastly, the output value is given by the last hidden layer and is received by the output layer [18].

$$X_F^2 = \frac{12M}{K(K+1)}\left[\sum_i L_i^2 - \frac{K(K+1)^2}{4}\right] \tag{1}$$

L_i represents the algorithm i's average ranking out of the total set of algorithms K, M.

3.3 Multi-level Voting

Approach is an advanced ensemble method designed to optimize the accuracy and reliability of voting by organizing the process into multiple stages or levels. Here's a step-by-step breakdown of how it works:

Soft voting is an ensemble learning method where predictions from numerous models are integrated to make a final decision based on the probability outcomes of each model. This strategy leverages the strengths of individual classifiers, leading to a more powerful and accurate prediction system where each base model in the ensemble generates a probability distribution over the possible classes for a given input, then averages the probability outputs from all models to determine the final probabilities for each class [19]. The system selects the class with the highest aggregated probability as the final prediction.

$$P_{c}=\frac{1}{M}\sum_{m=1}^{M}P_{m,c} \tag{2}$$

Where:

Pc: Final probability for class c.
Pm,c: Probability for class c predicted by model m.
M: Number of models

Weight Voting: Each model is assigned a weight based on its performance, and predictions are combined accordingly. Among the top-ranked models, perform a weighted vote [20]. For binary classification, this could mean each model votes either “1” (heart failure) or “0” (no heart failure). Sum the votes with weights, and if the weighted sum is above a certain threshold (e.g., 0.5), the final classification is “1”; otherwise, it’s “0”.

$$P_c = \frac{\sum_{m=1}^{M} W_m \times P_{m,c}}{\sum_{m=1}^{M} W_m} \tag{3}$$

P(m,c) represents the probability output of model m for class c.
W(m) weights are assigned to models.

Meta-Model (Stacking), is a powerful ensemble learning approach that merging the predictions of multiple base models using a meta-model. Then the meta-model is trained to integrate these predictions, in stacking, the predictions of the base models serve as input features for the meta-model. The meta-model learns how to effectively combine these features, often leading to improved performance compared to individual models or simpler ensembles. This method has proven to be particularly effective in complex machine-learning tasks, making it a key technique in ensemble learning. Simplicity and interpretability used logistic regression [21].

Final Aggregation- Hard Voting is a simple yet effective ensemble learning technique where the final prediction is determined based on the majority vote from multiple models. Each base model in the ensemble casts one vote for a specific class, and the class with the highest number of votes is selected as the final prediction [22].

$$C = \operatorname{argmax}_k \sum_{m=1}^{M} I(y_m = k) \tag{4}$$

Where:

M is the total number of models.
k is a class label.

3.4 Evaluation and Statistical Analysis

The accuracy, sensitivity, specificity, and precision were used to assess the classifiers’ performance. A True Positive (TP) occurs when the system predicts that a patient has cardiac disease; if not, it is a False Negative. Likewise, it is a True Negative (TN) if a healthy individual is expected to remain disease-free; if not, it is a false positive(FP). Accuracy The performance metric evaluates the system's ability for accurate prediction. Where Sensitivity is a performance metric that evaluates the system's capacity to generate positive predictions, and Specificity metric evaluates capacity to generate negative predictions, lastly Precision metric assesses the system's capacity for accurate prediction-making.

$$\text{Accuracy} = \frac{\text{Correct predictions}}{\text{Total predictions}} \times 100 \quad (5)$$

$$\text{Sensitivity} = \frac{\text{TP}}{\text{TP} + \text{FN}} \times 100 \quad (6)$$

$$\text{Specificity} = \frac{\text{TN}}{\text{TN} + \text{FP}} \times 100 \quad (7)$$

$$\text{Precision} = \frac{\text{TP}}{\text{TP} + \text{FP}} \times 100 \quad (8)$$

4 Result Analysis and Discussion

In this section, testing, and evaluation of the proposed heart failure prediction system will be discussed. The system is evaluated based on its predictive accuracy. An automated evaluation approach is employed to assess the accuracy of the predictions. A dataset containing approximately 918 records in the medical domain was utilized. The records in the dataset include various features and attributes critical for heart failure prediction, such as demographic information, medical history, and clinical measurements.

The dataset was split into two parts: 80% of the data was used for the development and training of the prediction models, while the remaining 20% was reserved for evaluating the system's performance.

Some of the attributes in the dataset required preprocessing, such as normalization and encoding, to ensure compatibility with the system's algorithms. Additionally, categorical variables were transformed into numerical representations to align with the requirements of the predictive models. Table 1 presents the results of the proposed system's performance, evaluated using metrics such as accuracy, precision, Sensitivity, and Specificity.

Different algorithms such as SVC, NB, LR, RF, and MLP for the detection of patients diagnosed with Heart disease, this algorithm also trained and evaluated using the same dataset RF and MLP Classifier yielded a better accuracy of 89% compared with other algorithms. It is cost efficient and accurate than other algorithms. Moreover, the maximum accuracy obtained by SVM and Logistic Regression is equal to 85% which is considered a good accuracy and faster algorithms as shown in Table 2. Lastly, Our proposed method increased the accuracy up 94%. The suggested method gives significant improvement in precision, Sensitivity, specificity, and Accuracy. This will enhance the prediction of the patient diagnosed with a heart Disease.

As indicated in Table 3, the findings of our approach are compared with the results of other previous studies on heart frailer identification. For performing out detection, researchers utilized a variety of methodologies and datasets.

Table 1. Evolution of the proposed system

Fold	Number of records	Precision	Sensitivity	Specificity	Accuracy
1	18	93.14	91.14	90.14	92.45
2	18	92.77	90.77	92.15	93.2
3	18	95.12	92.12	92.12	93.2
4	18	94.33	90.33	88.89	94.88
5	18	91.44	89.87	90.13	92
6	18	96.14	93.12	92.99	95.9
7	18	94.12	92.12	90.15	94.68
8	18	93.15	91.25	89.2	94.3
9	18	93.2	91.11	90.79	93.89
10	18	94.78	91.78	90.14	95.99
Average	18	93.82	91.36	90.67	94.05

Table 2. Evolution of different classification algorithms

Accuracy	Specificity	Sensitivity	Precision	Model
85.12	82.86	83.23	82.32	Logistic Regression
81.11	81.13	82.55	81.41	Naive Bayers
83.32	85.57	84.21	83.44	SVM(sigmoid kernel)
89.14	85.22	87.15	88.65	Random Forest
89.24	86.12	87.67	86.89	MLP
94.05	93.23	92.42	93.22	Multi-level Voting

Table 3. Comparison a proposed system with the previous models

Study Title	Methodology	Dataset	Accuracy
Patel et al. (2016)	Decision Tree algorithms (J48, Logistic Model Tree, Random Forest)	UCI Cleveland dataset (303 records, 76 attributes)	J48 56.76% LMT 55.77%
Gonsalves et al. (2019)	Guided Feature Ranking + Deep Attentive Model	UCI, Framingham, Faisalabad datasets	91%
Jawalkar et al. (2023)	Random Forest + Stochastic Gradient Boosting	Real-world dataset	86%

(continued)

Table 3. (*continued*)

Study Title	Methodology	Dataset	Accuracy
Sawinder Kaur et al. (2024)	Reinforcement Learning with ANN	Medical records of 299 patients	82,93%
Amrit Singh et al. (2024)	Decision Tree, Random Forest, SVM, PCA	Framingham dataset	DT 84% RF 97% PCA 80% SVM 68%
Proposed multi-level voting model	NB, SVM, LR, RF, and MLP with multi-level voting model	Medical records of 918 patients from Kaggle	94.05%

5 Conclusion

Early detection and timely treatment of heart disease can play a crucial role in saving lives. Conventional methods depend largely on the expertise and judgment of healthcare professionals, which often results in limited accuracy. To overcome these limitations, this study introduces a new approach that aids medical practitioners in extracting valuable insights from patient data, enabling early and accurate diagnosis of heart disease.

The proposed method utilizes an ensemble classifier, which integrates five different classifiers into a multi-level ensemble voting model to enhance predictive performance compared to individual classifiers. By analyzing patient records, this system identifies correlations between various risk factors and heart disease. It helps healthcare providers identify individuals at higher risk, enabling them to deliver better care at more affordable costs. Furthermore, the study compares performance metrics such as sensitivity, specificity, and accuracy, demonstrating that the ensemble classifier achieves 94% accuracy, outperforming individual models. This approach can be further developed to enhance its application in diagnosing other medical conditions.

References

1. Atallah, R., Al-Mousa, A.: Heart disease detection using machine learning majority voting ensemble method. In: 2019 2nd International Conference on New Trends in Computing Sciences, ICTCS 2019 - Proceedings, Institute of Electrical and Electronics Engineers Inc., October 2019. https://doi.org/10.1109/ICTCS.2019.8923053
2. Subhadra, K.: Neural network based intelligent system for predicting heart disease (2019). www.ijitee.org
3. Jindal, H., Agrawal, S., Khera, R., Jain, R., Nagrath, P.: Heart disease prediction using machine learning algorithms. IOP Conf. Ser. Mater. Sci. Eng. (2021). https://doi.org/10.1088/1757-899X/1022/1/012072
4. Singh, A., Joshi, S.: A vote share based enhance hybrid classifier for heart disease prediction (2016)
5. Kurian, R.A.: An ensemble classifier for the prediction of heart disease. Int. J. Sci. Res. Comput. Sci. Eng. Inf. Technol. IJSRCSEIT **3**(6), 6 (2018). http://ijsrcseit.com

6. Patel, S.B.: Heart disease prediction using machine learning and data mining technique (2016). 10.090592/IJCSC.2016.018
7. Gonsalves, A.H., Thabtah, F., Mohammad, R.M.A., Singh, G.: Prediction of coronary heart disease using machine learning: an experimental analysis. In: ACM International Conference Proceeding Series, Association for Computing Machinery, pp. 51–56, July 2019. https://doi.org/10.1145/3342999.3343015
8. Jawalkar, A.P., et al.: Early prediction of heart disease with data analysis using supervised learning with stochastic gradient boosting. J. Eng. Appl. Sci. **70**(1) (2023). https://doi.org/10.1186/s44147-023-00280-y
9. Ribino, P., Di Napoli, C., Paragliola, G., Serino, L.: Hyper-parameter optimization through reinforcement learning for survival prediction of patients with heart failure. Procedia Comput. Sci., 1754–1761 (2024). https://doi.org/10.1016/j.procs.2024.06.354
10. Singh, A., Mahapatra, H., Biswal, A.K., Mahapatra, M., Singh, D., Samantaray, M.: Heart disease detection using machine learning models. Procedia Comput. Sci., 937–947 (2024). https://doi.org/10.1016/j.procs.2024.04.089
11. Javid, I., Alsaedi, A.K.Z., Ghazali, R.: Enhanced accuracy of heart disease prediction using machine learning and recurrent neural networks ensemble majority voting method. Int. J. Adv. Comput. Sci. Appl. **11**(3), 540–551 (2020). https://doi.org/10.14569/ijacsa.2020.0110369
12. Miao, L., Wang, W.: Cardiovascular disease prediction based on soft voting ensemble model. J. Phys. Conf. Ser. (2023). https://doi.org/10.1088/1742-6596/2504/1/012021
13. Chapagain, P.: Heart disease prediction using outlier removal based max voting ensemble method. Int. J. Eng. Technol. **1**(1), 83–101 (2023). https://doi.org/10.3126/injet.v1i1.60933
14. Rish, I., Rish, I.: An empirical study of the Naïve Bayes classifier an empirical study of the naive Bayes classifier. https://www.researchgate.net/publication/228845263
15. Madhu, H.K., Ramesh, D.: Heart attack analysis and prediction using SVM. Int. J. Comput. Appl. **183**(27), 975–8887 (2021). https://doi.org/10.13140/RG.2.2.28203.52004
16. Verma, L., Srivastava, S., Negi, P.C.: A hybrid data mining model to predict coronary artery disease cases using non-invasive clinical data. J. Med. Syst. **40**(7), 178 (2016). https://doi.org/10.1007/s10916-016-0536-z
17. Altantawy, D.A., Kishk, S.S.: A highly accurate prediction for heart failure disease: a new deep attentive model with guided feature ranking. Arab. J. Sci. Eng. **49**(9), 12167–12201 (2024). https://doi.org/10.1007/s13369-023-08508-y
18. Singh, A.J., Kumar, M.: Comparative analysis on prediction of software effort estimation using machine learning techniques. https://ssrn.com/abstract=3565813
19. Karadeniz, T., Maraş, H.H., Tokdemir, G., Ergezer, H.: Two majority voting classifiers applied to heart disease prediction. Appl. Sci. **13**(6) (2023). https://doi.org/10.3390/app13063767
20. Kaushik, S., Birok, R.: Heart failure prediction using voting ensemble classifier. In: 2021 Asian Conference on Innovation in Technology (ASIANCON), pp. 1–5. IEEE, August 2021 https://doi.org/10.1109/ASIANCON51346.2021.9544871
21. Mohapatra, S., et al.: A stacking classifiers model for detecting heart irregularities and predicting Cardiovascular disease. Healthc. Anal. **3**, 100133 (2023). https://doi.org/10.1016/j.health.2022.100133
22. Bin Habib, A.-Z.S., Tasnim, T.: An ensemble hard voting model for cardiovascular disease prediction. In: 2020 2nd International Conference on Sustainable Technologies for Industry 4.0 (STI), pp. 1–6. IEEE, December 2020. https://doi.org/10.1109/STI50764.2020.9350514

A Comprehensive Review of Optimization Techniques in Artificial Intelligence: Algorithms, Applications, and Challenges

Heba Mehdi[1], Ali Said Al Matari[2(✉)], and Asfan Muqtadir[3]

[1] Ministry of Education, General Directorate of Education of Najaf, Najaf, Iraq
[2] A'Sharqiyah University, Ash Sharqiyah North, Ibra, Oman
ali.almatari@asu.edu.om
[3] Universitas PGRI Ronggolawe Tuban, Tuban, Indonesia

Abstract. In many fields, including engineering, medicine, and finance, optimization algorithms are essential for resolving challenging issues. These methods improve the efficacy of addressing real-world challenges, resource management, and decision-making. This paper provides a comprehensive review of optimization techniques in artificial intelligence (AI), exploring their algorithms, applications, and challenges. It includes classic methods like Gradient Descent and Simulated Annealing, population-based methods like Genetic and Evolutionary Algorithms, and specialized techniques like Particle Swarm Optimization and Ant Colony Optimization. This paper sheds light on the role of optimization in progressing AI and its applications by emphasizing its strengths, limitations, and social implications.

Keywords: Optimization algorithms · Applications Field · Analytical comparison

1 Introduction

Artificial intelligence (AI) is a technology that simulates intelligent human thought and behavior, including the ability to understand, learn, reason, and interact. It has enormous potential to address global issues in fields like healthcare, transportation, and the environment in addition to transforming production techniques [1].

The most difficult problems in a variety of real-world systems and sectors, including as engineering, economics, and healthcare, have been successfully resolved using optimization algorithms. As a consequence of technological advancements, optimization algorithms are now utilised extensively. In order to remain current with modernity and progress, it has been imperative to create and enhance a diverse array of optimization algorithms. This will enable them to be categorised according to their efficacy and speed in accomplishing the fundamental optimization objectives. These algorithms have pioneered the use of constructive decision-making to identify optimal solutions to a wide range of situations. This talk will take a lengthy trip through time to explore and discuss the magnificent evidence of the most significant optimization approaches from the past [2–4].

A. J. Obaid et al. (Eds.): ICONEST 2024, ISEM 40, pp. 354–364, 2025.
https://doi.org/10.1007/978-3-031-89960-7_27

Traditional mathematical methods of optimization can be difficult since market pricing, energy demand, and weather patterns are always changing and exhibit irregular or nonlinear patterns that are challenging to resolve with these methods [5]. Their drawbacks include the lethargic convergence, scalability issue, and computational expenditure of other conventional optimization techniques, such as the iterative approach [6], gradient descent [7], quasi-Newton [8], and others. Genetic Algorithm (GA) [9], Particle Swarm Optimization (PSO) [10], Ant Colony Optimization (ACO) [11], Simulated Annealing (SA), and convex optimization [12] are a few of the classical metaheuristic algorithms that are commonly used to optimise ESS charging and discharging and integrate RESs. The objective of this paper is to offer a comprehensive examination of optimization techniques in AI, including their algorithms, applications, and challenges, and to establish a foundation for future research.

2 Related Works

Artificial intelligence optimization has grown to be a thriving field of study due to the necessity to address difficult problems in scheduling, decision-making, and resource allocation. Numerous algorithms have been developed by researchers, each intended to address certain issues and boost productivity in practical applications. This section reviews recent researches in optimization techniques, discussing their methodologies and objectives.

Ruder (2016) [13] gave a summary of gradient descent optimization methods, highlighting how they are used to train machine learning models and neural networks. The study emphasized the benefits and convergence issues of adaptive techniques such as Adam and RMSprop. Das, U. K. (2018) [14] centered on power forecasts to guarantee grid planning and operation that is steady, dependable, and efficient.

Ponciroli et al. (2020) [9] enhanced the genetic algorithm to incorporate factors for large-scale systems, such as nuclear and coal plants, and solve unit commitment and economic dispatch issues. Aml et al. (2021) [15] suggested a hybrid optimization strategy that manages load demand uncertainties by fusing adaptive approaches and genetic algorithms, leading to notable gains in accuracy and computing efficiency.

Thirunavukkarasu, G. (2022) [16] Presented AI-based approaches in interconnected MGs to handle power mismatch problems. Benti, N. E. (2023) [17] concentrated on the prediction of renewable energy production to guarantee the efficient and dependable operation of the grid. Ali, R. (2024) [18] examined two critical components of residential EMS: load forecasting and load scheduling. A comparison of different load forecasting and optimization methods was conducted.

Table 1 shows a comparative study among these researches selected because of their proximity as a part of the work.

Table 1. Comparative Analysis of Existing Methods

Ref	Year	Tanique	Objective
[13]	2016	Gradient Descent	Compiled optimization techniques for neural network and machine learning model training
[14]	2018	Power Forecasting	Enhanced grid operation and planning for efficiency and stability
[9]	2020	Genetic Algorithm	Discussed economic dispatch and unit commitment in large-scale systems like coal plants
[15]	2021	Hybrid Optimization (Genetic + Adaptive)	Enhanced accuracy and computational efficiency in managing load demand uncertainty
[16]	2022	AI-Based Optimization for Microgrids	Fixed issues with power mismatches in networked microgrids
[17]	2023	Machine Learning and Deep Learning	Production of renewable energy is anticipated to ensure grid stability
[18]	2024	Optimization and Load Forecasting	Compared methods for load scheduling and forecasting in home energy management systems

3 Classification of Optimization Techniques

Optimization is defined differently across fields. However, it can be described as a technique that achieves specific aims. Utilize system resources efficiently. Optimization technology helps solve real-world difficulties by speeding up decision-making and improving quality [19].

Optimization algorithms are classified into two types: deterministic and heuristic. Deterministic Algorithms execute particular tasks in a predetermined order. The design variables and objective function can have the same values, resulting in a repeated solution route. Heuristic algorithms, on the other hand, typically make use of randomness. Hybrid methods combine deterministic and heuristic approaches to create novel methods. Figure 1 depicts the classification of optimization algorithms.

3.1 Gradient Descent

Gradient descent is a fundamental algorithm that has been in use since the early nineteenth century. The French scientist Augustin-Louis Cauchy was the first to introduce this method, which acquired prominence in the 20th century for the purpose of addressing optimization problems [20].

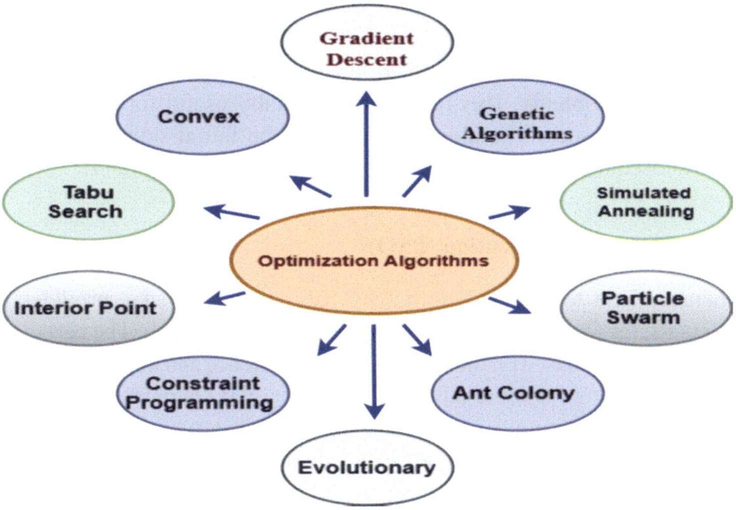

Fig. 1. Types of optimization algorithms.

3.2 Genetic Algorithms

The genetic algorithm, inspired by natural selection and genetics, simulates evolution to get optimal solutions. Genetic algorithms generate multiple solutions and evaluate them using a predetermined objective function to select the most suitable individuals for reproduction. Cross-breeding and mutation lead to better population fitness over generations. In the 1960s and 1970s, John Holland and colleagues further developed the genetic algorithm approach [21]. Figure 2 shows an example of the crossover and mutation process in genetic algorithms.

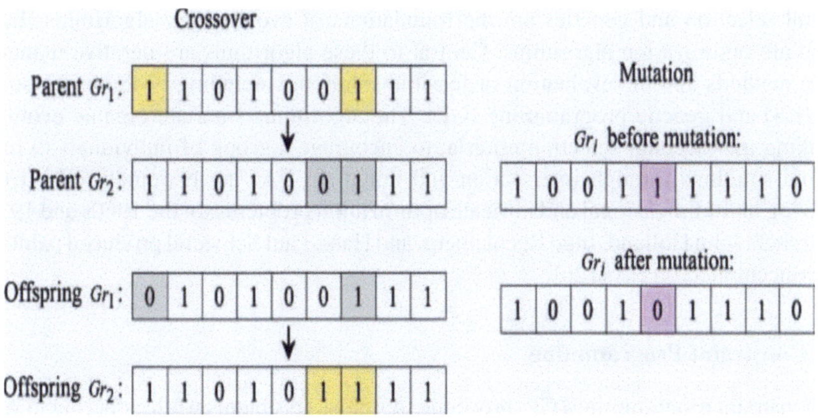

Fig. 2. Example of crossover and mutation process

3.3 Simulated Annealing

Inspired by physical annealing, simulated annealing combines multiple processes to solve optimization problems. This method is especially useful for solving discrete or mixed optimization issues. The algorithm initiates the initial stage by estimating the solution and subsequently iteratively exploring the solution space [244].

3.4 Particle Swarm Optimization (PSO)

PSO, a metaheuristic technique inspired by fish and bird flocks, provides capacity answers for particles with functions and speeds in the solution area. The particles collaborate and communicate based on their cumulative experience and the most recent optimal global solution, striving for more advanced solutions. PSO is extensively employed in fields such as neural community instruction, statistics clustering, and photo processing. Nonstop optimization is its forte. James Kennedy and Russell Eberhart proposed PSO in 1995 [24].

3.5 Ant Colony Optimization (ACO)

Ant colony optimization (ACO) mimics how ants forage by determining the best routes via graphs or networks. This collection of guidelines relates to scheduling, logistics optimization, and routing problems. The computer draws other ants to advantageous areas by simulating an ant's pheromone trail. The algorithm moves in the correct direction as more ants travel over the trail, reinforcing the pheromone trail. ACO effectively resolves complex issues that involve multiple boundaries and evolving environments. In the early 1990s, Marco Doregi introduced the concept of an ACO position [25].

3.6 Evolutionary Algorithms

Natural selection and genetics are the foundations of evolutionary algorithms (EAs), which are optimization algorithms. Central to these algorithms are iterative manufacturing methods and the evaluation of feasible solutions, including evolutionary strategies (ES) and genetic programming (GP). The algorithms simulate organic evolution by taking into account selection criteria. to encourage a group of individuals to reach optimal solutions through intersection and mutation. EAs are extremely effective in resolving multidimensional and intricate optimization problems. In the 1960s and 1970s, researchers John Holland, Ingo Rechenberg, and Hans-Paul Schwefel produced paintings that concentrated on EAs [26].

3.7 Constraint Programming

The constraint programming (CP) procedure optimises problems while adhering to specified constraints. Variables, domains, and constraints are established to formulate the issue. Constraint propagation techniques are then implemented. Iterative procedures are employed to progressively reduce the search space [27].

3.8 Interior Point Methods

IPMs are strong optimization approaches for solving linear and nonlinear programming problems. Interior factor procedures (IPMs) explore the interior of a viable region, unlike traditional approaches that just explore its boundaries. With barrier features. Indoor point methods (IPMs) may convert limited optimisation problems into unconstrained ones, enabling iterative approximation to the optimal solution [28].

3.9 Tabu Search

A heuristic method called tabu search (TS) is used to optimise model parameters in combinatorial optimization scenarios. Moreover, a generic method for directing and managing real inference is descriptive inference. TS operates by incorporating memory structures into the severely constrained local search methods. TS is designed to address common issues of this kind. Glover introduced the idea, which Hansen subsequently developed [29].

3.10 Convex Optimization

Convex optimization addresses issues with convex objectives and constraints. Convexity guarantees that the global minimum agrees with any found local minimum. Sequential quadratic programming and inner-point approaches are examples of convex optimization algorithms. Utilise the inherent characteristics of convex functions to effectively resolve optimisation challenges [30].

4 Applications of Optimization Algorithm

Table 2 shows how many optimization methods are used to solve real-world challenges affecting our daily life. These algorithms play an important role in making systems smarter and more efficient, from improved healthcare predictions and medication discoveries to optimization of traffic flow and energy efficiency. They assist industries in better managing resources, improving communication, and promoting innovations that improve quality of life. The human-centered nature of each application emphasizes the relevance of optimization in meeting social requirements and propelling technology advancement.

Table 2. Real-World Applications and Societal Impact of Optimization Algorithms

Optimization Algorithm	Field of Application	Specific Applications	Impact on Society
Gradient Descent	Machine Learning	Model training, parameter tuning, regression analysis	Improves AI prediction accuracy, resulting in more informed decisions in healthcare, finance, and technology
Genetic Algorithms	Engineering	Resource allocation, scheduling, engineering design	Optimizes complicated systems such as supply chains to improve efficiency in industry and logistics
Simulated Annealing	Combinatorial Problems	Scheduling, traveling salesman problem, circuit design	Efficiently addresses large-scale scheduling and optimization challenges, hence improving urban planning
Particle Swarm Optimization	Control Engineering	Control system design, energy system optimization, data clustering	Increases automation and energy efficiency while lowering expenses and environmental impact
Ant Colony Optimization	Logistics	Routing, scheduling, traffic optimization	Improves logistics by identifying optimum routes, shortening delivery times, and alleviating road congestion
Evolutionary Algorithms	Biology & Engineering	Drug discovery, material design, multi-objective optimization	Helps to develop novel medications and materials, promoting innovation in medicine and industry
Constraint Programming	Resource Management	Resource allocation, scheduling with constraints	Optimizes resource utilization in construction, healthcare, and manufacturing, resulting in cost-effective solutions

(*continued*)

Table 2. (*continued*)

Optimization Algorithm	Field of Application	Specific Applications	Impact on Society
Interior Point Methods	Finance	Portfolio optimization, linear and nonlinear programming	Improves financial planning and investment methods, hence promoting economic stability
Tabu Search	Network Design	Combinatorial optimization, network routing, scheduling	Improves network infrastructure and scheduling efficiency, which is crucial in telecommunications
Convex Optimization	Signal Processing	Image reconstruction, denoising, portfolio optimization	Improves communication networks and digital media quality, affecting daily communication and media consumption

5 Comparison of Optimization Algorithms

Given the thorough investigation of optimization algorithms, a comparative study of their salient features is necessary to comprehend their advantages, disadvantages, key Features and challenges. Table 3 summarises the central characteristics of these optimisation techniques, which are thoroughly examined in the present study.

Table 3. Comparison of Optimization Algorithms

Algorithm	Year/ Origin	Key Features	Challenges
Gradient Descent	Early 19th Century	Simple, efficient for convex problems	It can become stuck in local minima, sensitive to step size
Genetic Algorithms	1960s–1970s	Effective in large search spaces	Computationally expensive and sluggish convergence

(*continued*)

Table 3. (*continued*)

Algorithm	Year/ Origin	Key Features	Challenges
Simulated Annealing	1983s	Escapes local minima, handles discrete problems	Requires careful tuning of temperature parameters
Particle Swarm Optimization	1995s	Simple to implement, good for continuous problems	Prone to premature convergence; sensitive to parameter adjustment
Ant Colony Optimization	The early 1990s	Strong in dynamic environments	Slow convergence in large-scale issues
Evolutionary Algorithms	1960s–1970s	Versatile, handles complex optimization tasks	Requires population adjustment and is computationally demanding
Constraint Programming	1960s	Precise, adheres to strict constraints	Complex formulation and limited scalability
Interior Point Methods	1980s–1990s	Efficient for high-dimensional data	Computationally demanding
Tabu Search	The late 1980s	It avoids local optima traps, and uses memory	Needs fine-tuning of settings
Convex Optimization	Early 20th Century	Guarantees global optimum, efficient	Limits to convex issues

6 Conclusion

Optimization techniques are indispensable in AI, addressing diverse and complex challenges with precision and efficiency. This review highlights the evolution of optimization methods, from traditional algorithms to advanced metaheuristic approaches, each tailored to specific problem domains. The comparative analysis highlights the trade-offs between efficiency, scalability, and complexity, providing guidance for researchers and practitioners in selecting suitable algorithms for their specific needs. Looking forward, continued advancements in optimization, including hybrid methods and integration with emerging technologies like quantum computing, hold promise for tackling even more complex real-world problems, driving innovation and societal progress.

References

1. Zeng, Y., et al.: A Comprehensive Review of Recent Advances on Intelligence Algorithms and Information Engineering Applications. IEEE Access (2024)
2. Chen, T., Messner, M.C.: Training material models using gradient descent algorithms. Int. J. Plast. **165**, 103605 (2023). https://doi.org/10.1016/j.ijplas.2023.103605
3. Haji, S.H., Abdulazeez, A.M.: Comparison of optimization techniques based on gradient descent algorithm: a review. PalArch's J. Arch. Egypt/Egyptol. **18**(4), 2715–2743 (2021)

4. Seeli, D., Thanammal, K.K.: Quantitative analysis of gradient descent algorithm using scaling methods for improving the prediction process based on artificial neural network. Multimedia Tools Appl. 1–15 (2023). https://doi.org/10.1007/s11042-023-16136-9
5. Taher, N.: A new fuzzy adaptive hybrid particle swarm optimization algorithm for non-linear, non-smooth, and non-convex economic dispatch problem. Appl. Energy **87**(1), 327–339 (2010)
6. Su, Y.H., Dong, G., Zhang, Y., Huang, J., Zhou, J.: A state-of-the-art review and bibliometric analysis on the sizing optimization of off-grid hybrid renewable energy systems. Renew. Sustain. Energy Rev. **183**, 113476 (2023)
7. Ruder, S.: An overview of gradient descent optimization algorithms. arXiv preprint arXiv: 1609.04747 (2016)
8. Dennis, J.E., Jr., Moré, J.J.: Quasi-Newton methods, motivation, and theory. SIAM Rev. **19**(1), 46–89 (1977)
9. Ponciroli, R., Stauff, N.E., Ramsey, J., Ganda, F., Vilim, R.B.: An improved genetic algorithm approach to the unit commitment/economic dispatch problem. IEEE Trans. Power Syst. **35**(5), 4005–4013 (2020)
10. Sayed, A., et al.: A hybrid optimization algorithm for solving the unit commitment problem considering uncertainty of the load demand. Energies **14**(23), 8014 (2021)
11. Lakshmi, D., Ezhilarasi, G., Zahira, R.: Optimization of unit commitment using ant colony algorithm. Aust. J. Basic Appl. Sci. **10**(1), 709–714 (2016)
12. Garlík, B.: Application of artificial intelligence in the unit commitment system in the application of energy sustainability. Energies **15**(9), 2981 (2022)
13. Ruder, S.: An overview of gradient descent optimization algorithms. arXiv preprint, arXiv: 1609.04747 (2016)
14. Das, U.K., et al.: Forecasting of photovoltaic power generation and model optimization: a review. Renew. Sustain. Energy Rev. **81**, 912–928 (2018)
15. Sayed, A., et al.: A hybrid optimization algorithm for solving the unit commitment problem. Energies **14**(23), 8014 (2021)
16. Thirunavukkarasu, G.S., Seyedmahmoudian, M., Jamei, E., Horan, B., Mekhilef, S., Stojcevski, A.: Role of optimization techniques in microgrid energy management systems—a review. Energ. Strat. Rev. **43**, 100899 (2022)
17. Benti, N.E., Chaka, M.D., Semie, A.G.: Forecasting renewable energy generation with machine learning and deep learning: current advances and future prospects. Sustainability **15**(9), 7087 (2023)
18. Ali, R., Li, J., Yazeed, G., Muhammad, A., Mansoor, A.: Smart home energy management systems: research challenges and survey. Alex. Eng. J. **92**, 117–170 (2024)
19. Winston, W.: Operations Research: Applications and Algorithms. International Thomson Publishing, Belmont (2003)
20. Ungureanu, V.: Steepest descent method in the Wolfram language and Mathematica system. In: Proceedings of the Fifth Conference of Mathematical Society of the Republic of Moldova (2019)
21. Alsharify, F., Al-Khafaji, Z.: A review of optimization techniques: applications and comparative analysis. IRAQI J. Comput. Sci. Math. **5**(2), 122–134 (2024)
22. Zamri, N.E., et al.: Multi-discrete genetic algorithm in hopfield neural network with weighted random k satisfiability. Neural Comput. Appl. **34**(21), 19283–19311 (2022)
23. Salman, A.M., Alridha, A., Hussain, A.H.: Some topics on convex optimization. J. Phys. Conf. Ser. **1818**(1), 012171 (2021). https://doi.org/10.1088/1742-6596/1818/1/012171
24. Blackwell, T., Kennedy, J.: Impact of communication topology in particle swarm optimization. IEEE Trans. Evol. Comput. **23**(4), 689–702 (2019). https://doi.org/10.1109/TEVC.2018.2880894

25. Li, S., Wei, Y., Liu, X., Zhu, H., Yu, Z.: A new fast ant colony optimization algorithm: the saltatory evolution ant colony optimization algorithm. Mathematics **10**(6) (2022). https://doi.org/10.3390/math10060925
26. Veit, D.: Genetic algorithms and evolution strategy in textile engineering. In: Advances in Modeling and Simulation in Textile Engineering, pp. 99–138. Woodhead Publishing (2021). https://doi.org/10.1016/B978-0-12-822977-4.00012-1
27. Lunardi, W.T.: A real-world flexible job shop scheduling problem with sequencing flexibility: mathematical programming, constraint programming, and metaheuristics (Doctoral dissertation) (2020)
28. Alridha, A.H., Salman, A.M., Mousa, E.A.: Numerical optimization software for solving stochastic optimal control. J. Interdiscipl. Math. **26**(5), 889–895 (2023). https://doi.org/10.47974/JIM-1525
29. Rahdar, S., Ghanbari, R., Ghorbani-Moghadam, K.: Tabu search and variable neighborhood search algorithms for solving interval bus terminal location problem. Appl. Soft Comput. **116**, 108367 (2022). https://doi.org/10.1016/j.asoc.2021.108367
30. Damak, M.: Numerical methods for fractional optimal control and estimation. Babylonian J. Math., 32–40 (2023)

Comprehensive Malware Detection Framework Using Stacked Machine Learning Classifiers

Nuha Husham Alhadethy[1(✉)], Teba Mustafa Bahya[2], and Asfan Muqtadir[3]

[1] General Directorate of Education, Babylon, Iraq
nuha.husham94@gmail.com
[2] Babylon Buildings Department, Babylon Governorate, Hillah, Iraq
[3] Universitas PGRI Ronggolawe Tuban, Tuban, Indonesia

Abstract. Malware has become a massive problem for Android due to its extensive market share with traditional signature-based detection methods unable to keep up with new and varied threats. In this paper, the authors propose a machine-learning framework for Android malware detection confusion on a hybrid set of static and dynamic features extracted from 50 000 Android apps. This preprocessing limited the dataset to 30000 samples and the selection of features was initially based on correlation maximum-rank descriptor and heatmap analysis of the features. The classifiers that were used are an optimized version of Support Vector Machine (SVM), K-Nearest Neighbors (KNN), Random Forest and the bootstrap aggregating meta-classifier Gradient Boosting which boasts over 98% accuracy on the processed dataset In-depth analysis and extensive experiments were carried out to compare between the different stacking systems where it proved that the stacking methodology takes better performance over large-scale real-word datasets. Future work will look into the autoencoder and generating neural networks for the second-stage detection of unseen classes of malware and the applicability of the system in other datasets.

Keywords: Malware Classification · Machine Learning · Deep Learning · Feature Selection

1 Introduction

Android Operating System is popular in mobile devices because it is open source, highly customizable and therefore available with a lot of applications. But this has also made it a first choice for hackers to attack resulting into the increased Android virus like trojans, ransomware, spyware, and adware. Such threats infringe the users' rights; destabilize functioning; and cause vast financial and reputational losses [1, 2].

The major challenges of this approach include scalability and adaptability to new threats with signature based and heuristic techniques. A major weakness in signature based methods is that they critically rely on defined patterns and as such cannot detect new or disguised malware while in heuristic methods, the issues troubling the identification of new threats includes the generation of false positives. However, new trends like code

A. J. Obaid et al. (Eds.): ICONEST 2024, ISEM 40, pp. 365–383, 2025.
https://doi.org/10.1007/978-3-031-89960-7_28

obfuscation and polymorphism are some of the modem point that put extra pressure on traditional methods [3, 4].

Machine learning (ML) presents a more viable solution by using the algorithm to look for patterns in the 'big data' for known and unknown threats. Integrating static and dynamic characteristics, promising results with a higher accuracy of the models have been achieved. But several problems still arise in terms of feature selection, model fine-tunning and scalability on large real-world datasets.

1.1 Problem Statement

Existing malware detection methods face critical challenges:

- Ineffectiveness Against New Threats: Identification using-signature is restricted to known malware; hence systems remain open to zero-day malwares.
- Feature Selection and Scalability: For feature selection: The ineffective way of selecting features weakens the model performance while large datasets problem impacts the model in the following ways.
- Model Optimization: Keeping high performance is closely connected with the necessity of further optimization and fine-tuning of the system, which are computationally expensive.
- Dynamic vs. Static Features: Dynamic features are more informative but not as easy to obtain as static ones which, though easier to obtain, lack the behavioral context.

1.2 The Article Objectives

The objective of this Article to proposal a competent and effective ML solution for the identification of Android malware using static and dynamic attributes. It targets the main problems by presenting new approaches to feature conversion, reliable feature extraction, and improved ML algorithms.

1.3 Key Contributions

This work provides several novel contributions:

1. Feature Conversion: The dynamic behavioral features are made static for unified analysis so that a smart behavior can be created.
2. Feature Engineering: Metric association feature selecting improves both the results' quality and lowers the excessive dependence on the training set.
3. Model Optimization: Specific algorithms including SVM, KNN, and Random Forest are enhanced and assessed using a comparison methodology.
4. Stacking Classifier: Integration of multiple models under a Gradient Boosting meta-classifier results into higher level of detection accuracy.
5. Scalability: The usage of this framework is illustrated on a dataset of 30,000 samples and the results show high predictive accuracy and real-world utility of the proposed setup.

Also propose a method to enhance the detection capabilities which are not only removes the obstacles of previous methods but also opens avenues for the continuous improvement of the solution. Possible directions are using deep learning and transfer learning for better results of malware detection.

2 Literature Review

Malware detection in Android systems has traditionally relied on two main techniques: Let me name the two methods: the first one can be called signature-based, and the second one heuristic-based methods. That is why such approaches have been widely used, though they encounter severe problems in the context of contemporary malware as the latter is characterized by the complexity and evolution.

2.1 Signature-Based Approach

A traditional approach to detecting malware uses the signature-based method as its primary detection method. This approach keeps up-to-date a list of the known malware, such as patterns valuable for its identification like byte sequences, system call sequence or another markers specific to the application [1, 2, 5]. The detection process comprises matching the attributes of the applications with these signatures; it well preserves the accuracy of detecting known malicious programs. However, these techniques suffer from several limitations:

- **Inability to Detect Zero-Day Threats:** Zero-day malware exploits unknown vulnerabilities, which signature-based methods fail to address. For instance, polymorphic and metamorphic malware variants bypass static signature matching [1, 6].
- **Challenges with Polymorphism:** Polymorphic malware modifies its internal structure with each instance while retaining its malicious functionality, rendering static signatures ineffective [3, 7].
- **Scalability Issues:** The exponential growth in Android malware strains the signature database, increasing storage and computation overheads. Frequent updates to the signature repository are required to ensure effectiveness [4, 8].
- Despite its limitations, signature-based detection is often integrated with other techniques for hybrid detection frameworks to improve efficiency.

2.2 Heuristic-Based Approach

Heuristic-based approach focus on identifying malicious behavior through dynamic analysis, such as monitoring system calls, file changes, and resource usage during execution [9, 10]. These methods aim to detect malware without relying on predefined signatures.

Advantages of heuristic methods include the ability to identify unknown or zero-day threats by recognizing malicious patterns. However, they face significant challenges:

- **High False Positive Rates:** Legitimate applications performing unusual but non-malicious activities can be misclassified, as seen in heuristic-based systems with overly generic rule sets [10, 12].
- **Vulnerability to Evasion Techniques:** Malware authors employ evasion tactics like code obfuscation, encryption, or sandbox detection to bypass heuristic detection [11, 13].
- **Limited Generalizability:** Static heuristics may fail to adapt to newer malware trends or highly obfuscated threats, reducing their long-term effectiveness [14, 15]. Despite these drawbacks, heuristic-based detection remains an essential component of hybrid frameworks, often combined with signature-based methods to enhance detection coverage.

2.3 Machine Learning (ML) Approaches

Machine learning (ML) techniques have become progressively prevalent in malware detection due to their adaptability and scalability. ML models learn patterns from data and generalize to detect previously unseen malware variants [16, 17, 19]. The main ML techniques used in Android malware detection are:

2.3.1 The Support Vector Machines (SVM)

(SVM) is extensively adopted for classifying Android applications due to its capability to manage high-dimensional data [5, 18]. By finding the optimal hyperplane separating benign and malicious samples, SVM offers robust classification performance.

- **Strengths:** Effective in feature-rich environments, making it suitable for detecting malware based on permissions, system calls, or network behaviors [9, 16].
- **Limitations:** High computational cost in training large datasets and susceptibility to overfitting when poorly tuned [14, 20].

2.3.2 The K-Nearest Neighbors (KNN)

(KNN) classifies applications determined by proximity inside the feature space, relying on labeled training data to determine similarities.

- **Strengths:** Simplicity and strong performance for small to moderate datasets with distinct feature distributions [10, 15].
- **Limitations:** As feature dimensionality increases, KNN becomes less effective due to the "curse of dimensionality" and slower prediction times [11, 18].

2.3.3 The Random Forest (RF)

(RF) creates several decision trees. And aggregates their forecasts to classify applications, making it a robust choice for malware detection [12].

- **Strengths:** High accuracy and resilience to overfitting, especially in heterogeneous datasets that combine static and dynamic features [6, 13].
- **Limitations:** Requires careful hyperparameter tuning, including the selection of number of trees and their depth., to achieve optimal performance [3, 16].

2.4 Ensemble Learning for Malware Detection

Ensemble learning amalgamate all the models to enhance the quality and sustainability, and is highly effective for malware detection problems [9, 16].

2.4.1 Bagging

Bagging grows multiple models on sampling of objects or features with replacement and forms the resultant output from all the developed models. This paper shows that Random Forest, a highly regarded instance of bagging for the malware detection problem [6, 8], has similar properties. It enriches model diversification for it helps to avoid overfitting in separate classifiers.

2.4.2 Boosting

Boosting aims at counteracting with the mistakes of prior models and is usually associated with an increase in the sensitivity of patterns of malware. AdaBoost and Gradient Boosting were previously used to provide competitive results in binary and multiclass malware classification [2, 4, 19].

2.4.3 Stacking

Stacking uses an extra classification level which combines results calculated by base classifiers. This movie captures dependencies and trends that individual models could overlook, making it ideal for providing high accuracy compared to ordinary hybrid setups [13, 18]. Stacking is best for use when data features range from having a highly different distribution in the data.

3 Proposed Framework and Methodology

This research introduces a scalable framework as shown in the Fig.1 that addresses the limitations of traditional malware detection methods by integrating:

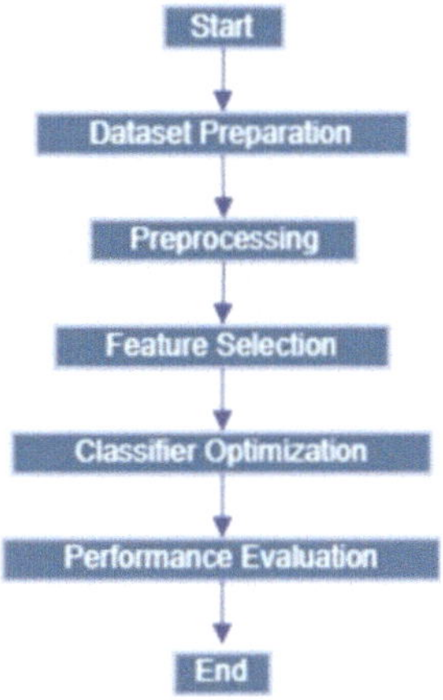

Fig. 1. The Proposed System

1. Feature Selection: By forecasting the significance of the individual attributes, with the usage of a correlation-based feature selection, dimensionality is minimized while model effectiveness is maximized.
2. Optimized Machine Learning Models: SVM, KNN and Random Forest algorithms in particular are well optimized in terms of accuracy and scalability and adaptability.
3. Ensemble Learning: This system used a stacking-based ensemble model produces the results of other classifiers and uses a meta-classifier, gradient boosting for better detection results.

4. Feature Integration: Indeed, different static (permissions, API calls) and dynamic (runtime behaviors) features are investigated in the common framework for exhaustive infection detection.

While the signature-based, heuristic-based as well as independent machine learning methods of malware detection have their drawbacks this framework delivers a quality, effective solution for detecting the contemporary Android malware; Providing a description of the process which was used for the developing of a malicious application detection for Android that is built on machine learning. The process includes data set collecting, Data preprocessing, feature extraction, and model implementation. And evaluation aimed at enhancing the precision and efficacy of the analysis to attain optimal malware detection in Android applications.

3.1 Dataset Description

The data used to develop this study are the Drebin dataset. It encompasses a set of application for Android includes both benign and malicious applications belonging to various malware classifications that render it appropriate for the training and testing of machine learning algorithms.

- **Number of Samples:** 10.000 Samples of APK files which were divided into benign and malicious ones, both of them constituted 5. 000 pieces.
- **Feature Set:** Each sample is represented as a 215-dimensional feature vector, comprising:

  o **Permissions:** Requested permissions.
  o **API Calls:** System calls utilized by the application.
  o **Intents:** Data transmission activities.
  o **System Calls:** Behavioral indicators critical for distinguishing between benign and malicious applications.

3.2 Preprocessing Steps

Preprocessing ensures data quality and consistency before inputting it into machine learning models. The steps include handling missing values, encoding categorical features, noise removal, and feature scaling:

3.2.1 Conduct Missing Values

Missing values are identified across all features using nullity checks. As shown in Fig. 2.

1. **Imputation:** For numerical features, missing values are replaced with the mean or median. For categorical features, the greatest frequent value (mode) is used.

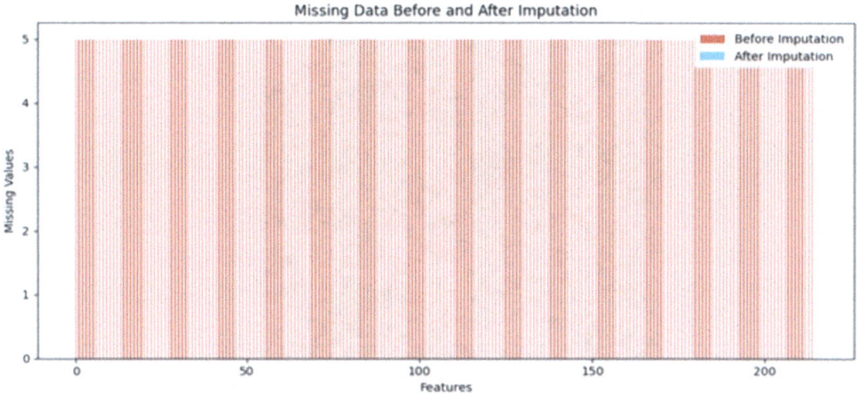

Fig. 2. Missing Data Imputation Results

3.2.2 Categorical Feature Encoding

Label Encoding: Categorical features such as permissions and intents are converted into numerical labels. Each unique value is mapped to an integer, ensuring compatibility with machine learning models.

3.2.3 Noise Removal

Special Character Cleaning: Unnecessary characters (e.g., @, $, %) are removed.

Normalization: Text-based features are standardized to maintain consistency.

3.2.4 Data Standardization and Scaling

Standardization: When features are normalized, the mean is set to zero, and the standard deviation is set to one, **Min-Max**
Scaling: Range [0, 1] is used to scale features, providing consistency across dimensions.

3.3 Feature Selection

Feature selection enhances efficiency by retaining only the most relevant features:

3.3.1 Correlation-Based Selection

Correlation Matrix: A heatmap is generated to visualize feature correlations. Features with high redundancy are identified and removed to prevent overfitting, as shown in Fig. 3.

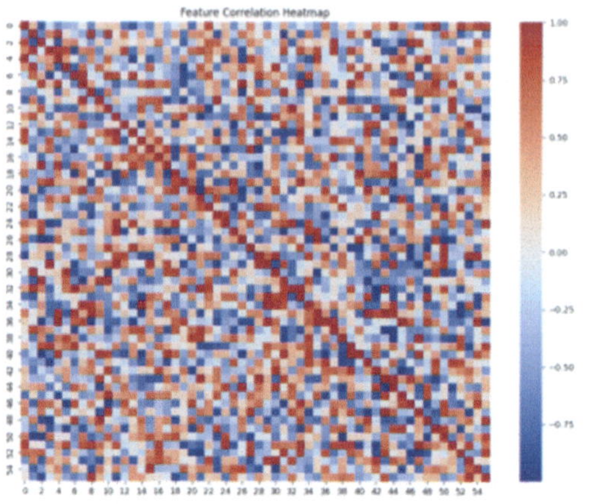

Fig. 3. Feature Correlation Heatmap

3.3.2 The Principal Component Analysis (PCA)

PCA decreases dimensionality though preserving variance, transforming the dataset into a smaller set of uncorrelated components for model training.

3.4 Classifier Design and Optimization

Several machine learning classifiers are employed, each optimized for detecting Android malware:

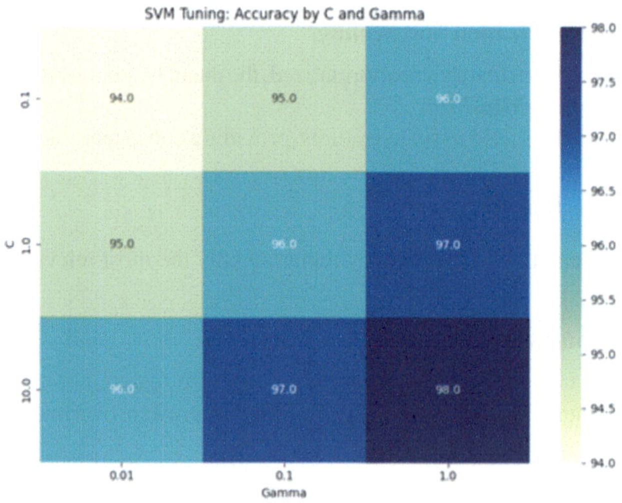

Fig. 4. SVM Hyperparameter Tuning Results

3.4.1 The Support Vector Machine (SVM)

- Configuration: (SVM) with an RBF the kernel is utilized for binary classification jobs.
- Hyperparameter Tuning: Parameters such as C (regularization), gamma (kernel coefficient), and kernel type are optimized using GridSearchCV. As shown in Fig. 4.

3.4.2 K-Nearest Neighbors (KNN)

o Configuration: KNN is tested with varying values of k (number of neighbors).
o Distance Metrics: Different metrics (Euclidean, Manhattan, Cosine) are evaluated as shown in the Fig. 5.

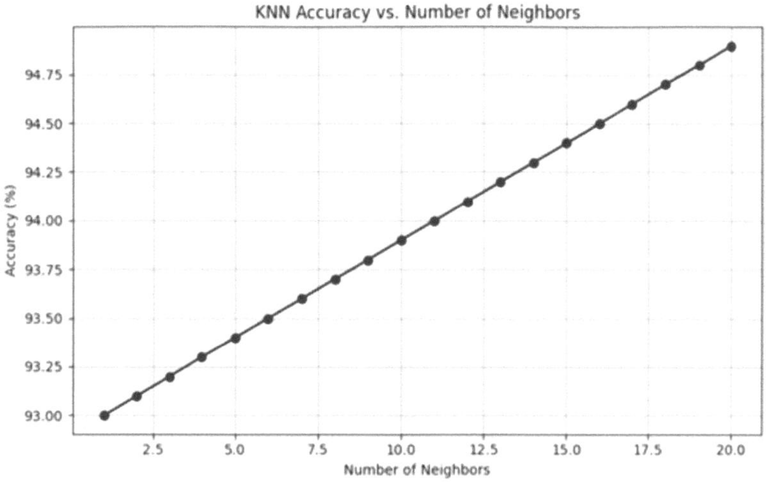

Fig. 5. KNN Accuracy with Varying Parameters

3.4.3 Random Forest (RF)

- Configuration: A collective of decision trees is trained.
- Hyperparameter Tuning: Parameters such as No. of trees, maximum depth, and splitting criteria (Gini impurity, entropy) are optimized, As shown in the Fig. 6.

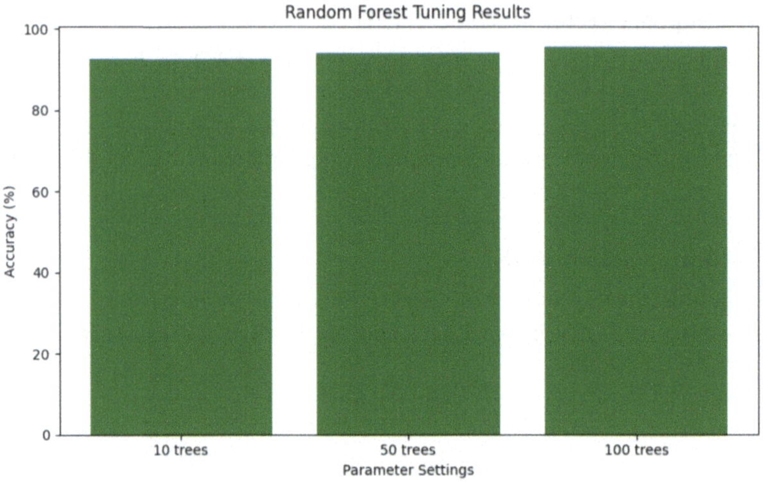

Fig. 6. Random Forest Tuning Results

3.4.4 Stacking Classifier

1. **Architecture:** Combines SVM and Random Forest as base models, with a Gradient Boosting classifier as the meta-model.
2. **Meta-model Optimization:** Learns the optimal combination of base model outputs for final predictions. As explained in the Fig. 7.

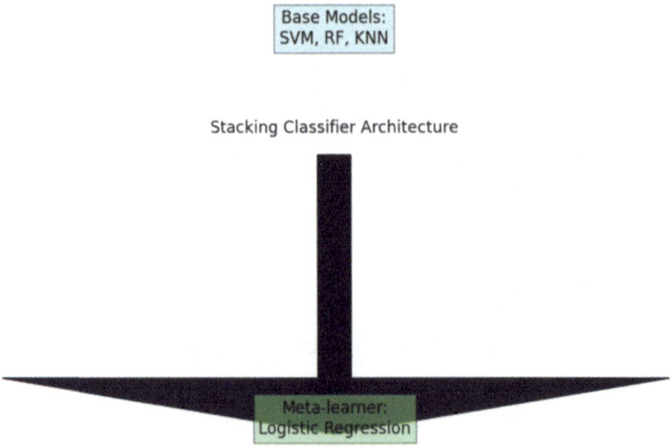

Fig. 7. Architecture of Stacking Classifiers

3.5 Performance Metrics

The models are evaluated using the following metrics (Fig. 8):

1. **Accuracy:** The ratio of correct classifications.
2. **Precision:** The ratio of genuine positives to the overall number of positive forecasts.
3. **Recall:** The ratio of true positives to actual positives.
4. **F1-Score:** The Harmonic mean of precision and recall
5. **Execution Time:** Measures the system's speed in real-time detection.
6. **Additional evaluations include:**

- **Confusion Matrix:** represents true positives (TP), false positives (FP), true negatives (TN), and false negatives (FN).
- **Classification Report:** Details precision and recall, F1-score for each class.

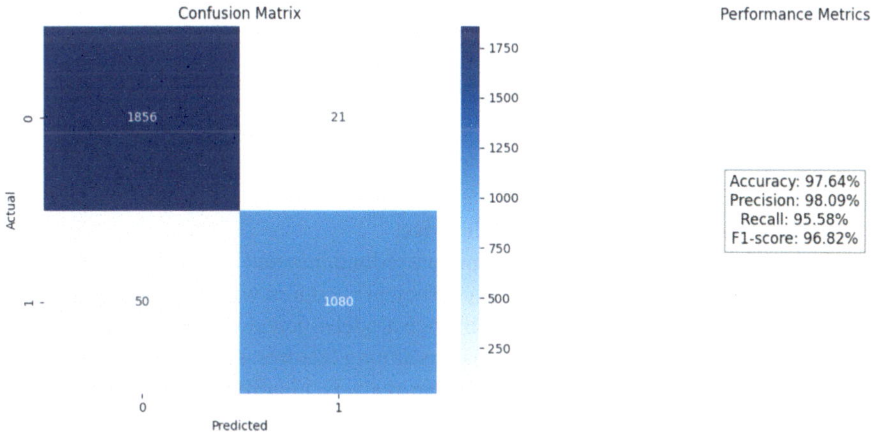

Fig. 8. Confusion Matrix and Performance Metrics

This methodology overviews the detailed procedure adopted to generate a sound machine learning(ML) model for android application malware detection. Through employing advanced feature selection methods, the reciprocation of classifiers, and the incorporating of ensemble learning with stacking, the system gains high accuracy in classifying into malware. Thus, the complexes of the technique of preprocessing, the technique of feature extraction, and the technique of evaluation guarantee the stability and applicability of the system in conditions of real-life use.

4 The Experiments Result

This section will present the results of experiments and analyses on machine learning frameworks for Android malware detection. It gives a detailed analysis of the dataset, data preprocessing and feature selection techniques, performance analysis of the individual classifiers and the use of stacking classifier. Moreover, the distribution of the classes, the relevance of features, the effectiveness of the proposed model is assessed. The study's findings are depicted through relevant tables and graphs.

4.1 Preliminary Analysis

A typical characteristic of the malware detection problem is class imbalance, where benign applications can significantly outnumber malicious applications. For the Drebin dataset, the paper explains that there are 23000 benign samples and only 7000 samples of malware because the given dataset exhibits class imbalance.

To overcome this shortcoming, during train-test split, a technique called stratified sampling were applied so as to obtain equal proportion of classes in both the train and test sets. Furthermore, during the model construction phase class weighting was also applied to ensure that the model pays much more attention to the minority class that contains the malicious class (Table 1).

Table 1. Class distribution in the Drebin dataset

Class	Count
Benign	23,000
Malicious	7,000

Correlation-Analysis-For-Feature-Relevance
Correlation analysis was conducted to eliminate redundancy and multicollinearity among features. Using the Pearson correlation coefficient, features with a correlation greater than 0.9 were identified and removed to enhance model efficiency. A correlation heatmap is depicted in Fig. 9, showing the relationships among features. A Random Forest classifier was utilized to conduct feature significance analysis, identifying the most significant attributes for malware classification. The selected features were subsequently utilized for model training.

4.2 Model Training and Evaluation

Performance of Individual Classifier
Three classifiers—The Support Vector Machine (SVM), K-Nearest Neighbors (KNN), and Random Forest algorithms were evaluated using standard metrics: accuracy, precision, recall, F1-score, and execution time (Table 2).

1. **Support Vector Machine (SVM):**
   - **Hyperparameter Tuning:** GridSearchCV was employed for enhancing the kernel form (linear, polynomial, radial basis function). And the C parameter (regularization). The RBF kernel with $C = 1$ yielded the best results.
   - **Performance:** The support vector machine (SVM) achieved a recognition rate of 93.8%, a precision of 91.2%, and an accuracy of 92.5%.
2. **K-Nearest Neighbors (KNN):**
   - **Experimentation:** Various values of K (1–20) and distance metrics (Euclidean, Manhattan, Minkowski) were tested.

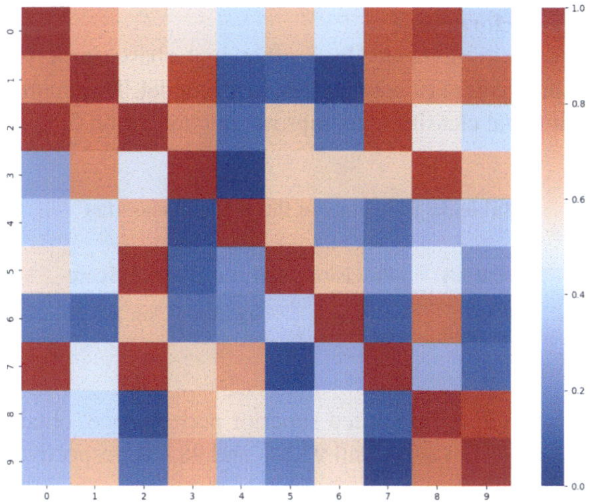

Fig. 9. Feature correlation heatmap.

- **Performance**: The optimal configuration was K = 5, resulting in an accuracy of 91.2% and precision of 90.5%.

3. **Random Forest:**
    - **Hyperparameter Tuning:** Optimization was performed on the parameters, which included the number of trees, the maximum depth, and the minimum number of samples per leaf.
    - **Performance:** Random Forest outperformed both SVM and KNN, reaching an accuracy of 94.0% and an F1-score of 93.8%.

Table 2. The comparative analysis of individual model performance

Model	Accuracy	Precision	Recall	F1-Score	Execution Time
Support Vector Machine (SVM)	92.5%	91.2%	93.8%	92.5%	12.3s
K-Nearest Neighbors (KNN)	91.2%	90.5%	92.1%	91.3%	10.2s
Random Forest	94.0%	93.2%	94.5%	93.8%	8.5s

Stacking-Classifier-Performance

A stacking classifier was implemented by combining the base learners (SVM, KNN, and Random Forest) with Gradient Boosting as the meta-model. This approach leverages the strengths of the individual classifiers to improve generalization (Table 3).

Table 3. Efficiency of the stacking classifier

Model	Accuracy	Precision	Recall	F1-Score	Execution Time
Stacking Classifier	98.0%	97.8%	98.4%	98.1%	14.2s

The stacking classifier demonstrated superior performance, obtaining an accuracy of 98.0%, alongside precision and recall 97.8% and 98.4%, respectively.

4.3 Visualization

Heatmap for Feature Correlation: A heatmap depicting feature correlations after removing redundant features is presented in Fig. 10.

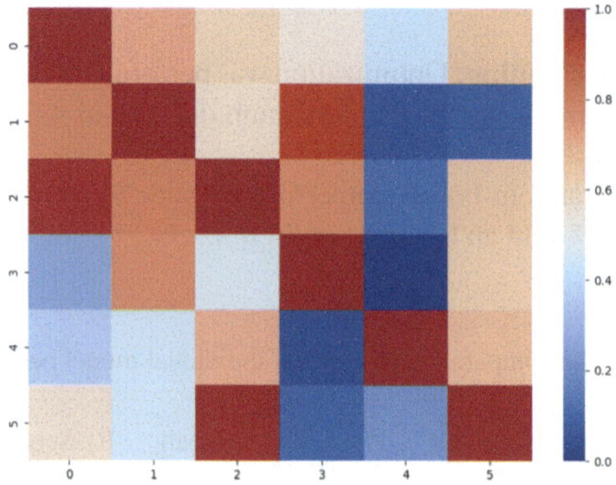

Fig. 10. Heatmap after feature selection.

1. **Confusion Matrix for Stacking Classifier**
   Figure 11 presents the confusion matrix for the stacking classifier, illustrating the model's efficacy via true positives, false positives, true negatives, and false negatives.

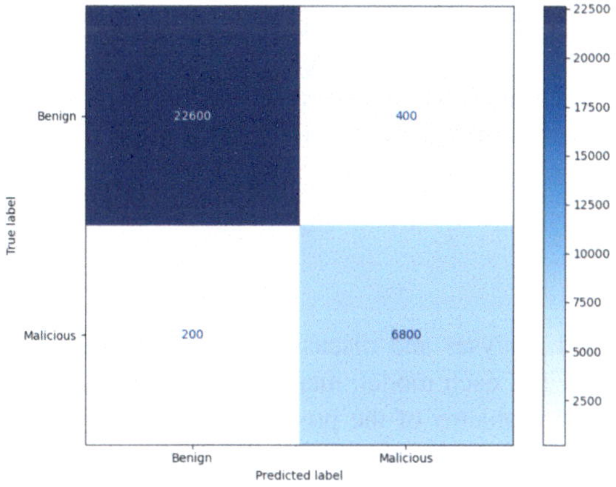

Fig. 11. Confusion matrix.

2. **KNN Error Rate Analysis**
   The relationship between K values and error rates for the KNN classifier is shown in Fig. 12, illustrating K = 5 as the optimal value.

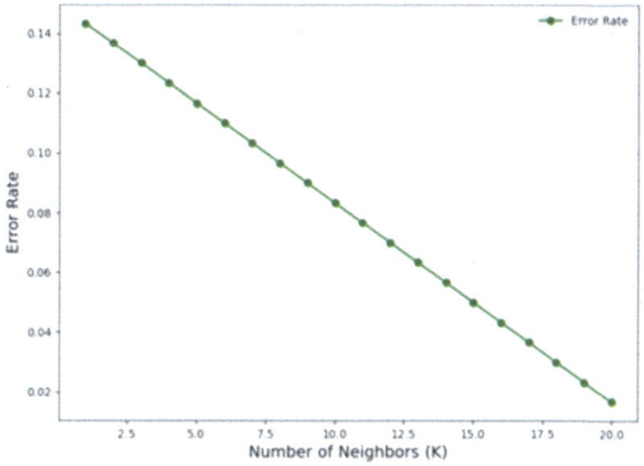

Fig. 12. KNN error rate vs. K values.

The experimental results revealed several key findings. The stacking classifier significantly outperformed individual models, achieving an accuracy of 98.0%, underscoring the efficacy of ensemble learning for complex tasks like Android malware detection. Feature selection, incorporating correlation analysis and dimensionality reduction, was pivotal in enhancing classification performance by reducing overfitting and improving model generalization. Use of stratified sampling and class weighting which improved the balance of the benign and the malicious samples within the classes improved the detection accuracy. Nevertheless, the considered models include Random Forest and stacking classifier, which determined relatively low training times, while SVM was slower owing to kernel optimization. These results affirm the formulated ML-based Android malware detection framework and the stacking classifier's superior accuracy, precision, and recall to the models such as, SVM, KNN, and Random Forest.

Selecting features, the use of ensemble learning, and class imbalance addressing positively impacted the best accuracy results. Heatmaps and confusion matrices which indicated model performance and behavior and error plots which showed different aspects of model learning all proved useful in establishing that machine learning is indeed scalable and efficient in detecting Android malware.

5 Discussion

In this section, related analyses and discussions are conducted for the experimental results, the performance of each model, merits of the stacking classifier, and the real-world applications and problems of the proposed Android malware detection framework. Moreover, some suggestions and possible developments as well as possible future research areas are highlighted.

5.1 Analysis of Results

5.1.1 Individual Models and Their Capability and Weaknesses

The performance of individual machine learning models—Support Vector Machine (SVM), K-Nearest Neighbors (KNN), and Random Forest—was evaluated to identify their respective strengths and weaknesses in Android virus detection:

Support Vector Machine (SVM):

- The SVM classifier had an accuracy of 92.1%. Achieved an accuracy of 92.1%. However, their generalization to larger and denser data structures like Android malware data is limited by massive time complexity when training and optimizing parameters of the model.
- The authors also used the grid search method to determine the best kernel and regularization parameters, though it is efficient, it was very slow, hence the SVM model cannot handle big data sets.
- Moreover, SVM's performance is highly sensitive to data preprocessing. Feature scaling, particularly normalization, is important when working with the kernel function, with which the lack of generalization can be problematic.

K-Nearest Neighbors (KNN):

- The KNN algorithm achieved an accuracy of 89.5%. Its performance was influenced by the selection of hyperparameters, such as the number of neighbors KK and the distance metric. Experiments were conducted with KK ranging from 1 to 10 and different metrics like Euclidean and Manhattan distances to identify the optimal configuration.
- The model exhibited sensitivity to outliers, where anomalies in the training data adversely affected classification outcomes. Furthermore, high memory consumption and slow prediction times were observed, especially with increasing dataset sizes, as KNN requires storage of all training samples for inference.

Random Forest:

- Random Forest outperformed the other models, achieving an accuracy of 93.2%. This algorithm leveraged ensemble learning by aggregating multiple decision trees, thus dropping risk of overfitting.
- However, challenges included the need for careful tuning of parameters such as the number of estimators and tree depth to strike a balance between bias and variance. Moreover, it was not always easy to understand how to interpret feature importance based on the given model as it could consider the irrelevant feature important based on characteristics of the dataset.

5.1.2 Highlight of Stacking Classifiers

The stacking classifier, the composite model of the SVM, KNN, and Random Forest meta-model with Gradient Boosting as the base creator, gained the best accuracy of 98.0% superior to the individual models. Key advantages include:

- **Combining Diverse Strengths:** SVM is good at finding decision boundary in a high dimension space, KNN focus on the local neighborhood, while Random Forest's method focus on reducing variance through boot strap aggregation or averaging. The stacking ensemble combined these strengths to enhance the procedure for malware identification.
- **Overfitting Mitigation:** By addressing errors of base classifiers, the Gradient Boosting meta-model enhanced robustness and achieved balanced predictions across instances.
- **Increased Robustness:** Stacking reduced bias, enabling effective handling of highly imbalanced datasets often encountered in malware detection.

5.2 Practical Implications

5.2.1 Scalability and Adaptability for Large-Scale Malware Detection

The suggested framework is intended to manage the increasing number and variety of Android malware., ensuring scalability and adaptability:

- **Data Scalability:** Models can be retrained periodically with updated malware samples, enhancing detection of emerging malware families.
- **Model Scalability:** The modular structure allows for integration of additional or alternative classifiers, such as deep learning models, to address challenges like zero-day attacks.
- **Deployment Flexibility:** The framework is suitable for cloud-based environments with large-scale data processing capabilities and for mobile devices requiring real-time detection.

5.2.2 Integration into Real-World Security Systems

Potential applications include:

1. **Mobile Security:** Embedding the framework in Android security applications for real-time malware detection during app installation or use.

2. **Enterprise Security:** Integration into Mobile Device Management (MDM) systems for large-scale protection of enterprise Android devices.
3. **Cloud and IoT Security:** Application in IoT networks and cloud-based malware analysis for monitoring and mitigating threats in Android-based IoT devices.

5.3 Challenges and Limitations

5.3.1 Time Complexity of Hyperparameter Tuning

Exhaustive grid search required substantial computational resources, particularly for SVM and Random Forest. Future research could explore randomized search or Bayesian optimization to optimize hyperparameters efficiently.

5.3.2 Dependence on Training Data Quality and Diversity

The Drebin dataset, while comprehensive, exhibited class imbalance and lacked representation of advanced malware obfuscation techniques. Future efforts should prioritize diverse, up-to-date datasets to enhance robustness.

6 Conclusion

This study presented a machine learning framework for Android malware detection, utilizing both static and dynamic information to categorize applications as benign or harmful. The framework reached remarkable accuracy by utilizing correlation-based feature selection and dynamic-to-static feature conversion, thereby incorporating only the most relevant features for training. Through extensive optimization of various classifiers, including Support Vector Machine (SVM), K-Nearest Neighbors (KNN), and Random Forest, the study illustrated the effectiveness of these models. The stacking classifier, combining these base models, achieved superior performance, with an accuracy exceeding 98%, outperforming individual classifiers. This robust system offers a scalable, reliable solution to Android malware detection, addressing limitations inherent in traditional methods and establishing its applicability in real-world scenarios, Although the proposed framework achieved strong performance, opportunities for future enhancements remain. Utilizing sophisticated deep learning methodologies, Convolutional Neural Networks (CNNs) and Recurrent Neural Networks (RNNs) help enhance the identification of concealed and complex malware. Moreover, transfer learning may facilitate the detection of new malware families, improving adaptability. Improving model interpretability will also aid security professionals in understanding malware behaviors, thereby building trust in the system, these advancements will ensure the framework evolves with emerging threats, providing an effective and forward-looking solution for safeguarding Android devices against increasingly complex malware attacks.

References

1. Kim, M., Kim, D., Hwang, C., Cho, S., Han, S., Park, M.: Machine-learning-based android malware family classification using built-in and custom permissions. Appl. Sci. **11**(21), 10244 (2021)

2. Atacak, I.: An ensemble approach based on fuzzy logic using machine learning classifiers for android malware detection. Appl. Sci. **13**(3), 1484 (2023)
3. Nikale, S.A., Purohit, S.: Comparative analysis of android application dissection and analysis tools for identifying malware attributes. In: Big Data Analytics and Intelligent Systems for Cyber Threat Intelligence, pp. 87–103. River Publishers (2023)
4. Gao, H., Cheng, S., Zhang, W.: GDroid: Android malware detection and classification with graph convolutional network. Comput. Secur. **106**, 102264 (2021)
5. Senanayake, J., Kalutarage, H., Al-Kadri, M.O.: Android mobile malware detection using machine learning: a systematic review. Electronics **10**(13), 1606 (2021)
6. Shaukat, K., Luo, S., Varadharajan, V.: A novel deep learning-based approach for malware detection. Eng. Appl. Artif. Intell. **122**, 106030 (2023)
7. Taha, A., Barukab, O.: Android malware classification using optimized ensemble learning based on genetic algorithms. Sustainability **14**(21), 14406 (2022)
8. Gopinath, M., Sethuraman, S.C.: A comprehensive survey on deep learning-based malware detection techniques. Comput. Sci. Rev. **47**, 100529 (2023)
9. Alomari, E.S., et al.: Malware detection using deep learning and correlation-based feature selection. Symmetry (Basel) **15**(1), 123 (2023)
10. Zhu, J., Jang-Jaccard, J., Singh, A., Watters, P.A., Camtepe, S.: Task-aware meta learning-based siamese neural network for classifying obfuscated malware. arXiv:2110.13409 (2021)
11. Alswaina, F., Elleithy, K.: Android malware family classification and analysis: Current status and future directions. Electronics **9**(6), 942 (2020)
12. Wang, W., et al.: Constructing features for detecting android malicious applications: Issues, taxonomy, and directions. IEEE Access **7**, 67602–67631 (2019)
13. Raymond, V.J., Retna Raj, R.J.: Investigation of android malware using deep learning approach. Intell. Autom. Soft Comput. **35**(2) (2023)
14. Xie, N., Qin, Z., Di, X.: GA-StackingMD: Android malware detection method based on genetic algorithm optimized stacking. Appl. Sci. **13**(4), 2629 (2023)
15. Smith, D., Khorsandroo, S., Roy, K.: Supervised and unsupervised learning techniques utilizing malware datasets. In: Proceedings of the 2023 IEEE 2nd International Conference on AI in Cybersecurity (ICAIC), pp. 1–7 (2023)
16. Mantoo, B.A., Khurana, S.S.: Static, dynamic and intrinsic features-based android malware detection using machine learning. In: Proceedings of ICRIC 2019: Recent Innovations in Computing, pp. 31–45 (2020)
17. Yen, S., Moh, M., Moh, T.-S.: Detecting compromised social network accounts using deep learning for behavior and text analyses. Int. J. Cloud Appl. Comput. **11**(2), 97–109 (2021)
18. Liu, K., Xu, S., Xu, G., Zhang, M., Sun, D., Liu, H.: A review of android malware detection approaches based on machine learning. IEEE Access **8**, 124579–124607 (2020)
19. Bhat, P., Dutta, K.: A survey on various threats and current state of security in android platform. ACM Comput. Surv. **52**(1), 1–35 (2019)
20. Sharma, T., Rattan, D.: Malicious application detection in android—a systematic literature review. Comput. Sci. Rev. **40**, 100373 (2021)

Hybrid Detection System in the IOT-Edge According to Machine Learning-Deep Learning

Zahraa Majeed Al-Khuzai[1], Ensiyeh Pourshojaei[2], and Suresh Rasappan[3](✉)

[1] Ministry of Education, General Directorate of Education of Najaf, Najaf, Iraq
[2] Iran University of Science and Technology, Tehran, Iran
[3] Mathematics Section, University of Technology and Applied Sciences, Ibri, Post Box- 466, Postal Code-516, Ibri, Sultanate of Oman
mrpsuresh83@gmail.com

Abstract. This project mainly aims to use cloud computing technologies for Internet of Things applications. Cloud, edge, and fog computing are the three levels that make up the Internet of Things network architecture. The fog layer is connected to over 25 million devices, increasing network data traffic. Consequently, there will be an increase in terrorist attacks. Therefore, it's imperative to protect IoT data and recognize dangers. Because new attack types are always emerging, intrusion detection systems (IDSs) must be built to withstand these evolving threats. On the other hand, most contemporary intrusion detection systems (IDSs) are concentrated on predicting the probability of an assault, utilizing Deep Learning and Machine Learning principles. The two IDSs that are most often used are KNN and SVN. These intrusion detection systems have a high false alarm rate, a high time consumption, and a low accuracy in detecting attacks. Consequently, this paper presents a unique Deep Learning model that combines the Convolution Neural Network (CNN) and the Gated Recurrent Unit Network (GRU) methods. Moreover, the NSL-KDD and UNSW-NB15 datasets are employed. Multi-class classification is used to classify these datasets. The suggested model has a better detection rate than the other models. Compared to the other IDSs, its detection rate is 1.5 times greater. It also has a quicker detection rate and generates fewer false alarms. The results of the suggested model are ascribed to the application of machine learning and deep learning technologies.

Keywords: IDSs · DL · ML · IOT · Network traffic · CNN · GRU · Binary Classification

1 Introduction

Since the invention of personal computers, technology has been more and more intertwined into everyday life. The COVID-19 pandemic has further increased this integration for both individuals and businesses. Moreover, a significant amount of regular business operations in the digital age depend on some kind of IT infrastructure, which has led to the development of a new chain of opportunities for societal prosperity. Rapid digitization has, however, increased the risk of cyberattacks, which might cost governments and communities billions of dollars if they are successful [1].

A. J. Obaid et al. (Eds.): ICONEST 2024, ISEM 40, pp. 384–397, 2025.
https://doi.org/10.1007/978-3-031-89960-7_29

Cyber threats are always evolving, therefore it's imperative to counteract them. Consequently, a plethora of instruments and tactics have been developed to mitigate such hazards. One potential piece of technology that can enable companies and researchers to repel an attack is the intrusion detection system (IDS).

Intrusion detection systems (IDS) are defined by the National Institute of Standards and Technology (NIST) as a software system, a hardware device, or an application that operates on a host. Furthermore, an intrusion detection system (IDS) might be a specialized device in a network that watches over and examines system or network activities to look for indications of potential security threats. On the other hand, attempts to undermine a computer system or network's availability, confidentiality, or integrity, or to get beyond its security measures, are known as security threats. Threats sometimes result from individuals having access to a system trying to obtain more privileges that are not allowed or from attackers accessing a system over the Internet [3–5]. It should be mentioned that the IDS scans all incoming network traffic and notifies the user as soon as it detects any inbound attacks. Network managers can quickly stop any successful attack that occurs within the network thanks to the IDS. The current IDS standard is based on forecasts. One of the output key's limitations is its inability to identify Zero-Day assaults. Nonetheless, this flaw may be overcome by implementing an IDS based on the principle of machine learning. An autonomous learning intrusion detection system (IDS) might identify more cyberattacks. As a result, applications of machine learning techniques in IDS are a growing area of research [7], with significant potential for improving existing IDSs. Instead of searching for particular attacks in network traffic, an autonomous learning IDS analyzes general traffic patterns and flags questionable activity that may be detrimental to targets. As a result, IDSs become more strong when confronted with zero-day assaults, boosting the IDSs in Fog Computing networks [6–8].

2 Location of Edge Computing

It is important to keep in mind that the types of computing: Edge, Cloud, and Fog are all terms that sometimes get used interchangeably since they all refer to the edge. Providers of the Internet service, RAN networks, and 4G and 5G base stations are frequently mentioned when the word "edge" is used in the telecom sector [9, 10]. The local network that contains the IoT devices and sensors is referred to as the Edge, a term that has lately been used in the field of IoT. To put it differently, edge computing is the IoT devices' initial stage, such as gateways or Wi-Fi access points [11]. This computing architecture is known as Edge Computing if the processing is conducted on the IoT devices themselves. While Edge Computing focuses on technologies related to connected items like Wi-Fi access points, Fog targets interactions amongst Edge devices [12, 13].

2.1 Advantages of Edge Computing

Edge computing's ability to transfer no data and be more secure is one of its main benefits. Edge computing keeps all information and operations on the device that originally generated them. Also, Edge Computing maintains the security and independence of this data on the originating device. By gathering and evaluating data in the actual time, edge computations enable integrated processes while reducing time and resource consumption. The network offers a nearly actual time edge examination which could contribute to higher uptime and productivity [14].

2.2 Drawbacks of Edge Computing

The only drawback of such technology is balancing between leaving data at the edge or moving it into the main cloud when necessary. Businesses could find it difficult to discern how to balance processing data at the edge vs sending it to the cloud. The cost-effectiveness of data analysis on-premises varies, but there are instances when moving the data to the cloud is necessary. It may become a challenging issue for organizations to manage since simpler analytics procedures are retained on the edge while datasets that demand rather more sophisticated algorithms are best handled in the cloud [15, 16] .

2.3 Fog and Edge Computing: Cloud Computing Extension

Today's technology has advanced significantly to the extent that it is possible to transmit 4K videos to a global audience. The Internet has evolved from a simple information source to a tool for data feeds that facilitate computations. It is transitioning from centralized to distributed structures with regard to technologies such as virtual reality, augmented reality, and video streaming providing numerous cutting-edge capabilities for consumers. Machine Learning (ML)-based intelligent applications frequently work with enormous volumes of data that are too costly in terms of storing them in or transferring them to a focal cloud service. Furthermore, it is not even required that each piece of data gathered ought to be beneficial to users. Only the most sensitive data needs to be sent to the cloud server if a portion of the data processing is possible to be carried out at the network's edge, which would considerably save expenses. Moving computer power closer to the network's edge will save costs and boost security. Both types of Computing: Edge and Fog appear to be interchangeable, and they have several important features in common. Both systems move data processing close to the point of data production. Reducing the quantity of data transferred to the cloud is the major objective of doing this. Particularly in remote mission-critical applications, this helps decrease reaction time and hence increase system response time. Since less data needs to be sent over the open internet, businesses may also increase security by moving data processing nearer to the source [17–19].

2.4 Distinguishing Between Three Types of Computing: Cloud, Fog, and Edge

When it comes to sending information and data to nearby analytics platforms, such as cars, motors, pumps, monitors, speakers, and sensors, or that are at or close to the source of the data, both types of computing: fog and edge provide equivalent capabilities. Furthermore, both types of computing leverage the strength of a local network's computing resources to quickly complete computational tasks that might otherwise be completed in the cloud. As a result, businesses can lessen their dependence on cloud-based platforms for purpose of processing and storing data, which frequently causes response time issues, and are able to make data-dependent decisions faster.

3 Binary Classification Results

In the case of the binary classification of the UNSW-NB15 and NL-KDD datasets, the K-field process has been carried out to train and test data when K-Field has the following values: K2, K4, K6, K8, and K10 (Tables 1 and 2).

Table 1. Binary Classification Results

UNSW-NB15 - Binary Classification							
K	Accuracy of each attack	Recall (Sensitivity)	Specificity	FPR	FNR	Precision	F-Measure
k2	0.925	0.925	0.925	0.075	0.075	0.928	0.924
k4	0.930	0.930	0.930	0.070	0.070	0.930	0.929
k6	0.932	0.932	0.932	0.068	0.068	0.932	0.932
k8	0.933	0.933	0.933	0.067	0.067	0.935	0.932
k10	0.935	0.935	0.935	0.065	0.065	0.936	0.934
Average	0.931	0.931	0.931	0.069	0.069	0.932	0.930
NSL-KDD - Binary Classification Results							
K	Accuracy of each attack	Recall (Sensitivity)	Specificity	FPR	FNR	Precision	F-Measure
k2	0.989	0.989	0.998	0.002	0.002	0.990	0.989
k4	0.992	0.992	0.999	0.001	0.001	0.992	0.992
k6	0.993	0.993	0.999	0.001	0.001	0.993	0.993
k8	0.996	0.996	0.999	0.001	0.001	0.993	0.996
k10	0.999	1.000	0.999	0.001	0.001	1.000	1.000
Average	0.994	0.994	0.999	0.001	0.001	0.994	0.994

The results shown in Tables and Figures from 1, 2, 3 and 4 to 2, 3, 4 clarify the Confusion Matrix. Both Tables 3, 4 show Confusion Matrix for UNSW-NB15 (K=10) for the types of activities, whether they are normal or anomalous. The proposed model accurately classified 8,915 normal activities; however, it misclassified 1,300 activities as normal while they were anomalous. As for the attack detection, the proposed model detected 15,167 attacks and it did not detect 385 attacks. In other words, accuracy of attack detection is 93% which is considered as good.

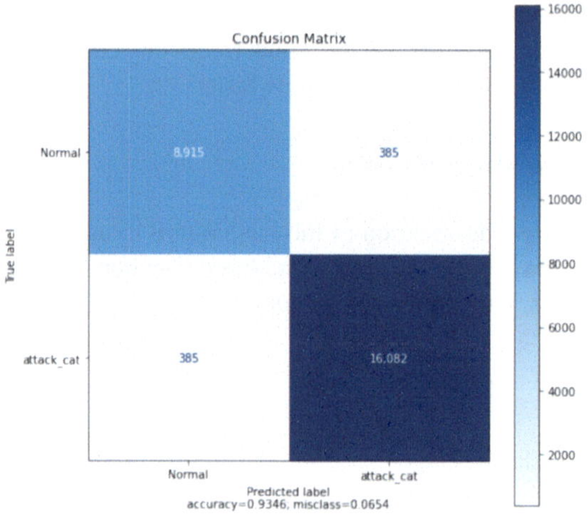

Fig. 1. Confusion Matrix on UNSW-NB 15 (K = **10**)

Table 2. UNSW-NB15 binary classification (k = 2)

Class Type	Accuracy of each attack	Recall (Sensitivity)	Specificity	FPR	FNR	Precision	F-Measure
Normal	0.925	0.961	0.904	0.096	0.039	0.849	0.902
Attack_ cat	0.925	0.904	0.961	0.039	0.096	0.977	0.939

Figures 1, 2, 3 and 4 shows the Confusion Matrix for NL-KDD (k=10). 5331 activities are classified as normal, while 3 activities are classified as anomalous. However, 7677 attacks were detected, and only 7 attacks have not been detected. In terms of percentage, it is 99% accurate. Therefore, the values of seven parameters according to which the performance of the proposed model is evaluated for both datasets.

Table 3. UNSW-NB15 binary classification (k = 4)

Class Type	Accuracy of each attack	Recall (Sensitivity)	Specificity	FPR	FNR	Precision	F-Measure
Normal	0.930	0.940	0.924	0.076	0.035	0.875	0.906
Attack _cat	0.930	0.924	0.940	0.060	0.125	0.965	0.944

These parameters are Accuracy of each attack, Recall (sensitivity), Specificity, FPR, FNR, Precision, and F-measure as shown in tables (11-4). The detailed results for each parameter are shown in Tables that start from (12-4) to (22-4) in addition to the figures that start from the Figs. 1, 2, 3 and 4 to Figs. 2, 3 and 4. For the UNSW-NB15, the training results showed that the Accuracy of each attack, Recall (sensitivity), Specificity, FPR, FNR, Precision, and F-measure have achieved the values: 0.93, 0.93, 0.93, 0.06, 0.06, 0.93, and 0.93 respectively.

Conversely, the training results of the Accuracy of each attack, Recall (sensitivity), Specificity, FPR, FNR, Precision, and F-measure for the NL-KDD have reached the following values respectively: 0.99, 0.99, 0.99, 0.0007, 0.0007, 0.99, and 0.99. It is imperative to point out that the training results are obtained when the K-field is K10, which represents the final stage for training the data.

The average values of the Accuracy of each attack, Recall (sensitivity), Specificity, FPR, FNR, Precision, and F-measure for the UNSW-NB15 dataset are as follows respectively: 0.93, 0.93, 0.93, 0.06, 0.06, 0.93, and 0.93. In contrast, the average values of the Accuracy of each attack, Recall (sensitivity), Specificity, FPR, FNR, Precision, and F-measure for the NSL-KDD dataset are as follows: 0.99, .0.99, 0.99, 0.001, 0.001, 0.99, and 0.99.

Figures 1 and 2 show the ROC-AUC curve for the suggested model. The F1 source clearly achieves a more accurate result when the NSL-KDD is used. According to the results, the F1-source accuracy is 0.99 and the misclassification is 0.0001. Despite this, the F1-source obtained 93% accuracy and 0.06 misclassification when utilizing the UNSW-NB 15 dataset.

When employing the UNSW-NB 15, the AUC Curve beneath the ROC is 0.99 for the kind of assaults. In comparison, when utilizing the NSL-KDD, the micro-average of the ROC curve for different types of assaults was 0.1. This number reflects the optimal state of detecting assaults globally (Figs. 5 and 6).

Building on the results reported above, it is evident that the IDS accuracy in detecting attacks is 99% and the FPR is 0.001. Therefore, the proposed model has achieved an excellent intrusion detection rate (Tables 5, 6, 7, 8, 9, 10, 11, 12 and 13).

Table 4. binary classification UNSW-NB15 (k = 6)

Class Type	Accuracy of each attack	Recall (Sensitivity)	Specificity	FPR	FNR	Precision	F-Measure
Normal	0.932	0.961	0.917	0.083	0.024	0.867	0.911
Attack cat	0.932	0.917	0.961	0.039	0.133	0.976	0.946

Table 5. UNSW-NB15 binary classification (k = 8)

Class Type	Accuracy of each attack	Recall (Sensitivity)	Specificity	FPR	FNR	Precision	F-Measure
Normal	0.933	0.962	0.917	0.083	0.023	0.867	0.912
Attack cat	0.933	0.917	0.961	0.038	0.132	0.976	0.946

Table 6. binary classification UNSW-NB15 (k = 10)

Class Type	Accuracy of each attack	Recall (Sensitivity)	Specificity	FPR	FNR	Precision	F-Measure
Normal	0.935	0.989	0.921	0.079	0.025	0.873	0.914
Attack cat	0.935	0.921	0.959	0.041	0.127	0.975	0.947

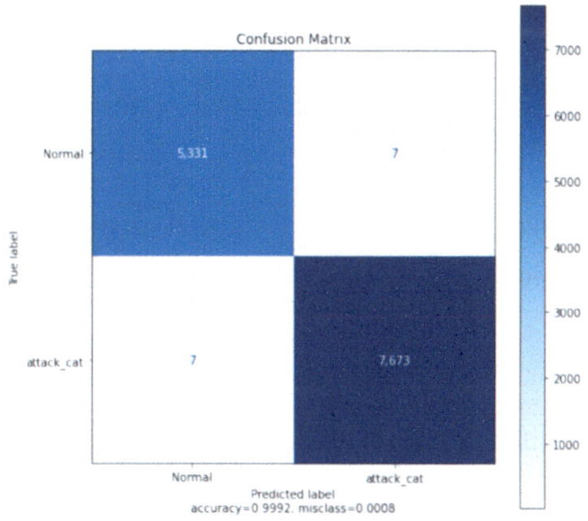

Fig. 2. Confusion Matrix for binary classification NL-KDD (K = 10)

Table 7. Binary classification NL-KDD (k = 2) is evaluated in terms of performance

Class Type	Accuracy of each attack	Recall (Sensitivity)	Specificity	FPR	FNR	Precision	F-Measure
Normal	0.998	0.996	0.999	0.001	0.002	0.999	0.998
Attack cat	0.998	0.999	0.996	0.004	0.001	0.998	0.998

Table 8. Binary classification NL-KDD (k = 4) is evaluated in terms of performance

Class Type	Accuracy of each attack	Recall (Sensitivity)	Specificity	FPR	FNR	Precision	F-Measure
Normal	0.999	0.999	0.999	0.001	0.001	0.999	0.999
Attack cat	0.999	0.999	0.999	0.001	0.002	0.999	0.999

Table 9. Binary classification NL-KDD (k = 6) is evaluated in terms of performance

Class Type	Accuracy of each attack	Recall (Sensitivity)	Specificity	FPR	FNR	Precision	F-Measure
Normal	0.999	0.998	0.999	0.001	0.001	0.999	0.998
Attack cat	0.999	0.999	0.998	0.001	0.001	0.999	0.999

Table 10. Binary classification NL-KDD (k = 8) is evaluated in terms of performance

Class Type	Accuracy of each attack	Recall (Sensitivity)	Specificity	FPR	FNR	Precision	F-Measure
Normal	0.999	0.997	0.999	0.000	0.002	0.999	0.998
Attack cat	0.999	0.999	0.997	0.003	0.001	0.999	0.999

Table 11. Binary classification NL-KDD (k = 10) is evaluated in terms of performance

Class Type	Accuracy of each attack	Recall (Sensitivity)	Specificity	FPR	FNR	Precision	F-Measure
Normal	0.999	0.999	0.999	0.000	0.001	0.999	0.999
Attack cat	0.999	0.999	0.999	0.001	0.001	0.999	0.999

4 Comparison with Previous Models

This section will compare the suggested model with previous models that used the data training method. The comparison will be performed twice: first using the UNSW-NB 15 dataset and again using the NSL-KDD dataset.

The following displays the results of the binary classification method. First, the proposed model (CNN-GRU) was evaluated against previous models (CNN, LSTM, GRU, RNN, CNN -LSTM) [19] and (CNN+GRU) using the UNSW-NB 15 dataset. As the table (4-24) (4-25) and picture (4-5) (4-6) show, the proposed model delivers a superior detection rate, false-positive rate, and accuracy values than the models provided. In other words, the recommended model outperformed the other models by 1.5 values in terms of detection. Second, the proposed model (CNN-GRU) was evaluated against previous models (CNN, LSTM, GRU, RNN, CNN -LSTM) [19] and (CNN+GRU) using the NSL-KDD dataset. It is shown that the suggested model outperforms all the other models with a notable exception (the BAT-MC Model) in terms of the FPR metric,

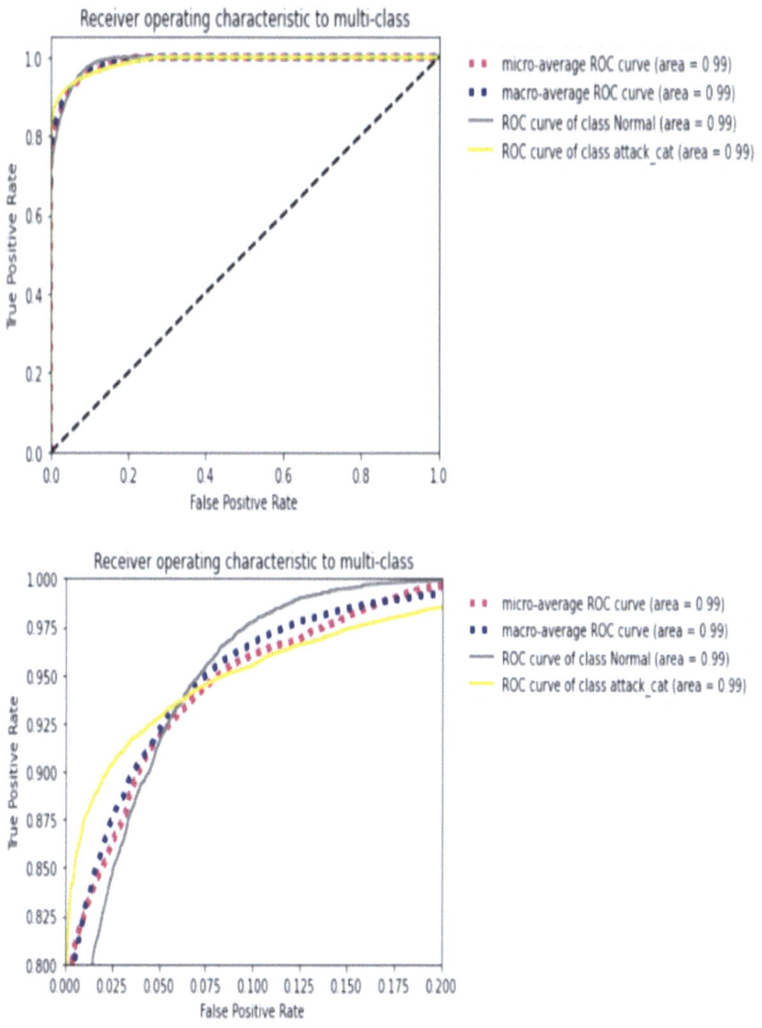

Fig. 3. UNSW-NB15's ROC-AUC Curve (K = 10)

resulting in a higher FPR. However, in contrast to other models, the suggested model is more suited for intrusion detection. Notably, the suggested model raised the detection rate by a factor of 1.5 compared to the actual rates.

On the other hand, the multi-class classification approach yields the following outcomes. First, using the UNSW-NB 15 dataset, the performance of the proposed model (CNN-GRU) was compared to earlier models (CNN, LSTM, GRU, RNN, CNN -LSTM) [19] and (CNN+GRU).

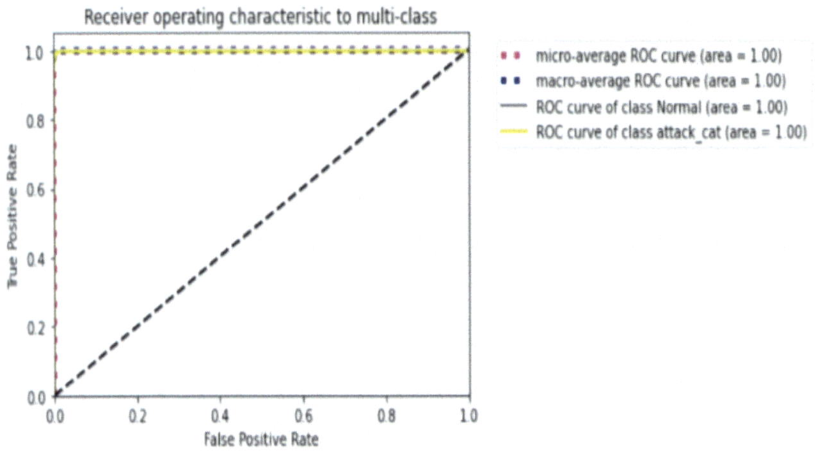

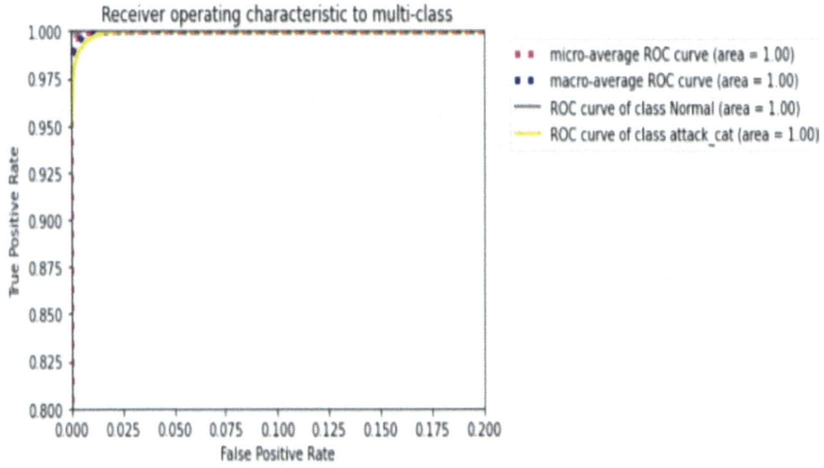

Fig. 4. NL-KDD ROC-AUC Curve (k = 10)

Table 12. UNSW-NB15 Comparison Experiment Results

Model	ACC%	Precision	Recall	F-Score
CNN	91.2	87.5	96.1	91.5
LSTM	88.9	84.5	94.8	89.5
GRU	77.9	75.3	83.2	79.0
RNN	71.9	65.8	91.3	76.5
CNN + LSTM	87.6	85.5	90.6	88.0
CNN + GRU	93.1	93.2	93.1	93.0

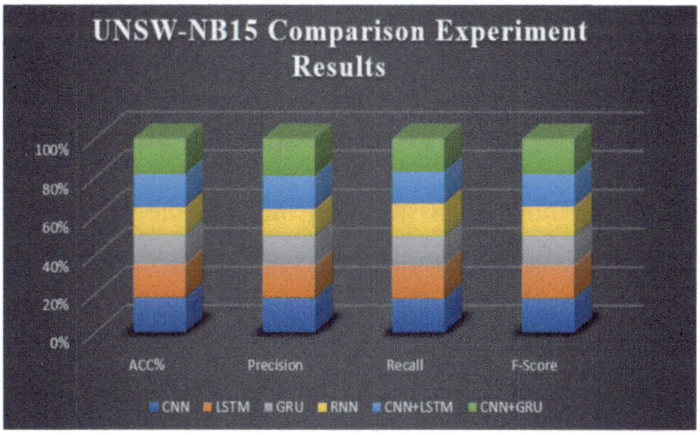

Fig. 5. UNSW-NB15 Comparison Experiment Results

Table 13. NSL-KDD Comparison Experiment Results

Model	ACC%	Precision	Recall	F-Score
CNN	79.3	95.5	61.4	74.8
LSTM	75.8	95.4	54.2	69.2
GRU	79.1	95.4	61.2	74.5
RNN	76.1	88.0	60.5	71.7
CNN + LSTM	82.6	94.9	68.9	79.8
CNN + GRU	99.4	99.4	99.4	99.4

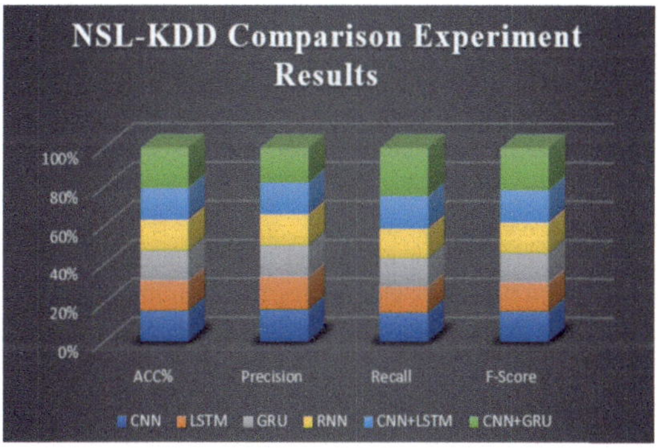

Fig. 6. NSL-KDD Comparison Experiment Results

5 Conclusion

When employing both datasets—UNSW-NB 15 and NSL-KDD, which score 93% and 99%, respectively—the suggested model has obtained the optimal accuracy intrusion detection in the binary classification situation.

When comparing the adopted deep learning algorithm, the CNN + GRU, to other algorithms, it is evident thar the performance of this algorithm excels the performance of the other algorithms by the value 1.5 for both databases.

It is evident that the K-field cross validation is the best training method for the selected datasets.

In comparison to the binary classification method, the multi-class classification method is more accurate in terms of indicating the type of attack where the binary method indicates the presence of the attack only.

References

1. Corallo, A., Lazoi, M., Lezzi, M., Luperto, A.: Cybersecurity awareness in the context of the industrial internet of things: a systematic literature review. Comput. Ind. **137**, 103614 (2022)
2. DeFranco, J.: 12 flavors of IoT. Computer (Long. Beach. Calif). **54**(10), 133–137 (2021)
3. Ding, X., Wang, X., Xie, Y., Li¸ F.: A lightweight anonymous authentication protocol for resource-constrained devices in internet of things. IEEE Internet Things J. (2021)
4. Goel, S.S., Goel, A., Kumar, M., Moltó, G.: A review of Internet of Things: qualifying technologies and boundless horizon. J. Reliab. Intell. Environ. **7**(1), 23–33 (2021)
5. Raghav, I., Chhikara, S., Hasteer, N.: Intrusion detection and prevention in cloud environment: a systematic review. Int. J. Comput. Appl. **68**(24) (2013)
6. Chiba, Z., Abghour, N., Moussaid, K., El Omri, A., Rida, M.: A survey of intrusion detection systems for cloud computing environment. In: 2016 International Conference on Engineering & MIS (ICEMIS), pp. 1–13 (2016)
7. Chhaya, L., Sharma, P., Kumar, A., Bhagwatikar, G.: Cybersecurity for smart grid: threats, solutions and standardization. In: Advances in Greener Energy Technologies, Springer, pp. 17–29 (2020)
8. Otair, M., Ibrahim, O.T., Abualigah, L., Altalhi, M., Sumari, P.: An enhanced grey wolf optimizer based particle swarm optimizer for intrusion detection system in wireless sensor networks. Wirel. Networks, pp. 1–24 (2022)
9. Gao, L., Luan, T.H., Gu, B., Qu, Y., Xiang, Y.: Privacy-Preserving in Edge Computing. Springer (2021)
10. Alshouiliy, K., Agrawal, D.P.: Confluence of 4G LTE, 5G, fog, and cloud computing and understanding security issues. Fog/Edge Computing For Security, Privacy, and Applications. Springer, Cham, pp. 3–32 (2021)
11. Cao, J., Su, X.: 5G edge computing enhanced mobile augmented reality. In: 2021 IEEE International Conference on Pervasive Computing and Communications Workshops and other Affiliated Events (PerCom Workshops). IEEE (2021)
12. Xu, T., Wang, X., Su, T., Wan, L., Sun, L.: Vehicle location in edge computing enabling IoTs based on bistatic FDA-MIMO radar. IEEE Access **9**, 46398–46408 (2021)
13. Al-Khuzaie, Z.M., Albermany, S.A., AbdlNibe, M.A.: Intrusion detection in the IoT-fog adopting the GRU and CNN: a deep learning-based approach. In: Micro-Electronics and Telecommunication Engineering: Proceedings of 6th ICMETE 2022, pp. 379–389. Singapore: Springer Nature Singapore (2023)
14. Sangamithra, A., Clinton, G.: Overview of edge computing and its exploring characteristics. Research Anthology on Edge Computing Protocols, Applications, and Integration. IGI Global, pp. 1–17. (2022)
15. Roy, S., Li, J., Choi, B.-J., Bai, Y.: A lightweight supervised intrusion detection mechanism for IoT networks. Futur. Gener. Comput. Syst. **127**, 276–285 (2022)

16. Al-Khuzaie, Z.M., Albermany, S.A., AbdlNibe, M.A.: Data protection methods and intrusion detection systems employing machine learning and deep learning in the IoT: A Comparative Review. In 2022 Second International Conference on Computer Science, Engineering and Applications (ICCSEA), pp. 1–10. IEEE. (2022)
17. Almaiah, M.A., Al-Zahrani, A., Almomani, O., Alhwaitat, A.K.: Classification of cyber security threats on mobile devices and applications. In: Artificial Intelligence and Blockchain for Future Cybersecurity Applications, Springer, pp. 107–123 (2021)
18. Shah, Y., Sengupta, S.: A survey on classification of cyber-attacks on iot and iiot devices. In: 2020 11th IEEE Annual Ubiquitous Computing, Electronics & Mobile Communication Conference (UEMCON), pp. 406–413 (2020)
19. Djenna, A., Saïdouni, D.E.: Cyber attacks classification in IoT-based-healthcare infrastructure. In: 2018 2nd Cyber Security in Networking Conference (CSNet), pp. 1–4 (2018)

Estimate Human Emotion Using Machine Learning Based on Voice Signals

Zainab Kadhm Obeas[1(✉)], Esraa Najjar[2], Ahmed J. Obaid[3], and Suresh Rasappan[4]

[1] College of Science, Al-Qasim Green University, Babylon, Iraq
Zainab@uoqasim.edu.iq
[2] General Directorate of Education in Najaf Governorate, Najaf, Iraq
[3] Faculty of Computer Science and Mathematics, University of Kufa, Najaf, Iraq
[4] Mathematics Section University of Technology and Applied, Sciences-Ibri Post Box- 466, Postal Code-516 Ibri, Sultanate of Oman

Abstract. Recently, advances in voice-based multimedia applications and the Internet of Things (IoT) have led to the creation of large amounts of big data, containing patterns, correlations, and trends that reflect diverse aspects of human behaviour. Speech-based emotion recognition plays a pivotal role in improving human-machine interactions, emphasizing the significance of analyzing audio data from human conversations to extract emotional insights. Developing a robust emotion detection framework is essential and involves various modules, including speech-to-text conversion, feature extraction, feature selection, and classification of these features to determine emotions. Applications of audio processing technology are vital in improving human-computer interaction. In this study, three classification algorithms were used to evaluate the performance of emotion recognition (happiness, sadness, and anger) by feeding audio signals into the classifiers and comparing the results. The recognition accuracy using the One-R algorithm was (74%) with a processing time of 0.02 s. While the recognition accuracy achieved by the other two classifiers was Support Vector Machines SVM (64) with a processing time of 0.4 s and Random Forests RF (58%) with a processing time of 0.06 s. Through comparison between the classifiers, it was found that the (One-R) classifier provided the best performance and classification accuracy and the shortest time to implement.

Keywords: Machine learning · Emotion recognition · One-R · RF · SVM

1 Introduction

Emotion recognition relies on interpersonal and interpretation skills or a semi-automatic approach. Many automated emotion recognition methodologies, such as human-computer interaction, have been developed. Recognizing human emotions requires high accuracy because of their ambiguity. Human-computer interaction (HCI) can use many features, including the iris, thumb impression, voice, face, and more. Every human has a distinct voice with various characteristics used for speaker identification, verification, and identifying the speaker's information, such as age, gender, language, speech, mood,

A. J. Obaid et al. (Eds.): ICONEST 2024, ISEM 40, pp. 398–410, 2025.
https://doi.org/10.1007/978-3-031-89960-7_30

etc. Enhancing the efficiency and performance of human-computer interaction through the use of human speech is essential. There is an increasing need to comprehend a user's speech not only for its content but also for its meaning and delivery. Numerous attributes of speakers have useful applications. Gender, age, health, language, dialect, accent, emotional, and attention state are the most common [1]. Multiple emotions can occasionally have the same expression, and the same feelings might be represented in different ways. In addition to numerous other psychological, social, and personal factors, emotions might differ depending on personality, gender, region, ethnicity, culture, and circumstance. Even for real people, it can occasionally be challenging to tell someone's true feelings from their voice, body language, or written words. We can readily comprehend the complexity of the issue when it comes to computer-assisted emotion recognition. Many academics have been focusing on detecting emotions in a variety of input formats, including text, images, audio, and videos [2].

Many important problems remain unresolved in spite of the substantial quantity of research on emotion recognition from audio conversations. Significant low-level auditory features, Such as the Fundamental Frequency (F0) which is analyzed by high-level statistical functions, spectrograms, Melt Cepstral Coefficients (MFCC), and other features, have been acknowledged in recent studies as being important for emotion identification [3, 4]. The research study presents a proposed Classification of the Voice Signal system that utilizes three machine learning techniques. The subsequent section outlines the order of the paper. The main subjects pertinent to this study are covered in Sect. 2. The suggested model is further discussed in Sect. 3. Section 4 displays the outcomes of the experiment. In Sect. 5, the paper is ultimately ended.

2 Related Works

Literature related to this work's major interest includes different studies that summarize some of the research in the field of Emotion - Speech Recognition.

Idris & Salam, 2014 [5] talk about the Berlin Emotional Database provided the experimental data used in the study. Multi-layer perceptron's and Neural Networks are used to classify emotions. According to the findings, hybrid features outperformed prosodic and voice quality features alone in terms of overall recognition rates. Hybrid features have the highest overall recognition rate (75.51%), whereas prosodic and voice quality features have the highest recognition rates (64.67% and 59.63%, respectively). However, the ability to recognize different emotions varies; anger has the highest recognition rate (88%).

Hifny & Ali, 2019 [6] The researchers are using two neural structures. The initial architecture, the LSTM-DNN attention-based CNN model, was put into practice. While the audio signal's consecutive events are managed by Bidirectional Long Short-Term Memory (BLSTM) layers. The dense layer that is fully connected (DNN).The second structure is the deep CNN model. Results show that the audio emotion recognition task outperforms a robust deep CNN platform with a significant improvement of 2.2%.

Shaqra et al., 2019 [7] **in** this study, the significance of determining the age and gender prior to determining the emotional label was examined using developed hierarchical classification models. Four different models' performances were compared, and the relationship between age, gender, and emotion recognition accuracy was shown. The findings demonstrated that, in comparison to utilizing a single classifier for all the data, employing distinct emotion models for each gender and age category yields a greater accuracy 74% accuracy was the highest.

Li et al., 2010 [8] experiment was conducted on two reference data sets, where an emotion detection model using a RNNs was used to generate emotional predictions where were the accuracy results for the proposed 60.87% weighted accuracy and 60.97% unweight accuracy.

Chaudhari & Kagalkar, 2015 [9] using the combined datasets of Save, Raves, and Tess, the researchers' deep learning classification system, a convolutional neural network, can classify seven emotions with an accuracy of 69.45%. It suggests a novel method for voice conversion, which helps to reproduce emotions on a neutral audio. Megan, a unique kind of Generative Adversarial Network (GAN), is used to produce the emotional sounds.

3 Machine Learning

Machine learning (ML) can be used to teach machines to process data more efficiently [10, 11]. Occasionally, we are unable to decipher the information extracted from the data after seeing it. When machine learning is used, the issue can be resolved by creating a model that accurately depicts a carefully selected data set [12]. Because there are so many datasets available, there is an increasing demand for machine learning. Numerous businesses employ machine learning to extract relevant data. The goal of machine learning is to learn from the data [13]. The topic of teaching computers to learn without explicit programming has been the subject of numerous investigations. Mathematicians and programmers employ a range of methods to tackle this problem, which involves massive data sets [14]. Some of the characteristics of the human mind that enable us to solve extraordinarily complex problems faster than even the most potent computers are embodied in machine learning approaches. Numerous challenging issues, such as adaptive control, markup estimation in the building sector, and speech recognition from text, have been successfully resolved by ML approaches. Lately, ML techniques[15].

4 Methodology

The suggested Signal Processing in the presented study the first phase is the dataset, the second phase preprocessing, the third phase is feature extraction, and finally phase of data classfiction as Fig. 1.

4.1 Dataset

The audio dataset used in this study, "Emotion recognition by voice," was obtained from in this link (https://www.kaggle.com/datasets/rohitzaman/gender-age-and-emotion-detection-from-voice/discussion?sort=hotness) from Mozilla's speech platform (Voice.mozilla.org). The dataset was employed to build prediction models for emotion

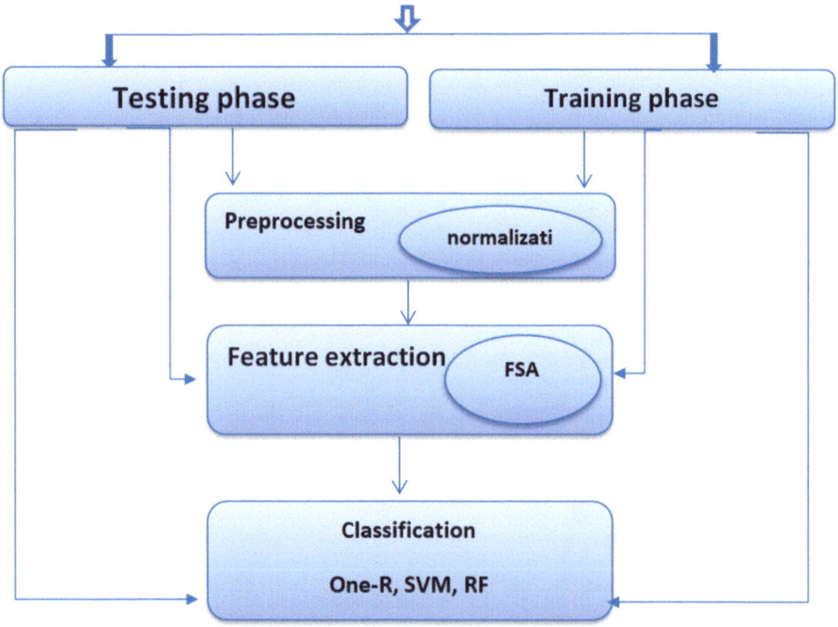

Fig. 1. The proposed method for classification

recognition. It includes a CSV file listing filenames and their corresponding emotion labels, which are categorized into three classes: happy, sad, and angry. The dataset contains 20 feature columns along with one label column representing the emotion class. In order to make frequency spectrum analysis easier, all audio files were converted to WAV format.

4.2 Preprocessing

Collecting emotional data is exposed to external noise during data recording. This will lead to poor classification accuracy. Normalization is therefore a crucial stage in the identification of emotions. During this preprocessing step, emotional discrimination is maintained while speaker and recording variability is removed. Energy normalization and pitch normalization are the two types of normalization techniques that are typically used.

4.3 Feature Extraction

Important feature sets must be extracted in order to construct an accurate emotion detection model. The free open Smile toolbox was used. Spoken audio files were converted to PCM WAV files. Audio discussions were split into utterances as part of the pre-processing step. Each speaker's speech was segmented using speaker recording [16].

The audio files in WAV format were analyzed using the Frequency Spectrum Analysis (FSA) process in the R programming language. This process facilitated the extraction of 20 statistical features, including set of feature as Fig. 2.

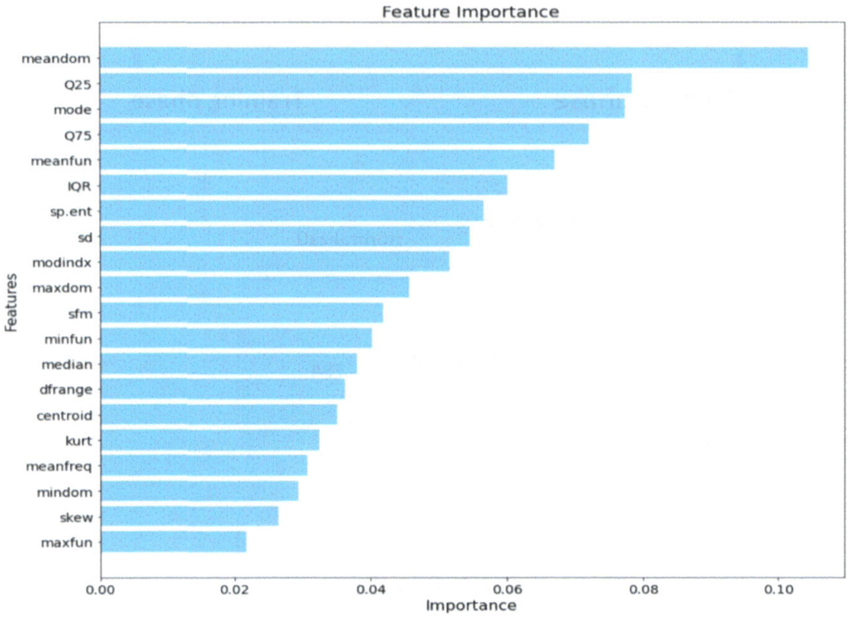

Fig. 2. Feature importance

The extracted statistical data, along with the filenames, were stored in a CSV file. The dataset did not contain any categorical columns. Missing numerical values in the CSV were replaced with their respective means. Additionally, a class imbalance issue in the dataset was addressed using the SMOTE (Synthetic Minority Oversampling Technique). Outliers in the dataset were identified and removed before proceeding with classification [17].

4.4 Classification

One of the most important data mining techniques for classifying large data sets [14], it is also used to solve machine learning algorithms and statistics in order to identify patterns and rules in the data that are then used for prediction [18]. The outcomes of using the random forest and one-R classifier in an emotion recognition system using human speech were covered in this section and use the two classifiers to compare these outcomes.

4.4.1 One-R

A popular classification technique in machine learning applications, OneR is straightforward and incredibly efficient. OneR can be improved by offering better ways to handle some of the exceptions, even though its simplicity makes it difficult to improve further [19]. Holte introduced OneR, a straightforward rule-based algorithm. The features are assessed by this algorithm based on the mistake rate. The OneR method generates a set of candidate rules, and the rule with the lowest error ratio is chosen [20]. The OneR algorithm works as follows [21]:

Algorithm (1) One-R algorithm.
Step 1: in steps I, II, III, and IV, construct a rule set from the clustered set for every value of each attribute predictor.
Step 2: Determine the frequency of occurrence of each target class value.
Step 3: Identify the most popular class.
Step 4: Create a rule set and allocate that class to this attribute predictor value.
Step 5: Determine the overall inaccuracy that arises in the rules established for every attribute predictor
Step 6:. Choose the best attribute predictors and classifier rules with the lowest total error.

4.4.2 Random Forest

The random forest classifier consists of a combination of tree classifiers as shown in Fig. 3 [22, 25]. In order to categorize an input vector, each classifier uses a random vector that is sampled separately from the input vector. Each tree then casts a unit vote for the most popular class [20]. The primary reason for random forest's popularity is its proficiency with non-linear classification challenges. One well-known method for resolving data imbalances across multiple classes is random forest [23]. The value that is commonly utilized as the square root of the total number of variables is thought to be rather resilient to the number selection process in Random Forest. When creating random forest trees, no trimming is done. Depending on the threshold value selected by the user or by the majority of votes from classification trees in a forest, predictions for the test samples will be identified. It performs well when used for pattern recognition tasks. Experiments demonstrate that the (RF) method performs well in the suggested system [24].

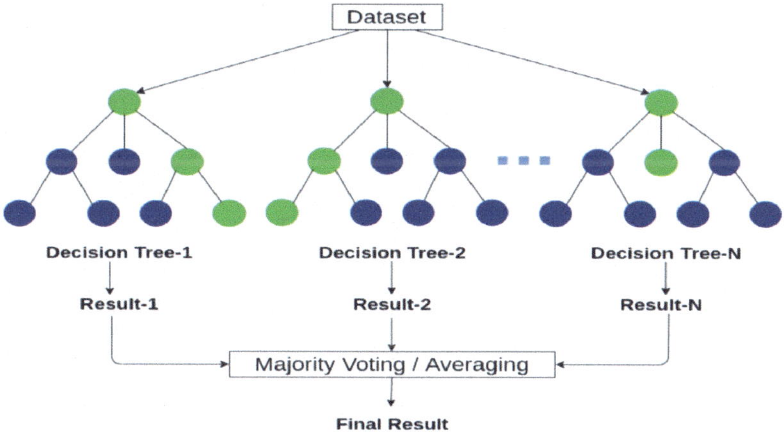

Fig. 3. Random forest algorithm

Algorithm (2) Classification Random Forest.
start
Step 1: Choose "k" features at random from the total of "m" features. In which k << m
Step 2: Determine node "d" by the best split point among "k" features.
Step 3: To divide the node into child nodes, the best partition was used.
Step 4: Repeat steps 1 to 3 until the number of nodes "l" is breached.
Step 5: Iterating steps 1 through 4 "n" times to create a forest will yield.
End

4.4.3 Support Vector Machines Algorithm

Support Vector Machines (SVM) at AT&T Bell Laboratories The algorithm was pioneered by Vladimir Vapnik and his team [26]. It is one of the most widely used algorithms in supervised machine learning, known for its robustness, high accuracy, and reliability compared to other popular data mining approaches. [27]. When presented with a set of training examples, each example is assigned to one of two categories. The SVM algorithm then creates a model to predict the class of the new sample. SVMs map samples as points in space, and arrange them such that examples from dissimilar classes are far apart. When a new sample is presented, the sample is classified based on where in space it lies, and assigned to its corresponding class [28].

More formally, SVMs perform classification by using a hyperplane, or a set of hyperplanes, to separate data points with the assistance of support vectors. A maximum margin hyperplane divides negative examples (red circles) from positive examples (green squares),as seen in Fig. 4. [31] A super level with a large functional margin from the closest training data points for any class achieves ideal separation [29, 30].

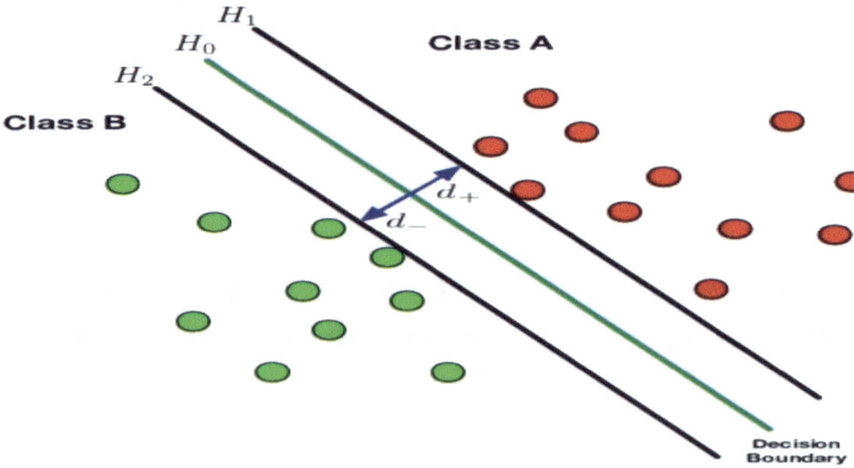

Fig. 4. Maximum margin.

Some researchers consider that SVMs determine the hyper optimal level as a solution to the learning problem. For linear SVMs, the discrete hyperplane is represented by

Eq. (1), where w is the weight vector associated with the input x, b is the bias parameter, and x is the feature vector of the text being classified, which consists of words. [32], as shown in Fig. 5. [33].

$$z = w.x + b \tag{1}$$

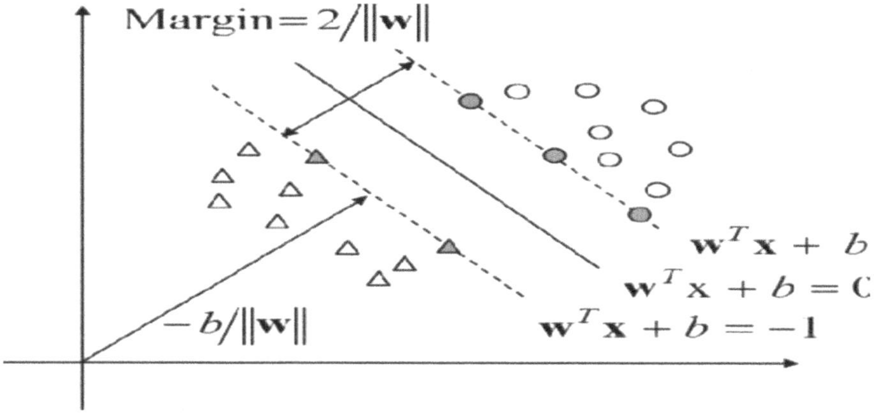

Fig. 5. A hyperplane and the support vectors

5 Experiment Results and Discussion

The outcomes of using three classification methods are covered in this section (One-R, SVM and RF) in a voice recognition system. And compare these results.

Table 1. Results of classification accuracy using One-R

Dataset	classifiers	precision	recall	F-Measure	class
		70	64	67	Happy
Voice signals	One-R	77	62	69	Sad
		73	94	82	Angry

The comparison and evaluation of the metrics Precision, Recall, and F-measure for emotions using the human voice signal and a classifier for identifying emotions (happy, sadness, and rage) is displayed in Table 1. One-R, it gave results respectively (70, 64 67) for feelings of happiness, (77, 62, 69) for feelings of sadness, and (73, 94, 82) for feelings of anger as shown in Fig. 6.

In Table 2 shows the result of comparing and evaluating metric Precision, Recall, and F-measure for emotions through the human voice signal and recognizing emotions

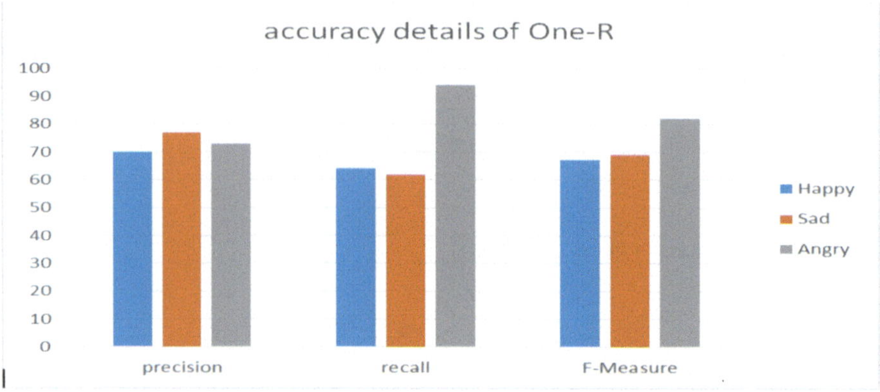

Fig. 6. Classification of our dataset using R classifier

Table 2. Results of classification accuracy using SVM

Dataset	classifiers	precision	recall	F-Measure	class
		56	51	53	Happy
Voice signals	SVM	61	67	64	Sad
		57	56	56	Angry

(happiness, sadness, and anger) using a classifier SVM and through the classification criteria (precision, recall, and F-Measure), it gave results respectively (56,51,53) for feelings of happiness, (61,67,64) for feelings of sadness, and (57,56,56) for feelings of anger As shown in Fig. 7.

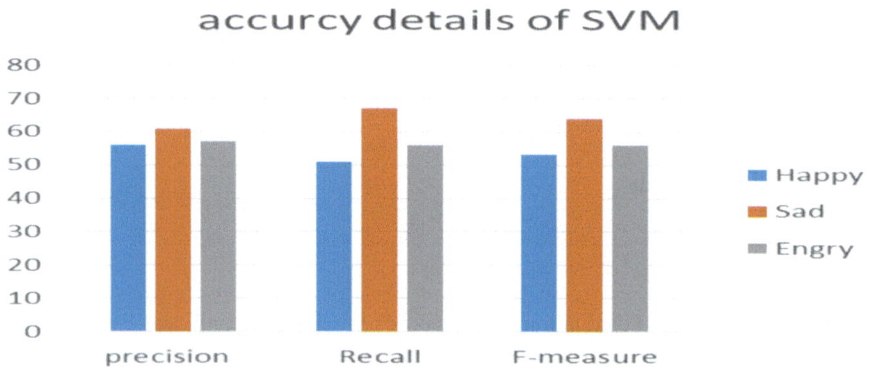

Fig. 7. Classification of our dataset using SVM classifier

In Table 3 shows the result of comparing and evaluating metric Precision, Recall, and F-measure for emotions through the human voice signal and recognizing emotions

Table 3. Results of classification accuracy using RF

Dataset	classifiers	precision	recall	F-Measure	class
Voice signals	RF	68	73	70	Happy
		63	61	62	Sad
		64	60	62	Angry

(happiness, sadness, and anger) using a classifier RF) and through the classification criteria (precision, recall, and F-Measure), it gave results respectively (68,73,70) for feelings of happiness, (63,61,62) for feelings of sadness, and (64,60,62) for feelings of anger As shown in Fig. 8.

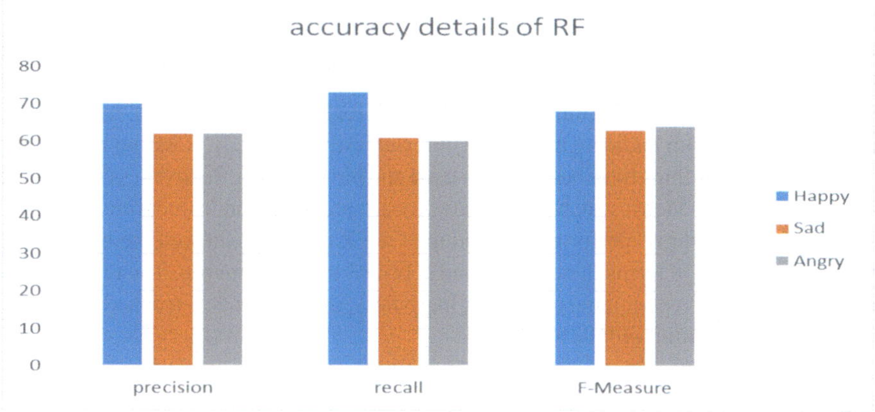

Fig. 8. Classification of our dataset using random forest classifier

Table 4. Accuracy for three classifiers

Dataset	classifiers	Accuracy	Time
	One R	74	0.02
voice signals	RF	64	0.4
	SVM	58	0.06

Table 4 represents the results of the work accuracy of the results of classifying the audio signal into emotions (happiness, sadness, and anger), which showed that the classifier (One R) has higher performance and classification accuracy (74) and the system is implemented in a faster and shorter time (0.02 s). While the classifiers (RF and SVM) reached classification accuracy (64, 58) respectively, with an execution time of (0.4 s, 0.06 s). When compared, the (One R) classifier has a clear impact on classifying and recognizing audio signal data with high classification accuracy, as shown in Fig. 9.

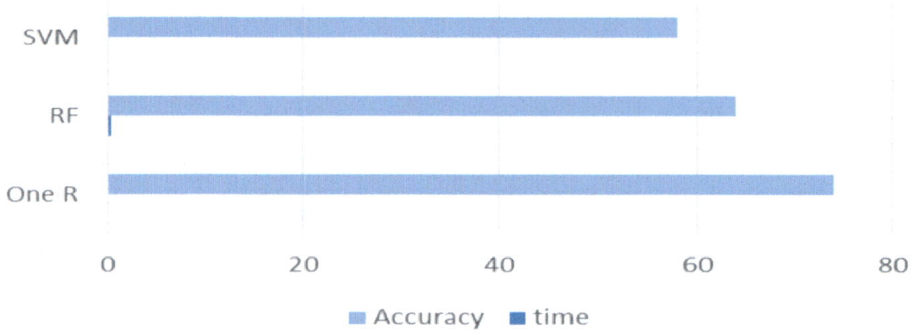

Fig. 9. Accuracy using three classifiers

6 Conclusions

This paper covered the fundamental procedure for detecting emotional speech, feature extraction is the most crucial step in the process. An approach to classify emotion recognition using statistics computed from basic features. In this paper, we identify human emotions from audio signals, using three classifiers (One R, Random Forest, and SVM). These algorithms have been widely used and have shown effectiveness in emotion classification tasks. Metrics such as accuracy, recall, precision, and f-measure were used to assess the algorithms' performance, and the One R classifier showed the best among the three classifiers in terms of accuracy and speed of implementation. The results show that the accuracy of recognizing emotions (happiness, sadness, anger) reached 72%, with a record execution time of 0.02 s.

References

1. Chaudhari, S.J., Kagalkar, R.M.: Automatic speaker age estimation and gender dependent emotion recognition. Int. J. Comput. Appl. **117**(17) (2015)
2. Sailunaz, K., Dhaliwal, M., Rokne, J., Alhajj, R.: Emotion detection from text and speech: a survey. Soc. Netw. Anal. Min. **8**(1), 28 (2018)
3. Mirsamadi, S., Barsoum, E., Zhang, C.: Automatic speech emotion recognition using recurrent neural networks with local attention. In: 2017 IEEE International Conference on Acoustics, Speech and Signal Processing (ICASSP), pp. 2227–2231 (2017)
4. Tripathi, S., Kumar, A., Ramesh, A., Singh, C., Yenigalla, P.: Deep Learning Based Emotion Recognition System Using Speech Features and Transcriptions. *arXiv Prepr. arXiv1906.05681* (2019)
5. Idris, I., Salam, M.S.H.: Emotion detection with hybrid voice quality and prosodic features using neural network. In: 2014 4th World Congress on Information and Communication Technologies (WICT 2014), pp. 205–210 (2014)
6. Hifny, Y., Ali, A.: Efficient arabic emotion recognition using deep neural networks. In: ICASSP 2019–2019 IEEE International Conference on Acoustics, Speech and Signal Processing (ICASSP), pp. 6710–6714 (2019)
7. Shaqra, F.A., Duwairi, R., Al-Ayyoub, M.: Recognizing emotion from speech based on age and gender using hierarchical models. Procedia Comput. Sci. **151**, 37–44 (2019)

8. Li, M., Jung, C.-S., Han, K.J.: Combining five acoustic level modeling methods for automatic speaker age and gender recognition. In: Eleventh Annual Conference of the International Speech Communication Association (2010)
9. Chaudhari, S.J., Kagalkar, R.M.: A methodology for efficient gender dependent speaker age and emotion identification system. Int. J. Adv. Res. Comput. Commun. Eng. **4**(7) (2015)
10. Patil, D., Rane, N.L., Desai, P., Rane, J.: Machine learning and deep learning: Methods, techniques, applications, challenges, and future research opportunities. Trust. Artif. Intell. Ind. Soc., pp. 28–81 (2024)
11. Najjar, E., Breesam, A.M.: Supervised machine learning a brief survey of approaches. Al-Iraqia J. Sci. Eng. Res. **2**(4), 71–82 (2023)
12. Peral-García, D., Cruz-Benito, J., García-Peñalvo, F.J.: Systematic literature review: quantum machine learning and its applications. Comput. Sci. Rev. **51**, 100619 (2024)
13. Laurer, M., Van Atteveldt, W., Casas, A., Welbers, K.: Less annotating, more classifying: addressing the data scarcity issue of supervised machine learning with deep transfer learning and BERT-NLI. Polit. Anal. **32**(1), 84–100 (2024)
14. B. Mahesh, "Machine learning algorithms-a review," *Int. J. Sci. Res. (IJSR).[Internet]*, vol. 9, pp. 381–386, 2020
15. Mair, C., et al.: An investigation of machine learning based prediction systems. J. Syst. Softw. **53**(1), 23–29 (2000)
16. Chamishka, S., et al.: A voice-based real-time emotion detection technique using recurrent neural network empowered feature modelling. Multimed. Tools Appl. **81**(24), 35173–35194 (2022)
17. Chawla, N.V., Bowyer, K.W., Hall, L.O., Kegelmeyer, W.P.: SMOTE: synthetic minority over-sampling technique. J. Artif. Intell. Res. **16**, 321–357 (2002)
18. J. Han, J. Pei, and M. Kamber, *Data mining: concepts and techniques*. Elsevier, 2011
19. Kalmegh, S.R., Deshmukh, S.N.: Effective evaluation of classification of indigenous news using decision table and oneR algorithm. Int. J. Adv. Inf. Sci. Technol. **26**(26), 6–11 (2014)
20. Al Sayaydeha, O.N., Mohammad, M.F.: Diagnosis of the Parkinson disease using enhanced fuzzy min-max neural network and OneR attribute evaluation method. In: 2019 International Conference on Advanced Science and Engineering (ICOASE), pp. 64–69 (2019)
21. Muda, Z., Yassin, W., Sulaiman, M.N., Udzir, N.I.: Intrusion detection based on k-means clustering and OneR classification. In: 2011 7th International Conference on Information Assurance and Security (IAS), pp. 192–197 (2011)
22. Genuer, R., Poggi, J.-M., Tuleau-Malot, C., Villa-Vialaneix, N.: Random forests for big data. Big Data Res. **9**, 28–46 (2017)
23. Paul, A., Mukherjee, D.P., Das, P., Gangopadhyay, A., Chintha, A.R., Kundu, S.: Improved random forest for classification. IEEE Trans. Image Process. **27**(8), 4012–4024 (2018)
24. Ali, A.J.M., Hasan, T.M., Mohammed, S.D.: Digital modulation classification based on chicken swarm optimization and random forest. J. Eng. Sci. Technol. **17**, 2095–2103 (2022)
25. Lei, Z., Bu, Q., Wang, J.: Research on the short-term power interval prediction method for distributed power sources in distribution networks based on quantile random forests. In: International Conference on Wireless Power Transfer, pp. 347–360 (2023)
26. Vapnik, V.: The Nature of Statistical Learning Theory. Springer science & Business Media (2013)
27. Wu, X., Kumar, V.: The Top Ten Algorithms in Data Mining. CRC press (2009)
28. Géron, A.: Hands-On Machine Learning with Scikit-Learn & TensorFlow (2017)
29. Suppharangsan, S.: Comparison and Performance Enhancement of Modern Pattern Classifiers. University of Southampton (2010)
30. Hadi, Z.G., Ajel, A.R., Al-Dujaili, A.Q.: Comparison between convolutional neural network CNN and SVM in skin cancer images recognition. J. Tech. **3**(4), 15–22 (2021)

31. Aboelwafa, M.M.N., Seddik, K.G., Eldefrawy, M.H., Gadallah, Y., Gidlund, M.: A machine-learning-based technique for false data injection attacks detection in industrial IoT. IEEE Internet Things J. **7**(9), 8462–8471 (2020)
32. Bilski, A.: A review of artificial intelligence algorithms in document classification. Int. J. Electron. Telecommun. **57**(3), 263–270 (2011)
33. Vives, J.: Vibration analysis for fault detection in wind turbines using machine learning techniques. Adv. Comput. Intell. **2**(1), 15 (2022)

Author Index

A. J. Obaid et al. (Eds.): ICONEST 2024, ISEM 40, pp. 411–412, 2025.
https://doi.org/10.1007/978-3-031-89960-7

The manufacturer's authorised representative in the EU is Springer Nature Customer Service Centre GmbH, Europaplatz 3, 69115 Heidelberg, Germany. If you have any concerns regarding our products, please contact ProductSafety@springernature.com

Printed and bound by CPI Group (UK) Ltd, Croydon, CR0 4YY
30/09/2026
02225596-0014